Yoshitaka Mori · Karl Lennert

Electron Microscopic Atlas of Lymph Node Cytology and Pathology

Translated by Klaus Küchemann

With 172 Figures

Springer-Verlag Berlin · Heidelberg · New York 1969

Dr. Y. Mori, Assoc. Professor, Department of Pathology, Division I., Kobe University
School of Medicine, Kusunokicho 7, Ikutaku, Kobe/Japan

Dr. K. Lennert, o. Professor, Direktor des Pathologischen Instituts
der Universität, D-2300 Kiel, Hospitalstr. 42

ISBN 978-3-642-86537-4   ISBN 978-3-642-86536-7 (eBook)
DOI 10.1007/978-3-642-86536-7

Preface

Each year sees the publication of hundreds of reports of experimental work on the lymphatic tissue, yet morphological studies of the cells involved can be counted on the fingers of one hand. Furthermore, anyone who tries to identify these cells by morphological criteria is accused of sophistry and hair-splitting, whereas it is accounted scientifically correct and unbiased to speak of "lymphoid cells", "blast cells" etc.

Not so many years ago things were different: there were too many names and too many classifications and everyone backed his particular fancy. People thought of cells in terms of rigid classes, nothing then being known about the transformability of mononuclear blood cells.

Today we must look for the middle way: cells should be named and defined according to morphological criteria but their transformation potential should be borne in mind. Once the cells are analysed and subdivided, it will be simple enough to set up proper classifications afterwards.

This book arose out of the conviction that there should be more criteria and more information available on the morphology of human lymphoreticular cells; previously such information had been restricted by the difficulty of the special hematological and cytochemical staining methods.

Another reason for compiling this book was a desire to aid practical diagnosis. Can the electron microscope facilitate the recognition of morbid changes in the lymph nodes and help us to understand them better? Will a knowledge of the ultrastructure perhaps help to explain the derivation of such controversial tumors as follicular lymphoma (germinoblastoma)? Do different cytoplasmic structures allow for a more selective chemotherapy of malignant tumors of the lymph nodes?

This Atlas presents some data which may help to answer these questions. There are still many gaps which can perhaps be bridged with the support of other methods, such as the use of a different fixative or electron microscopic demonstration of enzyme activity. As always, when problems are solved they generate fresh problems.

Despite the provisional nature of much of our material, we think we are justified in publishing it, indeed, that we have a duty to do so, because of the urgency of making basic information freely available — and here I mean no disrespect to the excellent Atlas by BERNHARD and LEPLUS. We propose to issue supplementary volumes to fill in the gaps and to devote our resources to solving the remaining puzzles.

There is one more reason — historical, this time — why we are publishing this work: we have been working for years with many standard staining and cytochemical methods to identify changes in sections and smears of lymph nodes. But we had come to the point where little in the way of fundamental new findings could be expected. At this stage we were joined by Professor YOSHITAKA

Mori, an experienced electron microscopist and an expert on the reticuloendothelial system; he grasped the opportunity to analyse the ultrastructure of the samples submitted to the Lymph Node Registry which had already been studied by our standard methods. This work was made possible by a grant from the Deutsche Forschungsgemeinschaft. The fact that shortly before the completion of this work Dr. Klaus Küchemann, former resident pathologist at the Philadelphia General Hospital under the late Professor W. E. Ehrich, became a member of our staff, encouraged us to publish our Atlas in English.

We therefore hope that intention and chance have proved a fertile union.

Kiel, April 1969   Karl Lennert

Acknowledgments

First we would like to express our appreciation to the „Deutsche Forschungsgemeinschaft" which by establishing and financing the Lymph Node Registry at the Deutsche Gesellschaft für Pathologie provided the basis for the opportunity of compiling this book.

We are most indebted to the many clinicians and pathologists who placed the interesting fresh specimens so generously at our disposal. Special credit is given to our colleagues of the Department of Surgery (Director: Professor Dr. LÖHR) and of the Hospital for Ear, Nose and Throat Diseases (Director: Professor Dr. MÜLLER) of the University of Kiel. Professor Dr. SCHNEIDER, Tübingen, and Dr. GRABENER, Rendsburg, contributed one case each of monocytic leukemia. Dr. MÜHLETHALER, Solothurn, has allowed us to use a specimen of histiocytosis X. Dr. ALTROGGE, Buxtehude, was kind enough to perform a second excision of a BURKITT's tumor, because the original specimen was fixed in formalin. Professor Dr. BOHLE, Tübingen, has supplied samples of histiocytosis X.

Some of the micrographs in this atlas were taken by other members of our staff: Fig. 51 by Doz. Dr. BLÜMCKE and Dr. NIEDORF; Fig. 33 by Dr. KAISERLING; Figs. 9—16—29—30 by Dr. MÜLLER-HERMELINK; Fig. 100 by Dr. NIEDORF. Greatful thanks are accorded to Doz. Dr. BLÜMCKE, Priv.-Doz. Dr. CAESAR, Priv.-Doz. Dr. LEDER and Dr. MÜLLER-HERMELINK for many suggestions offered during discussions on the various subjects.

Warmest thanks are also expressed to Dr. LUCY BALIAN RORKE, Chief, Department of Anatomic Pathology, Philadelphia General Hospital, and Miss THERESE JOYCE, Philadelphia/USA, for generous help in translation problems and for meticulous proof-reading.

The excellent drawings were prepared by our artist WOLFGANG VATER. Thanks are due to Miss ELKE CARSTENSEN for unremitting and skilled technical help. Miss ROSEL WEYER took proper care in the preparation of high quality magnifications of the micrographs. We also are grateful to Miss INGRID MOORS for efficient secretarial help in preparing the manuscript.

Finally, we express our gratitude to the publisher, in particular Dr. GÖTZE, for the generous and superb preparation of text and illustrations.

Contents

Contents

Key to Abbreviations

ag	azurophilic granule		n	nucleus
b	basement membrane		nb	nuclear dense body
bs	basophilic stem cell		ng	neutrophilic leukocyte
c	centriole		ni	nuclear inclusions
cap	capillary lumen		nl	nucleolus
co	collagen fiber		np	nuclear pockets
d	desmosome		nre	nonphagocytic reticulum cell
eg	eosinophilic leukocyte		p	phagosome
end	endothelium		pb	plasmablast
ep	epithelioid cell		pc	plasma cell
ery	erythrocyte		per	pericyte
f	cytoplasmic filaments		pmc	promyelocyte
fc	fibrocyte		ppc	proplasmacyte
g	glycogen		pr	protein deposits
G	Golgi apparatus		pre	phagocytic reticulum cell
gb	germinoblast		Rb	RUSSELL body
gc	germinocyte		re	reticulum cell
hi	immature sinus histiocyte		rer	rough endoplasmic reticulum
lbl	lymphoblast		rf	reticulum fiber
Lgh	LANGHANS' type giant cell		rt	retothelial cells
lp	lipid		ser	smooth endoplasmic reticulum
ly	lymphocyte		sg	specific granule
lys	lysosome		ssc	starry sky cell
m	mitochondria		StR	STERNBERG-REED cell
mc	mast cell		tf	tonofibrils
ml	melanin		tp	toxoplasma
my	myelin figure			

Material and Methods

The micrographs in this atlas were taken from 65 biopsy cases which were examined with an electron microscope during the last two years at the Institute of Pathology of the University of Kiel. Immediately following surgical removal, the fresh lymph nodes were fixed for two hours in buffered 1 % osmium tetroxide (Rhodin-buffer, pH 7.2). In a few cases prefixation with phosphate-buffered 6.5 % glutaraldehyde was employed. In addition, imprints of all lymph nodes were prepared. The imprints as well as cryostat sections of fractions of the lymph nodes were studied by enzyme-histochemical methods. Both a Pappenheim stain and the PAS reaction were done on the imprints. The remainder of the lymph node was fixed in formalin or in Zenker's formalin solution according to MAXIMOW, embedded in paraffin and stained by various procedures (Giemsa, H & E and reticulum stains).

Table 1

Case	Age (years)	Sex	Fixation	Diagnosis
1	50	m	formalin + OsO_4	Waldenström's disease
2	60	m	OsO_4	Reticulum cell sarcoma
3	71	m	OsO_4	Waldenström's disease
4	84	m	formalin + OsO_4	Reticulum cell sarcoma
5	67	m	formalin + OsO_4	Hodgkin's disease, rich in epithelioid cells
6	52	m	OsO_4	Thymic carcinoma
7	39	m	OsO_4	Metastasis of epidermoid carcinoma
8	67	m	OsO_4	Hodgkin's disease, mixed type
9	66	f	OsO_4	Chronic lymphocytic leukemia
10	60	m	OsO_4	Hodgkin's disease, nodular sclerosing type
11	30	m	formalin + OsO_4	Lymphosarcoma (? lymphocytic type)
12	42	f	OsO_4	Sarcoidosis
13	64	f	OsO_4	Chronic lymphocytic leukemia with hemolytic anemia
14	49	m	formalin + OsO_4	Hodgkin's disease, mixed type
15	23	m	OsO_4	Toxoplasmosis
16	48	m	OsO_4	Histiocytosis X
17	1.5	m	OsO_4	Histiocytosis X
18	46	f	formalin + OsO_4	Tuberculous lymphadenitis
19	65	f	OsO_4	Reticulum cell sarcoma
20	65	m	formalin + OsO_4	Reticulum cell sarcoma
21	54	f	OsO_4	Follicular lymphoma (germinoblastoma)
22	53	f	formalin + OsO_4	Follicular lymphoma (germinoblastoma)
23	33	m	OsO_4	Metastasis of mesothelioma
24	64	m	OsO_4	Hodgkin's disease, mixed type
25	73	m	OsO_4	Lipomelanotic reticulocytosis
26	69	m	OsO_4	Epithelioid cellular lymphogranulomatosis
27	83	m	OsO_4	Reticulum cell sarcoma
28	42	m	formalin + OsO_4	Reticulum cell sarcoma
29	68	m	OsO_4	Chronic lymphocytic leukemia
30	59	m	OsO_4	Nonspecific lymphadenitis

Material and Methods

Table 1 (Continued)

Case	Age (years)	Sex	Fixation	Diagnosis
31	5	f	OsO$_4$	Poorly differentiated leukemia, lymphoblastic type
32	52	m	OsO$_4$	Follicular lymphoma (germinoblastoma)
33	60	m	OsO$_4$	Nonspecific lymphadenitis
34	63	m	formalin + OsO$_4$	Follicular lymphoma (germinoblastoma)
35	23	m	OsO$_4$	Toxoplasmosis
36	32	m	OsO$_4$	Hodgkin's disease, mixed type
37	35	m	OsO$_4$	Nonspecific lymphadenitis
38	79	m	OsO$_4$ post mortem	Reticulum cell sarcoma
39	64	f	formalin + OsO$_4$	Reticulum cell sarcoma
40	39	m	OsO$_4$	Nonspecific lymphadenitis
41	13	m	OsO$_4$	Metastasis of neuroblastoma
42	45	m	OsO$_4$	Metastasis of malignant melanoma
43	59	m	formalin + OsO$_4$	Follicular lymphoma (germinoblastoma)
44	35	m	OsO$_4$	Nonspecific lymphadenitis
45	51	m	formalin + OsO$_4$	Reticulum cell sarcoma
46	25	m	OsO$_4$	Burkitt's tumor (European case)
47	55	f	OsO$_4$	Nonspecific lymphadenitis
48	59	m	OsO$_4$	Follicular lymphoblastoma (germinoblastoma)
49	25	m	OsO$_4$	Nonspecific lymphadenitis
50	42	f	OsO$_4$	Hodgkin's disease, mixed type
51	44	f	OsO$_4$	Lymphoepithelial carcinoma (SCHMINCKE-RÉGAUD)
52	30	f	OsO$_4$	Tuberculous lymphadenitis
53	12	m	OsO$_4$	Lymphosarcoma (lymphoblastic type)
54	25	m	OsO$_4$	Poorly differentiated leukemia, granulocytic type
55	45	f	OsO$_4$	Monocytic leukemia
56	64	m	glutaraldehyde + OsO$_4$	Monocytic leukemia
57	63	m	OsO$_4$	Poorly differentiated leukemia, granulocytic type
58	47	m	formalin + OsO$_4$	No diagnosis
59	13	m	OsO$_4$	Acute erythremia
60	70	m	OsO$_4$	Poorly differentiated leukemia, granulocytic type
61	34	m	OsO$_4$	Lymphosarcoma
62	65	m	formalin + OsO$_4$	Reticulum cell sarcoma
63	44	m	OsO$_4$	Nonspecific lymphadenitis
64	58	m	OsO$_4$	Reticulum cell sarcoma
65	18	m	OsO$_4$	Hodgkin's disease, mixed type
66	15	m	OsO$_4$	Toxoplasmosis

In addition to these freshly obtained lymph nodes, some nodes were collected from the formalin-fixed routine material which had been sent to the "Lymphknotenregister bei der Deutschen Gesellschaft für Pathologie" (Lymph Node Registry at the German Society of Pathology) in Kiel. This material was subsequently fixed in 1 % buffered osmium tetroxide for two hours after brief washing in buffer solution. The original formalin fixation is responsible for the fact that some of the pictures shown in this atlas are of poor quality. However, they are satisfactory for diagnostic purposes[1] as exemplified by some cases (e.g. Case 62). In addition to lymph nodes, we studied hyperplastic tonsils in collaboration with MÜLLER-HERMELINK. They were prefixed in formalin and postfixed in 1.0 % osmium tetroxide. By this method of fixation particularly instructive semithin sections were obtained.

---

1. HÜBNER and PAULUSSEN (1968).

For electron microscopy pieces of lymph node were embedded in Araldite. From each case sections 0.5 μ in thickness were stained with Mallory's Methylene blue-Azure II solution for light microscopy. After preparation of a semithin section suitable for orientation, slightly deeper cuts were examined with the Siemens Elmiskop I. Throughout this atlas, where the stain is not so designated, uranyl acetate and lead citrate stains were used. Diagnosis, age and sex of the patients are summarized in Table 1.

The So-Called Normal Lymph Node

I. The Cell Types
Figs. I—VII and 1—21

Considerable advancement in our knowledge concerning the function of the lymph node cells has come with improvement of new immunologic technics. However, classification and nomenclature of the cells remains a matter of controversy. It should be kept in mind that a wide variety of cells is encountered in the lymph node and that single cell types are also subject to certain variations. Not every cell can be rigidly classified. Although the fine structure of the lymph node has been reported by many authors[1], they have used many different classifications, especially with respect to the lymphocytic series and the reticulum cells. The terminology used in this atlas is essentially similar to that recommended by us in our handbook[2]. In a previous paper we have reported comparative studies of light and electron microscopic appearance of the cells[3].

1. Lymphocyte

The lymphocyte is the smallest cell in the lymph node usually being smaller than 8 μ in diameter. It has, as a rule, round, ovoid or frequently somewhat irregular cytoplasm. The nucleus occupies large portions of the cell and displays the most pronounced electron opacity of all lymph node cells. For the most part the nuclei are round or ovoid, but frequently they are deeply indented. Nuclear chro-

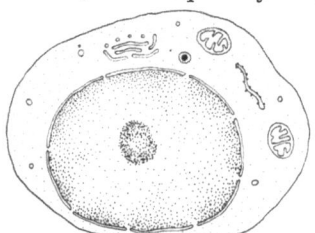

Fig. I. Lymphocyte.

matin congregates under the nuclear membrane and also somewhere within the nucleus. The nucleolus is rather inconspicuous and often not visualized. One or two nuclear dense bodies[4] are usually observed. There are always a few mitochondria unevenly distributed throughout the cytoplasm; they are round and measure about 0.5 μ in diameter. A small amount of finely vesicular smooth-surfaced and long

1. Tanaka (1957, 1958, 1963); Fresen and Wellensiek (1958, 1959); Matsuura (1958); Hosokawa (1960); Mori (1960); Sorensen (1960); Han (1961); Horie (1961); Clark (1962); Moe (1963, 1964); Bairati and Amante (1964); Bernhard and Leplus (1964); Yamori (1964); Miura (1965); Movat and Fernando (1965, 1966); Swartzendruber and Hanna (1965); Brooks and Siegel (1966); Müller-Hermelink and Caesar (1969).

2. Lennert (1961).

3. Lennert, Caesar and Müller (1966).

4. Weber, Waipp, Usenik and Frommes (1964); Brooks and Siegel (1967); Büttner and Horstmann (1967).

channels of rough-surfaced endoplasmic reticulum is sometimes seen within the cytoplasm. Occasionally, a few lysosomes with a limiting membrane are observed. Many free ribosomes are uniformly dispersed throughout the cytoplasm, but generally no polyribosomes can be made out.

Thus far we are unable to identify an electron microscopic equivalent of lymphocyte precursors, that is, cells which are referred to as lymphoblasts etc. Their light microscopic identity also remains controversial.

2. The Characteristic Cells of the Germinal Centers[1]

a) Germinocyte

Germinocytes can be found only in germinal centers. They range up to 10 μ in diameter and are ovoid or somewhat irregular in shape. Usually, the nucleus is ovoid, but it sometimes presents with slight indentations. The chromatin is focally concentrated at nuclear margins. However, the chromatin is

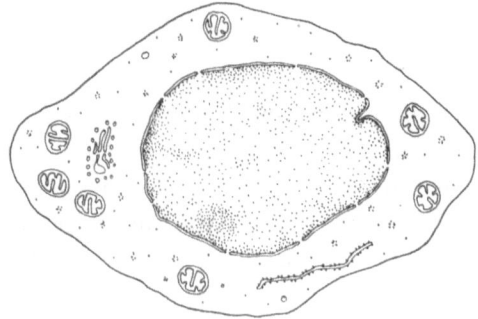

Fig. II. Germinocyte.

somewhat finer than in lymphocytes and, on the whole, the nucleus exhibits less opacity than does the lymphocytic nucleus. Generally, one small nucleolus is located in the center of the nucleus. The nucleolus is not sharply outlined against the rest of the karyoplasm and is occasionally found in an eccentric position. Some nuclei have nuclear pockets[2] (nuclear projections[3], nuclear pseudoinclusions[4]) containing cytoplasmic elements. There are several round mitochondria within the cytoplasm. The Golgi apparatus is small and is constituted of vesicular and cisternal components. Many ribosomes are disseminated throughout the cytoplasm, but on occasion they are arranged in clusters as polyribosomes. Endoplasmic reticulum, both smooth and rough, is uncommon.

b) Germinoblast

Germinoblasts measure about 12—14 μ in diameter and are of almost ovoid or somewhat irregular shape. Their nucleus is about 8 μ in diameter, and the chromatin is arranged in small aggregates at the nuclear membrane. Round large nucleoli with distinct nucleonemata occupy central or peripheral positions. Occasionally, nuclear pockets are found in some cells. They can be discerned, however, exclusively in germinocytes and germinoblasts, but never in lymphocytes. Polyribosomes are pro-

---

1. LENNERT, CAESAR and MÜLLER (1966); MÜLLER-HERMELINK and CAESAR (1969).
2. ROBERTSON and MACLEAN (1965); SMITH and O'HARA (1968).
3. See MOLLO and STRAMIGNONI (1967).
4. BLOOM (1967).

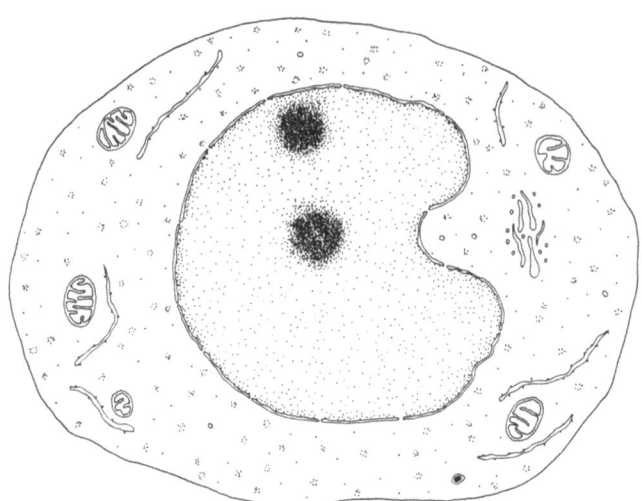

Fig. III. Germinoblast.

minently displayed in the cytoplasm and are much more pronounced than in germinocytes. Several large mitochondria can be distinguished in the cytoplasm, most of them located at one side of the nucleus. They are slightly irregular in shape and size, for the most part, however, being ovoid or tubular. A few lysosomes are also observed. A small amount of endoplasmic reticulum, both granular and agranular, can be seen on occasion. The Golgi apparatus is well developed.

c) Desmosome-connected Long-branching Reticulum Cell (see p. 9ff.)

3. Basophilic Stem Cell (Large Pyroninophilic Cell)

The basophilic stem cells apparently occur both in germinal centers and in the pulp, especially of the cortex of the lymph node. In no case could we detect a difference between the large basophilic cells of both areas. In germinal centers basophilic stem cells are frequently located peripherally near the lymphocytic mantle.

The basophilic stem cells are larger than 15 μ in diameter and have more cytoplasm than germinoblasts. The nucleus is also large and irregular and of medium electron density. Diffuse chromatin material is suspended in the nucleoplasm. There are two or three nucleoli characteristically located near the nuclear membrane. The cytoplasm is rich in diffuse polyribosomes. Mitochondria, 0.5 μ in diameter, are unevenly disseminated. Sparse endoplasmic reticulum studded with a few ribosomes can be seen. We never found the thread-like structures which have been described by PARKER, WAKASA and LUKES[1] in antigen-stimulated lymphocytes of lymphocyte cultures. The Golgi apparatus is well developed and more prominent than in germinoblasts. It consists mainly of vesicular and cisternal elements. Occasionally, around the Golgi zone, small lysosomes — similar to dense membrane-bound granules — dot the cytoplasm. However, other parts of the cytoplasm fail to show lysosomes.

1. 1968.

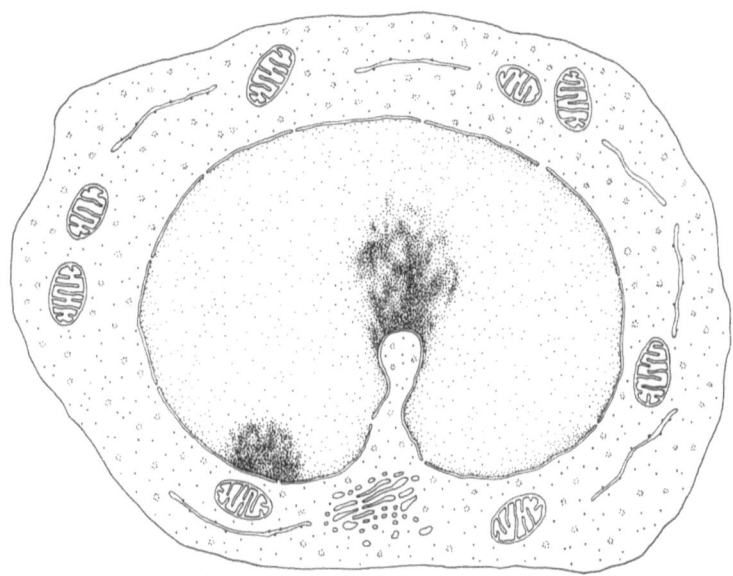

Fig. IV. Basophilic stem cell.

4. Plasma Cell Series

a) Plasma Cell

The ultrastructure of the plasma cell has already been reported by many authors[1]. The ovoid cell measures about 10 μ in largest diameter. The nucleus is round and eccentrically located. The chromatin is marginated and clumped near the nuclear periphery. The nucleolus is usually of medium size. A substantial amount of ergastoplasm (rough endoplasmic reticulum) is scattered throughout the cytoplasm. This correlates well with the striking basophilia of the cell observed with the light microscope.

The ergastoplasm contains occasional sac-like dilatations which are filled with varying amounts of light or dark amorphous material which most likely represents immunoglobulin. Distended ergastoplasmic tubules may occasionally reach several microns in diameter and in some instances may occupy virtually the entire cytoplasm. Large dilated ergastoplasmic tubules or sacs are the so-called Russell bodies. The Golgi apparatus is well developed and corresponds to the paranuclear clear zone seen with the light microscope. It consists of a lamellar membrane, small vesicles and somewhat dilated vacuolar elements. Mitochondria are large and almost globular in configuration. The mitochondria of plasma cells are surrounded by ergastoplasmic lamellae. This arrangement of mitochondria and ergastoplasm differs from that of reticulum cells, although in the latter cell types the ergastoplasm may occupy a prominent place in certain areas of the cytoplasm.

Sometimes one lipid globule is found in the cytoplasm. It is bound by a single limiting membrane and contains moderately electron-dense material. It is identical with the so-called lipochondrion seen in light microscopy.

1. BRAUNSTEINER and PAKESCH (1955); DOHI, HANAOKA and AMANO (1957); KAUTZ, DE MARSCH and THORNBURG (1957); WELLENSIEK (1957); STOECKENIUS and NAUMANN (1957); STOECKENIUS (1958); WELSCH (1960); BESSIS (1961); REBUCK and LOGRIPPO (1961).

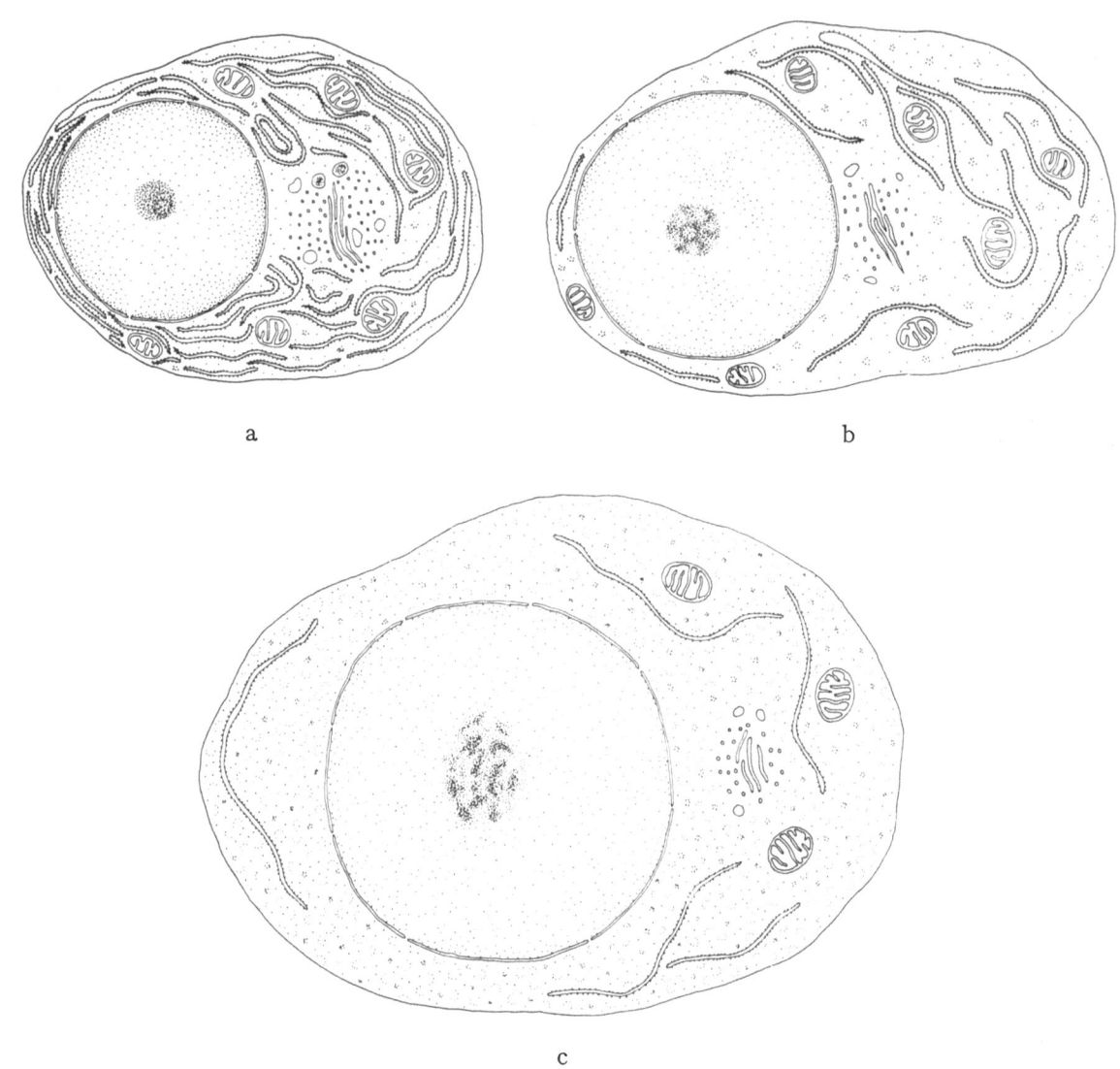

a

b

c

Fig. V a—c. Plasma cell, proplasmacyte, plasmablast.

b) Proplasmacyte

These cells are round and measure about 10—12 μ in diameter. Their nuclei may be either centrally or eccentrically placed. The nuclear chromatin is not as coarse as that of plasma cells and tends to accumulate in the center of the nucleus. One or two nucleoli are present. They tend to be larger and to have more distinct nucleonemata than those of the plasma cells. A varying amount of rough endoplasmic reticulum is disbursed throughout the cytoplasm and it shows a less regular lamellar arrangement. A conspicuous number of free ribosomes and polyribosomes is observed. The Golgi apparatus is almost as prominent as that found in plasma cells. A few lysosomes can occasionally be seen in the cytoplasm. Mitochondria are large, almost round and interspersed between the rough endoplasmic reticulum. Sometimes solitary lipid globules may be demonstrated.

c) Plasmablast

Plasmablasts are larger than proplasmacytes and exhibit a greater nuclear-cytoplasmic ratio. Their nucleus is usually round and provided with one or two large nucleoli. The chromatin material is diffusely suspended within the nucleoplasm and shows no clumping. Many ribosomes and polyribosomes are dispersed throughout the cytoplasm. Rough endoplasmic reticulum does not fill the entire cytoplasm as in proplasmacytes. In contrast to plasma cells and proplasmacytes the ergastoplasm is only occasionally arranged in parallel manner. It is usually branched and displays cistern-like distensions. This is also true of the Golgi apparatus. The mitochondria of the cells are as large as those of plasma cells.

In addition to plasmablasts with this characteristic picture, there are numerous transitional stages to proplasmacytes on the one hand and to an immature cell similar to the basophilic stem cell on the other hand. These intermediate types are distinguished by differences in the content of the endoplasmic reticulum.

Table 2

Cell type	Nucleus		Cytoplasm				
	Chromatin clumps	Nucleolus	Mitochondria (largest diameter)	Golgi apparatus	Endoplasmic reticulum	Ribosomes	Polysomes
Lymphocyte	+++	small	0.3—0.5 μ	small	+	++	+
Germinocyte	+	medium-size	0.3—0.5 μ	medium-size	+	+	++
Germinoblast	+	large	0.3—0.5 μ	large	+	+	+++
Basophilic stem cell	+	giant	0.4—0.5 μ	large	+	+	++++
Plasma cell	+++	small	0.5—0.6 μ	giant	++++	+	+
Proplasmacyte	+	medium-size	0.5—0.6 μ	large	+++	++	++
Plasmablast	+	large	0.5—0.6 μ	medium-size	++	+++	+++

5. Reticulum Cell

At the present time, the precise definition and classification of the reticulum cell is still confusing. Introduction of the term "histiocyte"[1] has merely replaced the term "reticulum cell", but has left the basic problem untouched.

Discussion regarding the nature of reticulum cells has been revived by the important new papers of NOSSAL and associates[2], who correlated submicroscopic morphology and function. They agreed with the concept of MARUYAMA and HANAOKA[3] in distinguishing 2 main types: 1) macrophages with active antigen ingestion and digestion, and 2) long-branching desmosome-connected reticulum cells that retain the antigen on the cell surface. They do not show phagocytosis. Since length and width of the cellular processes cannot always be estimated, we have combined those long-branching reticulum cells with all reticulum cells without phagocytosis: *nonphagocytic reticulum cells*. However,

1. GALL (1958).
2. NOSSAL, ABBOT and MITCHELL (1968); NOSSAL and LUMMUS (1968).
3. 1966.

the presence of two different types of nonphagocytic reticulum cells must be emphasized: first, the long-branching reticulum cell united by desmosomes and, second, the nonphagocytic reticulum cell devoid of desmosomal connections. The first type with desmosomal connections seems to be specific for follicles and particularly for germinal centers, while the second type, distinguished by the lack of desmosomes, is present in the pulp.

A second type is represented by *phagocytic reticulum cells*. They may also be referred to as *"macrophages"*. However, macrophages are not exclusively derived from reticulum cells, but may also stem from monocytes. Those monocytogenic macrophages cannot be distinguished from reticulogenic macrophages. The derivation of fully developed macrophages cannot be determined from their morphologic appearance. Applying strict criteria macrophages are identical with the "histiocytes" of GALL's classification. They may be designated mature histiocytes as contrasted with the immature histiocytes. *Immature histiocytes* appear in large numbers in the sinuses in the various kinds of inflammation of the lymph node (e.g. toxoplasmosis). They should be strictly separated from the retothelial cells and macrophages of the sinus. They are dealt with more thoroughly in the chapter on immature sinus histiocytosis.

For the description of *monocytes* the reader is referred to page 15.

a) Nonphagocytic Reticulum Cell

This cell type possesses irregular polygonal cytoplasm with long slender cytoplasmic processes extending between neighboring cells. It is rather difficult to survey all parts of the cell in one picture, since the processes are narrow and tangled. Frequently the cells are grouped around blood vessels and bear a definite morphological resemblance to adventitial cells. The cytoplasm, exclusive of the extensions, measures about 10 μ in largest diameter. Many small extensions of cytoplasmic processes can be seen between adjoining cells. In primary and secondary follicles (including germinal centers) they are linked together by distinct desmosomes which appear to be absent in the pulp. The nucleus shows striking variation in size and shape, being ovoid, lobulated or irregularly indented. There is a comparatively minor degree of electron density of the nucleus as compared to the other cells of the lymph node, since slender fine chromatin clumps line up along the nuclear membrane. This arrangement corresponds to the clear nucleus of this cell type as it appears in light microscopic images.

One or two small or occasionally moderately large nucleoli may be visualized, as well as a few nuclear dense bodies. Intracytoplasmic organelles are variable in different cells, but generally the smooth endoplasmic reticulum is quite obvious. Large numbers of small vesicles are scattered throughout the cytoplasm. Rough endoplasmic reticulum is often prominently displayed. Mitochondria are ovoid or long and rod-shaped and smaller than those of the lymphocytes and plasma cells. The Golgi apparatus is fairly prominent and divided into several groups. They comprise lamellar, vesicular and granular components. Usually several lysosomes are encountered in the cytoplasm with electron-dense granules, myelin figures and multivesicular bodies. Some cells contain fine threads arranged parallel to the longitudinal nuclear axis. Free ribosomes and polyribosomes are spread diffusely in the cytoplasm in variable numbers.

By means of its long processes this cell type produces a lattice in lymph follicles (including germinal centers) and pulp. Distinct long cytoplasmic processes of the cell are occasionally found around hyperplastic germinal centers (see also under follicular lymphatic hyperplasia). It is believed that at least some of the cells merely adsorb and retain antigen on their cytoplasmic membranes.

The nonphagocytic reticulum cell is often in direct contiguity to reticulum consisting of collagen fibers and electron-dense amorphous or fine filamentous materials. However, it has not yet been determined whether the cell is capable of producing collagen fibers.

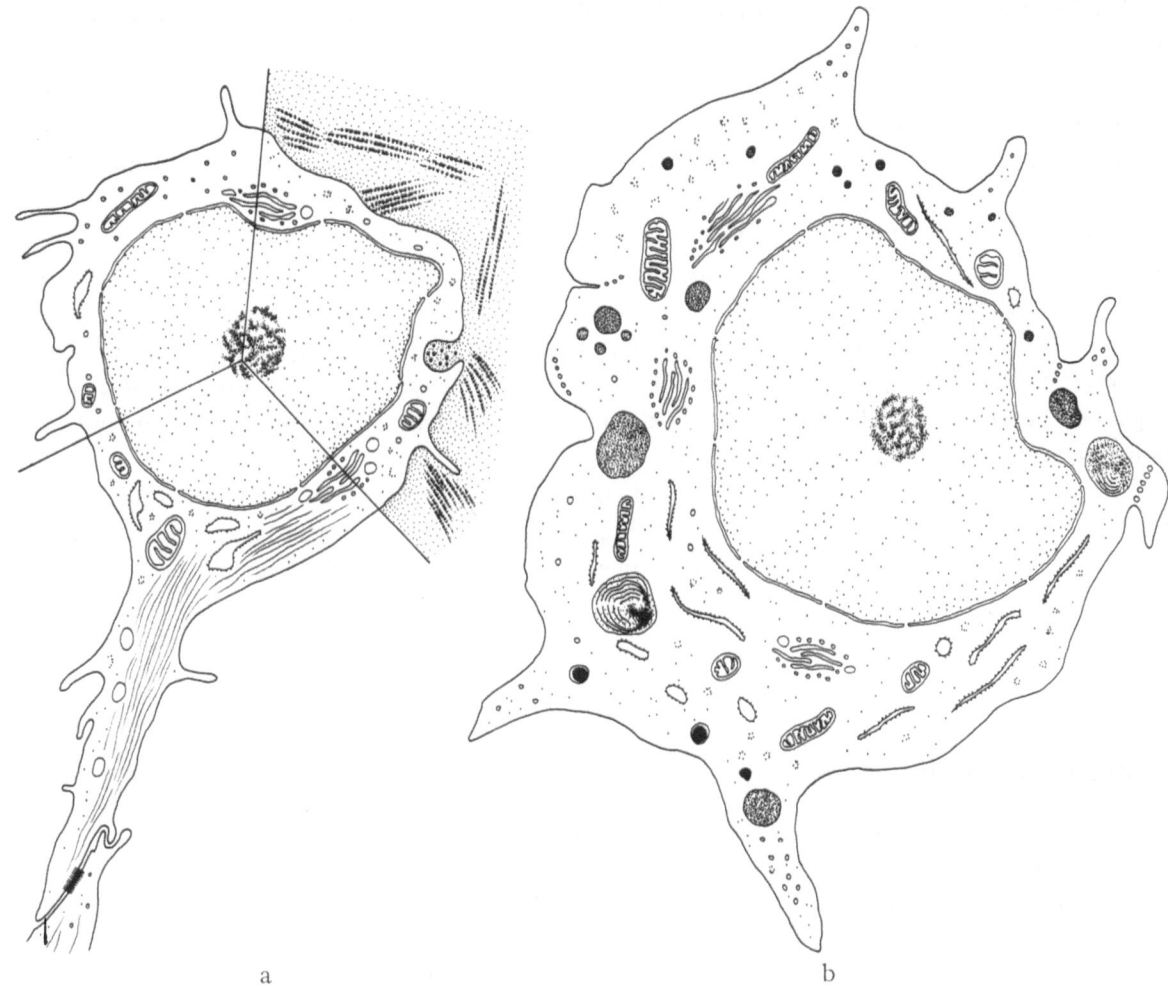

a   b

Fig. VIa and b. Nonphagocytic and phagocytic reticulum cell.

Some nonphagocytic cells often exhibit pronounced electron density and are called "dense reticulum cells", especially in hyperplastic germinal centers or in neoplastic tissues. It is not clear whether such cells represent a specific separate cell type or degenerative products[1]. They are rich in free ribosomes.

b) Phagocytic Reticulum Cell

Phagocytic reticulum cells may become the largest cells in the lymph node after ingestion of foreign bodies, and they sometimes measure more than 20 μ in diameter. They are subject to great variations in shape, showing many pseudopodal processes extending outward from their cytoplasm.

1. STOECKENIUS (1957, 1958); IZARD (1967); IZARD and DE HARVEN (1968).

They may occasionally be fairly long and wide, branching between adjacent cells. The enormous size of the cells, due to the many processes, enables them to contact foreign materials easily. The nucleus is round, ovoid or slightly folded, with one or two large nucleoli. Nucleoplasmic granules within the nucleus are dispersed diffusely with slight margination near the nuclear periphery. The chromatin structure is more delicate than in the nonphagocytic reticulum cell. The many mitochondria are unevenly distributed in the cytoplasm. They are round, ovoid or rod-shaped. The Golgi apparatus is prominently displayed and consists of lamellar, vesicular and vacuolar components. Sometimes the cell contains several Golgi groups. A large number of finely vesicular profiles of smooth endoplasmic reticulum are apparent throughout the cytoplasm. Rough endoplasmic reticulum is also found in most of the cells. A variety of lysosomes and lysosome-like particles can be seen more frequently than in nonphagocytic reticulum cells. Moreover, varying numbers of phagosomes are observed. Their character is not the same in all the cells. In some of them the original nature of the engulfed material can be recognized (neutrophilic leukocytes, erythrocytes, plasma cells etc.). However, some phagosomes merely contain amorphous light or dark dense granules. Lysosomes and residual bodies also show various compounds, i.e. myelin figures, multivesicular bodies, fine granular or dilated vesicular elements. Ribosomes are found especially at the periphery of the cell. Fine intracytoplasmic filaments are common in this cell type. On the whole, cytoplasmic matrices exhibit somewhat reduced electron density in contrast with nonphagocytic reticulum cells.

Although morphologic differences of the two types of reticulum cells can be discerned with relative ease, some cells may take on an intermediate pattern. This may be attributed to tangential sectioning which does not permit correct identification of those cells. Nevertheless, the authors have been unable to conclude whether or not transitional stages between the two cell types really exist, or whether these "intermediate" forms should be considered as only slight modifications of two separate cell lines (as suggested by MÜLLER-HERMELINK and CAESAR[1]).

6. Retothelial Cells

The term "retothelial" includes two types of sinus cells:

1. the endothelial sinus cells (littoral cells) lining the sinus wall, and

2. morphologically identical cells traversing the lumen of the sinus by means of long processes that are connected with the littoral cells.

Reticulum cells of the pulp of lymph nodes and of follicles are no longer classed with the retothelial cells[2]. Furthermore, the term "retothelial cell" is not applied to macrophages that are thought to arise from reticulum cells or monocytes. There may be an occasional exception in which a retothelial cell is presumed to round off and to become a macrophage.

The ultrastructure of the sinus wall is dealt with separately owing to structural variations in different areas (marginal, intermediate and medullary) of the lymph node. Retothelial cells exhibit variations in fine structures depending on their location.

a) Retothelial Cells of the Marginal Sinus

On the *capsular side* of the marginal sinus these cells are of relatively simple structure. Their cytoplasm covers the connective tissue of the capsule which is limited by a typical basement membrane

---

1. 1969.
2. LENNERT (1961).

Table 3

Cell type	Cytoplasm		Nucleus		Cytoplasmic organelles		
	Shape	Processes	Chromatin	Nucleolus	Lysosomes	Phagosomes	Endoplasmic reticulum
Non-phagocytic reticulum cell	irregular	long and slender	moderately coarse	medium size, sometimes large	+/++	−	+/++
Phagocytic reticulum cell	irregular, polygonal	short and wide	fine	medium size, sometimes large	++/+++	+/++	++/+++
Macrophage	irregular, ovoid	fine	fine	medium size	+++	+++	+++
Retothelial cell	fusiform	fine, rare	fine	small	+/++	(+)	+/++
Blood monocyte	round, ovoid	fine	fine, marginal condensation	small	+/++ specific granules	+	++/+++

that consists of amorphous or finely filamentous moderately electron-dense material. The cells are united by desmosomes. Their cytoplasm is usually thin except in the region of the nucleus. The nucleus shows moderate electron opacity with widespread nucleoplasm. Sometimes a small nucleolus and one or two nuclear dense bodies (sphaeridien[1]) are encountered. In the cytoplasm, a finely vesicular smooth endoplasmic reticulum is fairly well developed, and large numbers of pinocytotic vesicles may gather at the cytoplasmic periphery. The mitochondria are small and ovoid or long and rod-shaped. Scanty rough endoplasmic reticulum is observed only if the lymph nodes take up and process increased amounts of material.

The cells on the *inner side* of the marginal sinus are attached to a structure which corresponds to argyrophilic fibers as seen with the light microscope. It is composed of collagen fibers and amorphous or finely filamentous, moderately electron-dense material. It is thought to be equivalent to the basement membrane mentioned above, but does not represent a real basement membrane by itself. The nucleus of these cells is identical to the nucleus of the cell on the capsular side, notwithstanding the slightly thinner cytoplasm of the latter cell. Intracytoplasmic organelles are fairly prominent. They consist of finely vesicular or vacuolar smooth endoplasmic reticulum, rough endoplasmic reticulum and some lysosomes. Occasionally, the tubules of the rough endoplasmic reticulum are found to have saccular distensions. The mitochondria are small, round or ovoid. Most of the retothelial cells contain fine filamentous elements in the cytoplasm arranged along the longitudinal axis. The cells are also attached to neighboring cells by desmosomes.

The *intraluminal cells* of the marginal sinus were previously referred to as intrasinusoidal reticulum cells. However, we could not distinguish these cells from the retothelial cells bordering the sinus lumen. The difference is noteworthy only as far as the relation between cells and basement

1. BÜTTNER and HORSTMANN (1967).

membrane structures is concerned. In contrast to the retothelium lining the wall of the sinus, the intraluminal cells encase parts of the described "basement membrane" – like material in infoldings of the cytoplasmic membrane. The distinction is made on the basis of the fact that the cells are associated with the filamentous framework in the lumen, whereas the retothelium of the sinus wall covers both the inner and outer surface, only the latter representing a true basement membrane.

b) Retothelial Cells of the Intermediate and Medullary Sinus

In contrast to the retothelium of the marginal sinuses, intracytoplasmic organelles are most abundant in the cells of the medullary sinuses. A large amount of vesicular smooth endoplasmic reticulum and also fairly long channels of rough endoplasmic reticulum are seen. However, they lack further distinguishing characteristics.

In the authors' opinion, all retothelial cells are of the same basic type. Variations noted in fine structure appear to be related to the degree of activity which depends on the location of the cell in question.

7. Mast Cell

Mast cells[1] measure approximately 10 μ in diameter. They usually have a round nucleus with a small centrally placed nucleolus. The characteristic granules are found within the cytoplasm. They are round and measure from 0.5—1.0 μ in diameter. They are filled with varying amounts of light, dark and

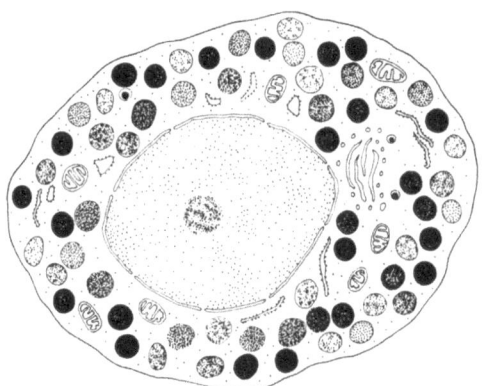

Fig. VII. Mast cell.

intermediate granular or amorphous homogeneous substances. The limiting membrane of the granules is apparently fragile and often cannot be seen. Mitochondria are ovoid or somewhat elongated and measure about 0.5 μ in diameter. They are interspersed between the specific granules. A few free ribosomes can also be made out in the cytoplasm. Paucity of smooth and rough endoplasmic reticulum is noted. The Golgi apparatus is well developed and consists of lamellar and vesicular components.

---

1. STOECKENIUS (1956); HIBBS, BURCH and PHILLIPS (1960); HIBBS, PHILLIPS and BURCH (1960); GUSEK (1962).

8. Other Cell Types

The blood monocyte measures 12—20 μ in diameter. Its rounded or kidney-shaped nucleus is eccentrically placed. Frequently, pronounced nuclear indentations are discernible along with marginal condensation of chromatin. An inconspicuous nucleolus may be found in apposition to the nuclear membrane. The nuclear-cytoplasmic ratio is slightly reduced. A few isolated relatively small spherical or ovoid bodies (? lysosomes) are scattered throughout the cytoplasm. Predominantly rod-shaped mitochondria are present in moderate numbers. They are interspersed along with abundant small (up to 0.4 μ in diameter) electron-lucent vesicles, occasional tubules of ribosome-associated endoplasmic reticulum and great numbers of free ribosomes and polysomes. The Golgi complex is inconspicuous and, for the most part, is constituted of small vesicles. It is frequently located in the area of the nuclear indentation. Especially after fixation in glutaraldehyde, small bundles of filaments may be identified in the cytoplasm. The cellular surface exhibits many short stubby pseudopodia.

Neutrophilic, eosinophilic and basophilic polymorphonuclear leukocytes may also occur in lymph nodes. The fine structure of their cytoplasm conforms to that of the corresponding cells of the peripheral blood. They have been described in detail by Low and FREEMAN[1], BERNHARD and LEPLUS[2] and many others. For this reason they are neglected in this book. This is also true of fibroblasts and smooth muscle cells of the capsule. A detailed description of the endothelial cells of the blood vessels is given in the chapter on blood vessels.

1. 1958.
2. 1964.

II. The Main Structural Units of the Lymph Node

Figs. VIII—IX and 22—31

The parenchyma of the lymph node consists of nodular and diffuse lymphoid tissue which are distinguished by "follicles" and "pulp". Located between these are the sinuses which are lined by retothelial cells, and the blood vessels.

1. The Follicles

According to EHRICH[1], the follicles of the lymphoid tissue may be divided into primary, secondary and tertiary follicles (nodules). The primary follicles mainly consist of small lymphocytes. These may be transformed into secondary follicles, which are distinguished by the presence of germinal centers. At the present time, the interfollicular cortical tissue is frequently referred to as tertiary follicles. However, these cannot be differentiated from the cortical pulp (paracortical zone).

The electron microscope reveals a few nonphagocytic long-branching reticulum cells in the *primary* lymphoid follicles in addition to small lymphocytes. Reticulum cells form a spongelike meshwork by means of long slender cytoplasmic processes which encase lymphocytes. There is no distinct boundary between the follicles and the surrounding lymphatic pulp. Occasionally, nonphagocytic reticulum cells with cytoplasmic extensions along the sinus wall are encountered.

The *germinal centers* of the *secondary* follicles consist of a number of germinocytes, germinoblasts, basophilic stem cells and reticulum cells of both the nonphagocytic (long-branching) and phagocytic varieties. The lymphocytic cuff does not include germinocytes and germinoblasts, but a few basophilic stem cells and, on occasion, a plasma cell are seen. Germinocytes can be differentiated from lymphocytes at the periphery of the follicle owing to the more pronounced electron opacity of the lymphocytic nucleus as compared with the nucleus of germinocytes. Phagocytic reticulum cells are provided with more cytoplasm and contain a conspicuous number of phagosomes and residual bodies. Light microscopically they are spoken of as macrophages with tingible bodies. Long slender cytoplasmic processes of the nonphagocytic desmosome-connected reticulum cells often surround the germinal center, as if forming a boundary between the outside and inside of the germinal center[2]. The nonphagocytic reticulum cells produce a lattice by means of their long cytoplasmic processes. However, the fine processes positioned between the other cell types do not always originate from the reticulum cells. The germinocytes, germinoblasts and also basophilic stem cells are often furnished with fine short cytoplasmic processes, and it is difficult to distinguish the processes of the reticulum cells from those arising from the other cell types by their fine structure alone. Occasionally, some plasma cells and their precursors are encountered in germinal centers. All cells of the germinal centers may display interlacing cellular extensions forming a lattice together with processes of the reticulum cells. This feature contrasts with the smooth cellular outlines in the lymphocytic rim where the cell membranes are devoid of cytoplasmic processes.

The *tertiary* follicles contain numerous small lymphocytes and frequently a striking number of reticulum cells, particularly those of the phagocytic type. There may also be many basophilic stem

1. 1946.
2. MÜLLER-HERMELINK and CAESAR (1969).

cells, but germinoblasts and germinocytes do not occur. The tertiary follicles are the chief sites of postcapillary venules, where the recirculation of the lymphocytes and their transformation into basophilic stem cells takes place. This is found, for example, in lymph nodes which participate in reactive inflammatory hyperplasia caused by skin homografts or contact dermatitis (delayed type of hypersensitivity).

2. The Pulp

While it is difficult to distinguish the diffuse lymphoid tissue of the cortex from tertiary follicles, the medullary pulp can be more clearly defined.

Lymphocytes and reticulum cells are observed in the cortical pulp (paracortical zone), but there are no distinguishing features in the fine structural characteristics between the pulpal cells and those of primary and tertiary follicles. The medullary cords are composed of lymphocytes and reticulum cells, frequently with plasma cells in addition. The reticulum cells in the cords are usually of the phagocytic variety. However, not every reticulum cell contains ingested material. Macrophages without obvious phagolysosomes can usually be differentiated from "nonphagocytic reticulum cells" by the cytologic criteria outlined in the chapter on reticulum cells. Desmosomal connections of reticulum cells could not be identified. Some reticulum cells are in apposition to reticulum fibers. However, it is still doubtful whether the reticulum cells are capable of producing these fibers.

3. The Lymph Sinuses

The composition of the sinuses varies depending on functional demands. The structure of the marginal sinus is simple, whereas that of the medullary sinus is much more complex. On the capsular side the marginal sinus is delineated by a distinct basement membrane which consists of three layers. The inner (often the follicular) side of the marginal sinus and neither side of the medullary sinus are bound by a basement membrane; they are lined by retothelial cells enwrapping fine fragmented threads containing a few collagen fibers. They correspond to the argyrophilic fibers seen with the light microscope. This membrane-like structure is comparatively compact in the marginal sinus, while it appears much looser in the medullary sinus. The intraluminal retothelial cells form a cellular network in the sinus similar to that of the bordering retothelial cells, but their interlacing cytoplasmic extensions enwrap reticulum fibers. The cellular lattice is much more prominent in the medullary sinuses than in the marginal sinuses. The ultrastructure of each cell type has already been described. Morphologic variations of stimulated retothelial cells will be discussed in the chapter dealing with sinus catarrh.

4. The Blood Vessels

The blood vessels of lymph nodes are ultrastructurally similar to those in other organs. The capillary endothelium occasionally contains a few fenestrations and rests upon a distinct basement membrane. Only the *postcapillary venules* exhibit characteristic features. Their inner lining is crowded with high endothelial cells that are tightly bound together by close interdigitations. Chromatin

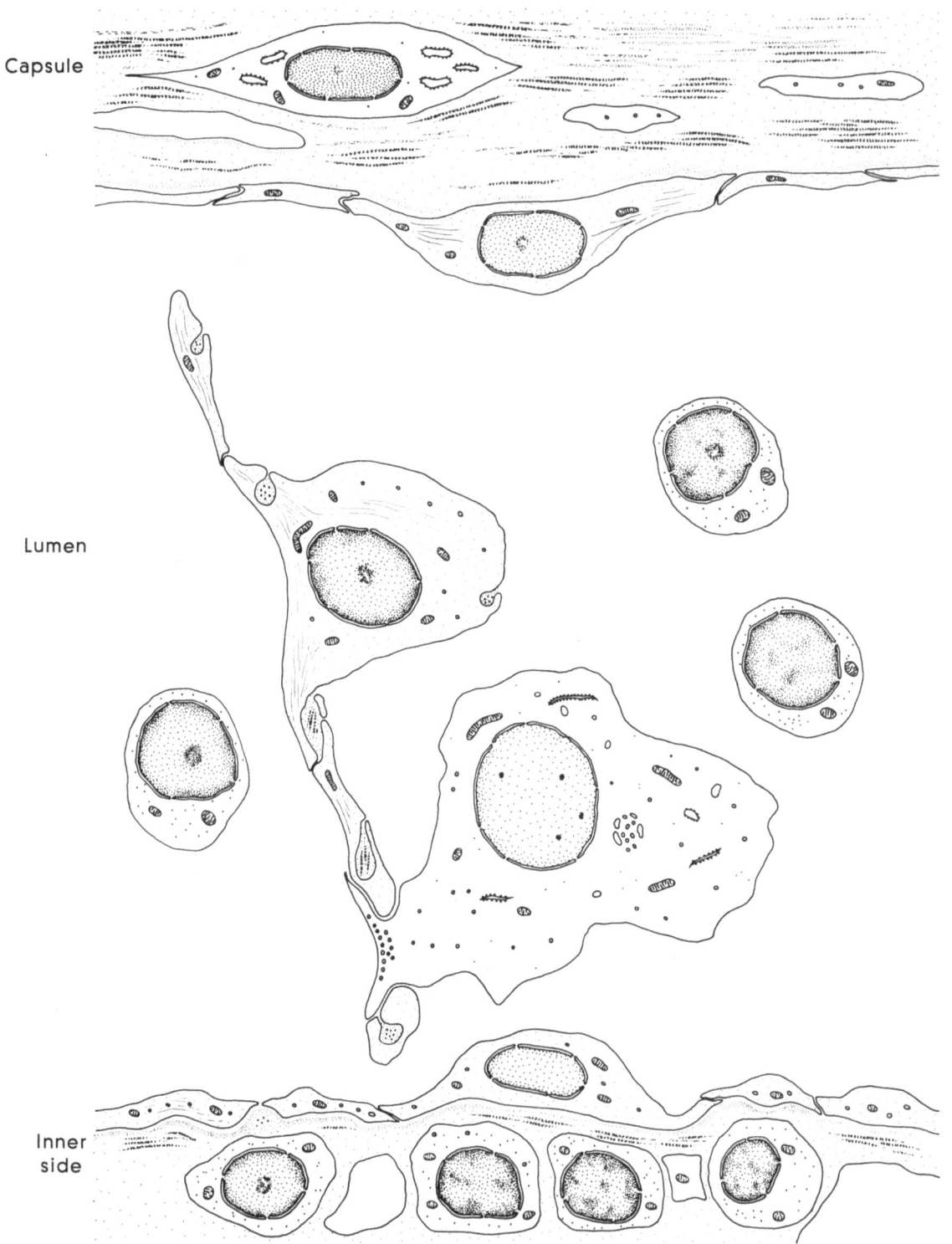

Capsule

Lumen

Inner
side

Fig. VIII. Schematic drawing of marginal sinus with littoral and traversing cells.

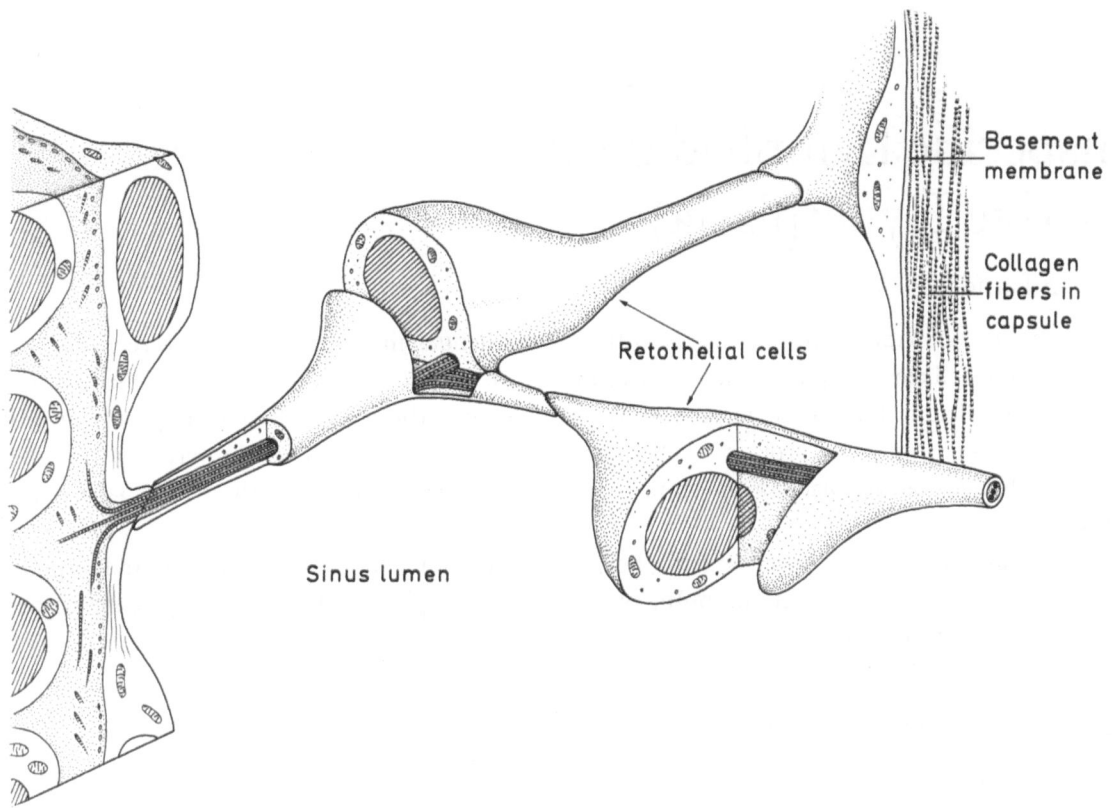

Fig. IX. Three-dimensional diagram of a marginal sinus with particular emphasis on traversing reticulum fibers.

is uniformly dispersed throughout the oval nucleus. The cytoplasm is typically filled with a conspicuous number of phagolysosomes. These variously sized phagolysosomes are of diverse composition. In some of them multivesicular bodies, myelin figures or electron-opaque granules are seen. Mitochondria are usually somewhat elongated, exhibiting distinct shelf-like ridges or cristae. The Golgi apparatus is prominently displayed and contains numerous minute vesicles. Fine tubular smooth endoplasmic reticulum is scattered throughout the cytoplasm and pinocytotic vesicles gather especially at the basal area of the cell. A medium amount of somewhat distended rough endoplasmic reticulum is also present in the cytoplasm. Frequently lymphocytes can be seen passing through the cytoplasm of the endothelium[1].

1. Gowans and Knight (1964).

Nonspecific Lymphadenitis (Reactive Hyperplasia)

Figs. X—XI and 32—45

The lymph node is capable of responding to different challenges in several ways. The reaction may take place within follicles, pulp and sinuses. In follicles we find production and enlargement of germinal centers (follicular lymphatic hyperplasia). In the pulp there is a diverse hyperplasia with numerous basophilic stem cells, a plasma cell hyperplasia and a reticulum cell hyperplasia (reticulocytosis). The changes taking place in the sinuses are categorized as sinus catarrh and immature sinus histiocytosis.

Five reaction types will be discussed in the present chapter. They include follicular lymphatic hyperplasia, diverse hyperplasia of the pulp, plasma cell hyperplasia (plasmacytosis), sinus catarrh and immature sinus histiocytosis.

1. Follicular Lymphatic Hyperplasia

Follicular lymphatic hyperplasia is distinguished by a more or less pronounced formation and enlargement of germinal centers.

In the germinal center, germinocytes, germinoblasts and basophilic stem cells are found. Occasionally, small groups of plasma cells and their precursors are also encountered, primarily in the intermediate zone and at the periphery of the germinal center. Commonly, in the hyperplastic germinal center; large macrophages with tingible bodies, the "starry sky cells", are of a high order of frequency. Their abundant cytoplasm contains much phagocytized material, residual bodies and lysosomes. The cytoplasm of nonphagocytic reticulum cells is increased in amount and yields more lysosomes than usual. Interspersed between the germinal center cells are occasional amorphous electron-dense materials equivalent to protein deposits as they appear by light microscopy.

2. Diverse Hyperplasia of the Pulp

In nonspecific inflammation of the lymph nodes, advanced hyperplasia of the cortical and medullary pulp is common. Most of the hyperplastic pulp is essentially accounted for by lymphocytes varying both in size and basophilia, in addition to basophilic stem cells, plasma cells, proplasmacytes, plasmablasts and phagocytic reticulum cells. In the cortical pulp (paracortical zone) basophilic stem cells often predominate. In other cases one may primarily find plasma cells and plasma cell precursors. The cytoplasm of reticulum cells is augmented and phagosomes and lysosomes appear in greater numbers than in nonstimulated lymph nodes. Macrophages with tingible bodies are not as frequent in the pulp as in germinal centers. The reticulum cells tend to be encircled by proliferating plasma cells, suggesting a functional interrelationship between antigen capture and antibody formation[1].

1. BESSIS and THIÉRY (1961).

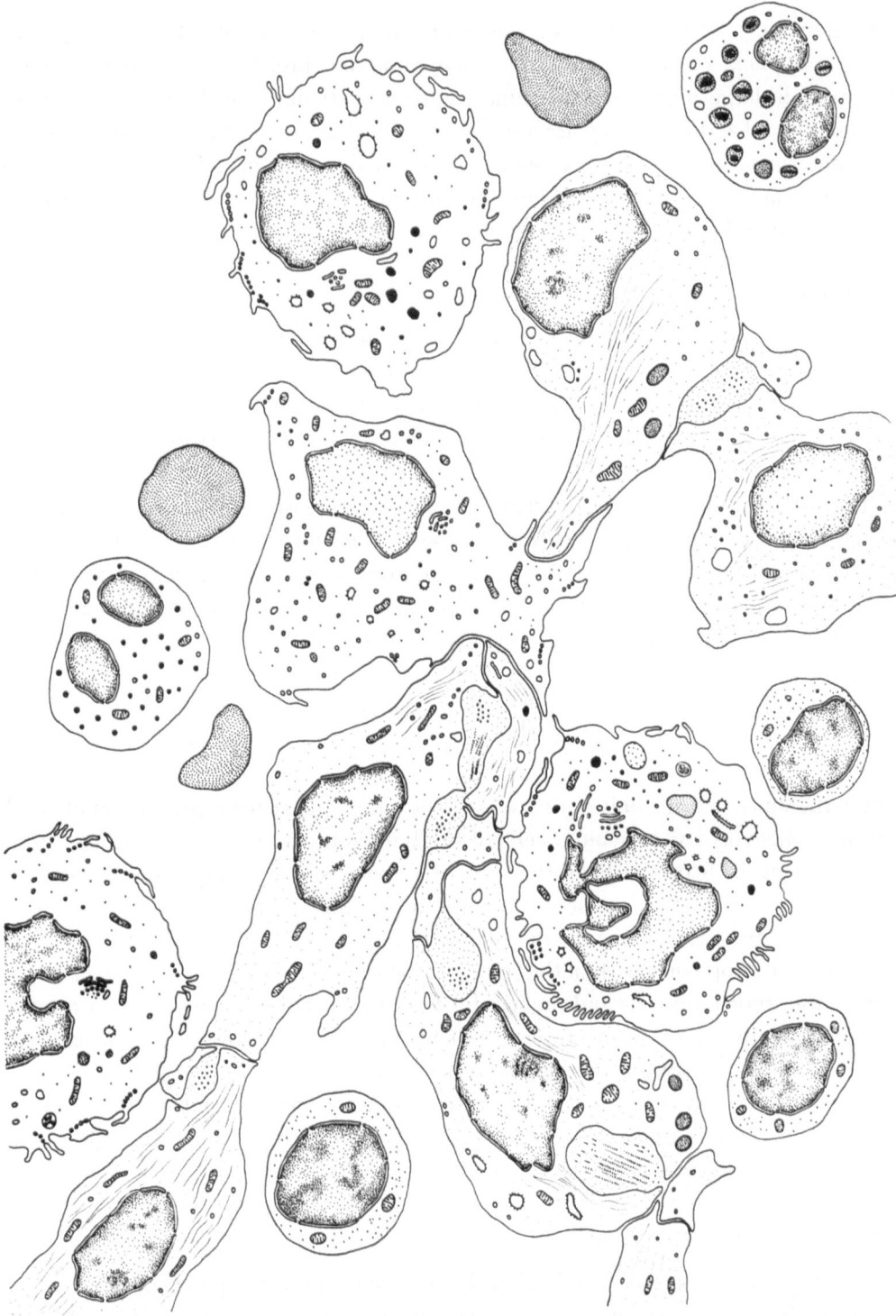

Fig. X. Sinus catarrh. This diagram depicts retothelial cells traversing the sinus lumen. Macrophages, granulocytes, lymphocytes and erythrocytes are also seen.

3. Plasma Cell Hyperplasia (Plasmacytosis)

Plasma cells may be found anywhere in the lymph node. Most frequently they are increased in the medullary pulp, next in the cortical pulp, where numerous precursors are usually added. Finally, they may appear in large numbers in germinal centers, particularly in hyperimmune reactions, for instance in rheumatoid arthritis. They are always recognized with ease by their ergastoplasm. Russell bodies make one suspicious of secretory retention in (? older) plasma cells. The plasma cells finally disintegrate. They are diminished in size and become intensely "stainable" due to pyknotic nuclei. At this stage they may be ingested by macrophages either of the germinal centers or sinuses.

4. Sinus Catarrh

According to our definition, acute sinus catarrh is distinguished mainly by distension of the sinus in combination with swelling of its retothelial cells and infiltration of granulocytes.

In contrast, a striking increase in the number of retothelial cells and macrophages in the sinus characterizes chronic sinus catarrh, which is equivalent to the sinus histiocytosis of the American literature.

In sinus catarrh, retothelial cells are furnished with an increased amount of cytoplasm and many more intracytoplasmic organelles than are normally seen. The cytoplasm is crowded with a considerable amount of rough and smooth endoplasmic reticulum, and the cytoplasmic membrane forms delicate villous folds. Lysosomes and residual bodies with a variety of components are augmented. These findings are more distinct in the intraluminal retothelial cells, especially those of the medullary sinuses; they are presumed to result from the position of the cells which exposes them to more extensive and frequent stimulation by the lymph stream. Also, the pronounced reactivity of the bordering retothelial cells of the medullary sinuses is in keeping with their location (see Chapter II/3 on lymph sinuses).

The cells in the sinuses are not only activated but they also appear to increase in numbers. They may originate from monocytes and reticulum cells of the lymphoid parenchyma. Both cell types may be interspersed between the littoral cells of the sinus. This suggests an immigration, at least of the reticulum cells, into the sinus. The monocytes, on the other hand, may be contained in the inflowing lymph and emigrate from the sinus into the lymphoid parenchyma.

Large rounded macrophages with active phagocytosis and numerous intracytoplasmic organelles arise not only from reticulum cells and monocytes but also probably from detached retothelial cells. The cytoplasmic membrane of the macrophages shows many fine extensions. These usually contain a great number of residual bodies of variable ultrastructure.

The two cell types appearing in sinus catarrh, that is, enlarged retothelial cells and macrophages, can be differentiated only with great difficulty by the light microscope, but distinction is easier with the electron microscope.

5. Immature Sinus Histiocytosis

In this condition retothelial cells are only slightly enlarged. The usual pinocytotic vesicles and lysosomes and intracytoplasmic filaments are seen. Occasional macrophages may be identified in the sinus lumen. The amount of vesicular smooth and rough endoplasmic reticulum appears slightly increased. This condition is characterized by an abundance of immature histiocytes[1]. They are of ovoid or some-

---

1. LENNERT (1961).

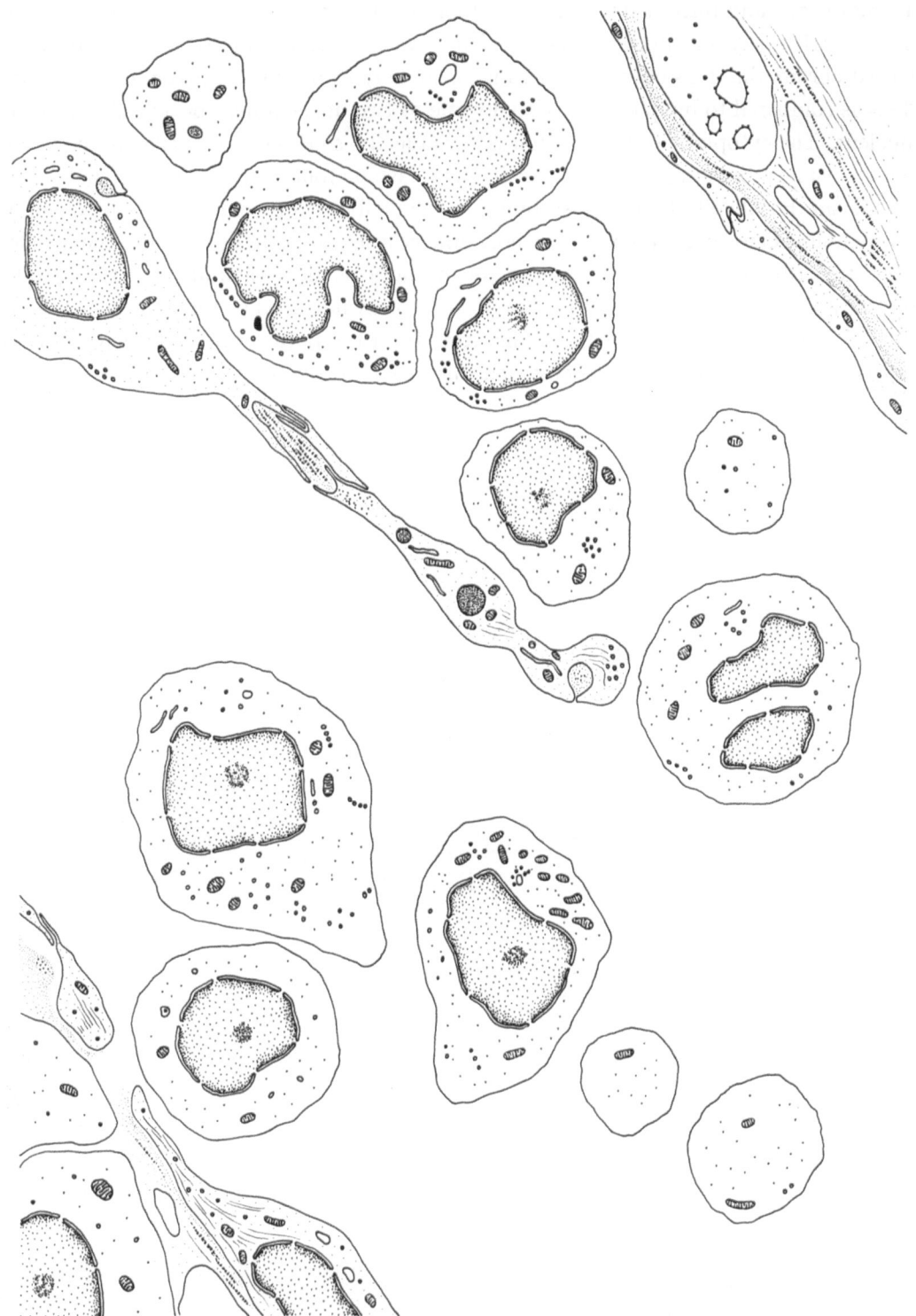

Fig. XI. Diagram of immature sinus histiocytosis. Sinus lumen containing immature histiocytes.

what irregular shape and furnished with more cytoplasm than lymphocytes. Their nucleus is ovoid or often irregular with deep indentations. The nucleoplasm is spread diffusely with slight margination of condensed chromatin. This chromatin arrangement is reflected by a greater electron lucidity of the nucleus as compared to lymphocytes. Usually one or two nucleoli may be visualized in the center of the nucleus. The cytoplasmic composition is variable. In some cells vesicular profiles resembling smooth endoplasmic reticulum, often arranged in linear fashion (pinocytotic vesicles), make up the bulk of the cytoplasm. Focal collections of lysosomes, occasionally with myelin figures, are common. Dark granules resembling dense core vesicles (catecholamine vesicles) are usually evident. Some cells possess quite a few free ribosomes. A few strands of rough endoplasmic reticulum are often present. The mitochondria are generally ovoid or rod-shaped and somewhat smaller than those of lymphocytes. This kind of histiocyte often appears to be migrating from the pulp through the retothelial cell wall of sinuses. However, the type of cell from which they originate is still a matter of dispute.

Specific Lymphadenitis

Figs. XII and 46—65.

This term is applied to reactive alterations of the lymph node producing characteristic histological features from which we can draw certain limited conclusions concerning etiology. Most of the specific inflammations of lymph nodes are distinguished by the presence of epithelioid cells. They may appear in large or small clusters. The large clusters (tubercles) are prone to scarring and/or hyalinisation. There may also be caseous necrosis. Areas with small epithelioid cell clusters do not tend to produce fibers or undergo necrosis. Tubercles may appear in sarcoidosis as well as in tuberculosis. Small foci of epithelioid cells are particularly evident in Piringer's lymphadenitis, which, in most cases, is due to toxoplasmosis of the lymph node. In addition, focal collections of epithelioid cells may occur in infectious mononucleosis, in syphilis (stage I or II) and in Hodgkin's disease.

The following text deals only with tuberculosis, sarcoidosis, and toxoplasmosis, which are the most important and most frequent specific inflammatory lesions of lymph nodes in Europe.

1. Tuberculosis

In tuberculosis of lymph nodes, the tubercles are morphologically similar to those in other organs. They consist of epithelioid cells and Langhans' giant cells surrounded by a peripherally increasing amount of fibroblasts. These produce collagen fibers. Hyalin deposits are frequently found at the periphery. An extraordinary accumulation of lymphocytes is not demonstrable in these lymph nodes. One can often see a small zone of necrobiosis with degenerating epithelioid cells and with protein/collagen deposits in what appear to be the central areas of the tubercles.

The *epithelioid cell* ranges up to 20 μ or more in diameter. Its cytoplasmic membrane displays irregular projections. At points of contact between neighboring cells the invaginations of one cell often interdigitate closely with evaginations of another. The nucleus is situated eccentrically and measures about 10 μ in diameter. For the most part the nuclear membrane exhibits irregular indentations and the nuclear chromatin is uniformly dispersed. A few rather inconspicuous nucleoli may be visualized. Numerous mitochondria and a large amount of smooth endoplasmic reticulum fill a large portion of the cytoplasm. Ovoid or elongated rod-shaped mitochondria tend to line up at the periphery of the cytoplasm. They display denser matrices than the cytoplasm and contain mitochondrial cristae in orderly array. Many endoplasmic vacuoles may be encountered in the center of the cells. In typical epithelioid cells they comprise electron-lucent materials. But in some cells, presumed to be somewhat younger epithelioid cells, electron-opaque material is enclosed within the vacuoles. A small amount of rough endoplasmic reticulum tends to concentrate at the periphery of the cell. Several Golgi apparatuses occupy central positions. They consist of lamellar, vacuolar and vesicular components. YAMORI[1] has divided epithelioid cells into two groups: typical epithelioid cells and "pre-epithelioid cells". The latter do not yet reveal distinct interdigitations, they have a paucity of rough

---

1. 1969.

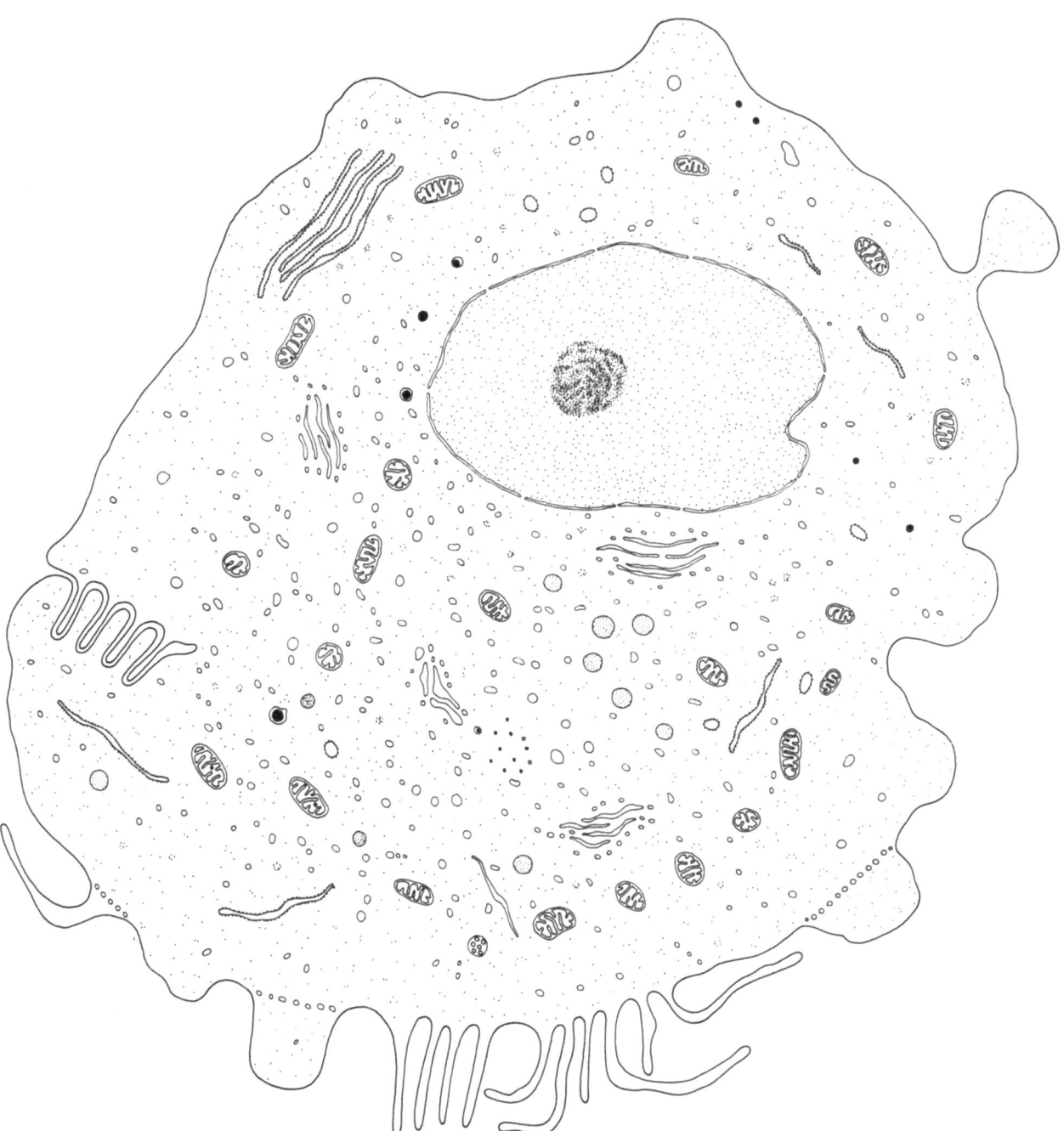

Fig. XII. Epithelioid cell.

endoplasmic reticulum and ribosomes are greatly reduced in number. Large phagosomes are lacking in contrast to macrophages which contain these structures. Our studies show that such preepithelioid cells appear in small numbers in tuberculosis, sarcoidosis and HODGKIN's disease. In toxoplasmosis, however, they outnumber the (mature) epithelioid cells which are more common in the three conditions just mentioned.

Table 4. *Electron microscopic criteria of macrophages and "mature" epithelioid cells*

	Macrophages	Epithelioid cells
Size	10—15 μ	very large, 15 μ
Cellular membrane	prominent extensions	fine extensions
Smooth endoplasmic reticulum	intermediate in amount	prominent
Rough endoplasmic reticulum	scanty	occasionally increased
Golgi apparatus	single	multiple
Free ribosomes	scanty	almost lacking
Phagosomes	large, common (diagnostic criterion)	small, rare

Electron microscopic studies of tuberculosis in the lymph nodes of rabbits were reported by MAT-SUURA[1]. He provided information about the ultrastructure of epithelioid cells, furnishing evidence for their reticulum cell origin. Although the ultrastructure of the epithelioid cells has also been described by other authors[2], the origin of the cell is still a matter of dispute. Some authors support their monocytic origin[3], but others suggest a reticulum cell origin. We were unable to elicit the true nature and origin of the epithelioid cells from our human biopsy material.

The ultrastructure of the *Langhans' giant cells* closely resembles that of the epithelioid cells except for the multiple nuclei. Several Golgi apparatuses occupy central positions. Some cells are also dotted by numerous electron-dense granules. Interdigitating cellular membranes can be made out between irregular cytoplasmic processes. Schaumann bodies show lamellar calcium deposits. Their electron density is variable, but outer layers are generally much denser than inner layers (Fig. 53).

2. Sarcoidosis

Tubercles are also found in sarcoidosis. In this condition epithelioid cells conform ultrastructurally to the same pattern as those in tuberculosis. Most of the cells seem to have a smaller amount of electron-transparent vesicular endoplasmic reticulum and a greater number of electron-dense granules. These differences may signify an earlier developmental stage of the epithelioid cells.

3. Toxoplasmosis (Piringer's Lymphadenitis)[4]

In toxoplasmosis epithelioid cells are arranged in small groups ("kleinherdige Epitheloidzellreaktion") without tubercle formation. Prominent follicular lymphatic hyperplasia, immature sinus histiocytosis[5] and a diverse hyperplasia of the pulp ("bunte Pulpahyperplasie") are also noted. The epithelioid cells in toxoplasmosis possess a large amount of vesicular electron-lucent smooth endoplasmic reticulum and long rod-shaped mitochondria. Large vacuolized elements are

---

1. 1958.
2. GUSEK (1962); YAMORI (1964).
3. LEDER (1967a).
4. PIRINGER-KUCHINKA (1953); LENNERT (1961).
5. LENNERT (1959, 1961).

common in the central region of the cytoplasm. Large phagosomes with cell debris are often encountered. In the germinal centers huge phagocytic reticulum cells, so-called "starry sky cells", are frequent. They contain numerous phagosomes and residual bodies of variable size and composition. Some of them yield myelin figures, others contain finely granular material of variable electron density, and still others are furnished with coarse amorphous electron-opaque material. Toxoplasma gondii[1], proliferating type, may be difficult to identify unless a longitudinal cut is shown in the section, but the organism can sometimes be recognized reliably by its bizarre large mitochondria, small round nuclei and double-layered cytoplasmic membrane.

Many activated and nonactivated lymphocytes, basophilic stem cells and reticulum cells accumulate in the hyperplastic pulp. Large nucleoli are often prominently displayed in reticulum cells, and some of them are provided with phagosomes containing abundant cytoplasm. These are considered to be an intermediate type between reticulum cells and epithelioid cells. Plasma cell proliferation takes place predominantly in the hyperplastic pulp. One may find quite a few plasmablasts, proplasmacytes and plasma cells.

The details of immature sinus histiocytosis have already been discussed in the chapter on non-specific inflammations.

1. LUDVIK (1956, 1960); BRAUNSTEINER, PAKESCH and THALHAMMER (1957); GERNHAM, BAKER and BIRD (1962); OLISA (1963); KIKKAWA and GUEFT (1964); WILDFÜHR (1964); HULDT (1966); OGINO and YONEDA (1966).

Hodgkin's Disease[1]

Figs. 66—77.

At the present time Hodgkin's disease is subdivided into several histologic types depending on the prevailing cells and the type of sclerosis. The histologic appearance may be correlated with a characteristic clinical behavior, proving the justification of this subgrouping. Of the four types defined in Rye (1966), we present one illustration each of the so-called mixed type and nodular sclerosing type of LUKES. One case of epithelioid cellular lymphogranulomatosis[2] is added. This type, to our knowledge, is not identical with the L&H type of LUKES and his collaborators[3]. On the contrary, it represents a previously undescribed, rather rapidly progressive, malignant systemic disease, whose main feature is a diffuse increase of epithelioid cells arranged in multiple small groups.

The decisive diagnostic criterion in all cases of Hodgkin's disease is the demonstration of Sternberg-Reed giant cells. However, certain light microscopic differences have been documented in the nodular sclerosing type[3] and in the epithelioid cell type[2]. These light microscopic differences have been confirmed by electron microscopy. Nevertheless, the recognition of typical Sternberg-Reed cells still remains the most important aid for diagnostic purposes. Therefore, we shall first turn our attention to this cell.

The typical *Sternberg-Reed cell* is multinucleated, although there may be only one nucleus. Such cells are called Hodgkin cells. The two types are mainly differentiated by the number of nuclei and the corresponding amount of cytoplasm. However, most nuclei of the multinucleated cells are more bizarre and irregular in shape.

The Sternberg-Reed cell is larger than 15 μ in diameter, revealing irregular cytoplasmic outlines, occasionally with prominent processes. The nucleus is large and bizarre. The nucleoplasm is less electron-opaque than in the other cells of the lymph node. Sometimes the chromatin is marginated near the nuclear membrane. Two or three very large nucleoli are seen in the nucleus, occasionally at its periphery. Round, ovoid or long rod-shaped mitochondria, 0.3—0.5 μ in largest diameter, are common. Finely vesicular smooth endoplasmic reticulum is disseminated in the cytoplasm, which is meagerly supplied with rough endoplasmic reticulum. Some cells contain several electron-dense lysosomes. The ribosome and polysome content is variable. In some cells there are only uniformly disseminated free ribosomes, while in others a prominence of polysomes, occasionally in focal collections, may be demonstrated. This finding is concordant with the varying degree of basophilia of the cytoplasm as seen by light microscopy. Sometimes an intermediate form between an enlarged reticulum cell and a Sternberg-Reed cell is noted. A few cells have fine intracytoplasmic filaments in close association with the nucleus. Similar filaments can also be recognized in reticulum cells, monocytes and leukemic cells. To date, there are no data concerning the nature and the function of the filaments.

1. ANDRÉ, DREYFUS and BESSIS (1955); FRAJOLA, GREIDER and BOURONCLE (1958); OGAWA (1962); BERNHARD and LEPLUS (1964); FORTEZA BOVER, BAGUENA CANDELA, FORTEZA VILA, BARBERA GUILLEM and BAGUENA CANDELA (1968); FORTEZA BOVER, PRIETO GARCIA, BARBERA GUILLEM, BAGUENA CANDELA and FORTEZA VILA (1968).

2. LENNERT and MESTDAGH (1968).

3. LUKES, BUTLER and HICKS (1966).

1. Mixed Type

In a case of the "mixed type" (Case 24) (our "classical lymphogranulomatosis") we found a pronounced number of Sternberg-Reed cells with extremely large nucleoli within bizarre giant nuclei. The cytoplasm was equipped with numerous free ribosomes and a small amount of vesicular endoplasmic reticulum. The other cells (lymphocytes, eosinophils etc.) retained the usual appearance. In other cases there were fewer Sternberg-Reed cells, otherwise no major difference was noted by light and electron microscopy.

2. Nodular Sclerosing Type (Lukes)

In the nodular sclerosing type (Case 10) typical Sternberg-Reed giant cells can rarely be found, but atypical reticulum cells with comparatively large nucleoli may be demonstrated. Their cytoplasm is of more or less intermediate character between that of the Sternberg-Reed giant cell and the epithelioid cell. These cells have a tremendous number of small vesicular elements in the cytoplasm.

3. Epithelioid Cellular Lymphogranulomatosis

The case of epithelioid cellular lymphogranulomatosis depicted in this atlas (Case 26) does not show sufficiently distinctive histological features to be regarded as absolutely representative of this subdivision of HODGKIN's disease. Therefore, the following description is presented with some reservation. There is a preponderance of epithelioid cells. No typical Sternberg-Reed cells are found, but small numbers of atypical reticulum cells with comparatively large nucleoli are seen. The epithelioid cells appear similar to those found in tuberculosis or sarcoidosis. They have abundant vesicular and vacuolar endoplasmic reticulum containing amorphous material of variable electron density. The Golgi apparatus is divided into several groups. Characteristically, in this case the number of plasma cells and their precursors is increased. They frequently contain nuclear "inclusions". Some of these inclusions are nuclear pockets with cytoplasmic components as seen in follicular lymphoma (germinoblastoma), others represent proteinaceous deposits as seen in Waldenström's disease or in multiple myeloma.

Another case of HODGKIN's disease (Case 5), which was studied after initial formalin fixation, revealed significant numbers of epithelioid cells that failed to take on the characteristic pattern of small epithelioid cell groups. Here, too, they presented with the classical structure seen in tuberculosis and sarcoidosis, but they did not, according to our definition, meet the criteria of macrophages or phagocytic reticulum cells.

Malignant Tumors of the Lymphoreticular Tissue

Figs. 78—115.

Although numerous excellent studies of the morphology of malignant lymphomas have been undertaken (the last one by RAPPAPORT[1]), and though the atlas of BERNHARD and LEPLUS[2] reports many cases that have been examined electron microscopically, we are far from having solved the problems of classification. This is particularly true if one tries to draw a sharp line between lymphosarcoma and reticulosarcoma. On the other hand, our light microscopic observations support the view that the malignant follicular lymphoma is a genuine tumor type, arising from germinal center cells. Hence we have coined the term "germinoblastoma". Of the lymphosarcomas we have merely investigated the lymphoblastic type, consisting of large cells. The electron microscopic studies conducted so far raise the question whether the so-called reticulum cell sarcoma in reality represents several tumor types. The position of Burkitt's tumor is still open to question, since the origin of the tumor cells is still unknown, and the same can be said of Waldenström's macroglobulinemia.

1. Germinoblastoma (Follicular or Nodular Lymphoma, M. Brill-Symmers)

Three of our five cases are considered in this chapter. Their ultrastructure proved to be identical in spite of variations in cellular size. The cells of the follicles vary in size from one case to another, but they are virtually uniform in a single case. The large nucleus is extremely irregular, exhibiting deep indentations. The nucleoplasm is dispersed diffusely, has few chromatin clumps and is of medium electron density. A nucleolus may be observed in some cells, but it is not prominent. In the cytoplasm a significant amount of finely vesicular endoplasmic reticulum is noted in addition to many long rod-shaped mitochondria. Occasionally, there may be some ergastoplasm. Some cells contain lysosomes. The cytoplasmic membrane of the cell possesses irregular processes, and many small pinocytotic vesicles are included. Bizarre nuclear pockets, including cytoplasmic components, are common. They always project into the cytoplasm, having arisen from the nuclear margin, and are surrounded by both a double layered nuclear membrane and a small complement of nucleoplasm. They are equipped with various cytoplasmic elements such as vesicles, granules, lamellar membrane, vacuoles and sometimes even mitochondria.

The nuclear pockets cannot be regarded as the characteristic feature of the cell in this disease, because they may be found in several other conditions as well as in other cells[3]. It is noteworthy, however, that the nuclear pockets are extremely frequent in Brill-Symmers' disease.

The fact that all three cases examined present the same electron microscopic picture and reveal numerous nuclear pockets similar to those seen in normal germinocytes and germinoblasts lends

---

1. 1966.
2. 1964.
3. LEDUC and WILSON (1959); CLARK (1960); BERNHARD and GRANBOULAN (1963); BRANDES, SCHOFIELD and ANTON (1965); FRASCA, AUERBACH, PARK and STOECKENIUS (1965); BERNHARD (1966); BUCCIARELLI (1966); BLOOM (1967); MOLLO and STRAMIGNONI (1967).

some support to our concept that the follicular lymphoma is a specific tumor of the germinal center. The presence of long-branching desmosome-connected reticulum cells[1] in all three cases of germino-blastoma may be considered even stronger evidence that this interpretation regarding the genesis of this neoplasm is justified. The long-branching reticulum cells appearing in this tumor could be demonstrated by both light and electron microscopy.

2. Lymphosarcoma

The so-called lymphoblastic lymphosarcoma is found mostly in children. The following description is based on a study of one pediatric case (Case 53). The tumor cells measure about 6—7 μ in diameter and reveal a high nuclear-cytoplasmic ratio. The cellular surface is smooth. Pseudopodal processes are usually absent. The nucleus is ovoid or somewhat irregular in shape, but occasionally has deep indentations of the nuclear membrane. A few inconspicuous nucleoli are usually observed, generally in apposition to the nuclear membrane. In the cytoplasm, mitochondria gather at one side of the nucleus. They are round or rod-shaped and less than 0.5 μ in largest diameter. There is a paucity of smooth endoplasmic reticulum, and rough endoplasmic reticulum is barely visible in the cytoplasm. The Golgi apparatus is fairly prominent and the centrioles are distinct. Free ribosomes and poly-ribosomes are disseminated in the cytoplasm. Occasionally, there is a meager supply of lysosomes. A few fat droplets or glycogen granules are sometimes discerned.

This cell definitely corresponds to the undifferentiated leukemic cell often seen in children which contains PAS-positive cytoplasmic granules ("paraleukoblastic" or "lymphoblastic" leukemia).

3. Reticulum Cell Sarcoma

On the basis of light microscopy several classifications of the so-called reticulum cell sarcoma have been proposed by various authors, but it is difficult to clearly identify the various subtypes with the electron microscope. On the basis of our studies we separate these into three types:

Type I: Virtually without endoplasmic reticulum

One case (Case 19) was placed in this category. The cells are completely undifferentiated. They are of irregular polygonal shape. The nuclear-cytoplasmic ratio is high and the nucleus has an irregular margin with deep indentations. The nucleoplasm is almost uniformly dispersed and contains fine chromatin clumps. One or two fairly large nucleoli are demonstrated, and the nucleoplasm appears somewhat more pale than in lymphocytes. Fine pseudopodal processes protrude from the cellular membrane. Tremendous numbers of free ribosomes and polyribosomes are disseminated in the cytoplasm. Mitochondria are round or rod-shaped. Vesicular smooth endoplasmic reticulum is scanty but some rough endoplasmic reticulum may be found. The Golgi apparatus is moderately prominent and is composed mainly of vesicular structures. Lysosomes are identified in the cytoplasm.

---

1. Lennert (1969); Lennert and Niedorf (1969).

Type II: With smooth endoplasmic reticulum

We have studied two cases (Cases 2, 64) of this type. This tumor is made up of two cell types: cells with much vesicular or vacuolar smooth endoplasmic reticulum and completely undifferentiated cells. The moderately differentiated cells are provided with an electron-dense nucleus which is denser than that of the first type. The nucleoplasm is finer than that of normal reticulum cells and shows indistinct chromatin clumps. A small nucleolus is usually visible. Some finely vesicular and moderately dilated vacuolar smooth endoplasmic reticulum is observed in the cytoplasm. The mitochondria are round, ovoid or rod-shaped. A few lysosomes with a distinct limiting membrane can sometimes be seen. In general, the Golgi apparatus is small and comprises vesicular, vacuolar and lamellar membranous structures. The observation of disseminated free ribosomes and polyribosomes is common.

Type III: With rough endoplasmic reticulum

This is the commonest type in our material which includes two definite and four questionable cases. The latter could not be placed into this category with reasonable confidence because of improper fixation.

The cells are generally larger than those of the other types. Their nucleus is large and round or slightly indented. Usually it contains a large nucleolus and small chromatin clumps. In the sparse cytoplasm long ergastoplasmic tubules are often demonstrated. Mitochondria are generally round and large. Numerous polyribosomes are disseminated in the cytoplasm. Little vesicular smooth endoplasmic reticulum may be present.

In all three types some differentiated reticulum cells are also intermingled with the tumor cells. However, it is not clear whether the differentiated reticulum cells are tumor cells or not.

Our studies thus far do not permit a final decision as to whether the three tumor types mentioned — all or in part — deserve to be classified as reticulum cell sarcomas. In particular, the cells of the third type sometimes resemble plasmablasts. Perhaps this type should really be placed with stem cell lymphomas, as has already been proposed by GALL and MALLORY[1]. Furthermore, the interpretation of type I is open to question due to the lack of specific features of its cells (? basophilic stem cells, ? undifferentiated reticulum cells).

4. Burkitt's Tumor

Electron microscopic observations of Burkitt's tumor have been reported by several authors[2]. BERNHARD and LAMBERT[3] subdivided this tumor into three types on the basis of ultrastructure. Most of their cases belong to the undifferentiated lymphoid cell sarcoma. Their studies are based on the African cases. In contrast, we investigated a European case, which showed all the characteristics of BURKITT's tumor by light microscopy of histologic sections and imprints. Cytochemical investigations (Sudan red etc.) were also done on this case.

1. 1942.
2. EPSTEIN and HERDSON (196?); ⌐ ARD and LAMBERT (1964); EPSTEIN, BARR and ACHONG (1965); ACHONG and EPSTEIN (1966); DORI); OETTGEN (1967); SEMAN, ROSENFELD, JASMIN and CAMAIN (1968); BERARD, O'CONOR, THOMAS and)69).
3. 1964.

Most of the tumor cells measure about 8 μ in diameter and show an almost smooth cell surface. The nucleus is large and of variable shape. Some nuclei are round or ovoid, but others show pronounced indentations. Usually coarse chromatin clumps occupy positions at the nuclear margins. However, we were unable to demonstrate nuclear pockets as have been described by ACHONG and EPSTEIN[1]. One or two nucleoli are commonly found. In the cytoplasm large mitochondria, about 0.5 μ in diameter, are concentrated mainly at one pole of the cell. They are usually round or ovoid but may be somewhat elongated and rod-shaped. The Golgi apparatus is inconspicuous and consists of finely vesicular components. Scanty vesicular smooth endoplasmic reticulum and long serpentine threads of rough endoplasmic reticulum are often evident in the cytoplasm. In some cells a few electron-dense, irregularly shaped, lipid vacuoles are noted. Some cells show a few electron-dense lysosome-like granules.

Scattered among these tumor cells are a few irregularly shaped cells of somewhat different appearance. They have long cytoplasmic processes extending between the tumor cells mentioned above. Their nucleus shows dense electron opacity and deep indentations of the nuclear membrane. Sometimes there are also large nucleoli. The cytoplasm is filled with strikingly deformed large mitochondria and numerous polyribosomes. A few lipid vacuoles and lysosome-like electron dense granules may also be encountered in these cells. Between the tumor cells first described and the darker cells numerous intermediate forms can be found. The latter cells give the impression of being degenerating tumor cells.

Numerous phagocytic reticulum cells containing diverse kinds of ingested cellular debris ("starry sky cells") are interspersed amidst the tumor cells. Occasional plasma cells are seen in addition. In our case, there was no evidence of a virus or mycoplasma in the tumor, though many authors have reported their presence in tumor cell cultures as well as in biopsy material.

5. Macroglobulinemia of Waldenström

Electron microscopic studies of Waldenström's macroglobulinemia have already been reported by several authors[2]. The most comprehensive work on this subject was performed by BESSIS [1961][3]. In general, he found a cellular ultrastructure resembling that of lymphocytes, while some of his cases also displayed ergastoplasm, as is found in plasma cells.

We have studied two cases, one after only osmium fixation. This case (Case 3) was composed of pleomorphic small cells and large cells with abundant cytoplasm. The small cells exhibit a high nuclear-cytoplasmic ratio. Their nucleus is ovoid, often with deep indentations. The nucleoplasm, fixed with osmium tetroxide, consists of small chromatin clumps and shows a lesser degree of electron density than does the nucleoplasm of lymphocytes. Characteristically, the outer nuclear membrane often extends outward. In the sparse cytoplasm several large, rod-shaped mitochondria, 0.5 μ in diameter, usually accumulate at one side of the nucleus. Lysosomes with a distinct limiting membrane and/or multivesicular bodies can often be demonstrated. An inconspicuous amount of vesicular, sometimes dilated, vacuolar endoplasmic reticulum is noted. This is usually studded with relatively

---

1. 1966.
2. BESSIS (1961); BRITTIN, TANAKA and BRECHER (1963); KUHN (1967); MALDONALDO, BROWN, BAYRD and PEASE (1966a, b).
3. 1961.

few ribosomes. Often the Golgi apparatus is more prominent than in lymphocytes. Some free ribosomes and polysomes are disseminated uniformly throughout the cytoplasm.

The small cells outnumber the large cells which present characteristic features. Their nucleus often proves to be irregular and stains less intensely than that of the small cells. The large cells are also equipped with many dilated cisternae with ribosomes containing moderately electron-dense amorphous substances. These most likely represent macroglobulin. The Golgi apparatus of the cell is quite prominent. It is composed of many vesicles with electron-dense contents and dilated cisternae. Free ribosomes are commonly seen and some lysosomes are also encountered.

In both cell types intranuclear inclusions enclosing moderately dense materials can often be found. Identical inclusions have already been reported by several authors in Waldenström's disease as well as in multiple myeloma cells. Kuhn[1] pointed out that the inclusions might originate from nuclear bodies. We found that the inclusion is mainly visualized in close proximity to perinuclear cisternae, suggesting an origin from the protein secreted at the outer nuclear membrane. Protein deposits are sometimes observed in intercellular spaces. These deposits usually show a somewhat greater degree of electron density than those in the cytoplasm or in intranuclear inclusions. However, according to other reports, intracytoplasmic and intranuclear protein deposits vary in electron density from one case to another and also in different cells of a single case.

The nuclear inclusions are also seen in large numbers in semithin sections examined with the light microscope and in autopsy material. The inclusions impart a greenish tinge in Goldner stains, whereas in plasmacytomas they appear red[2].

Our second case — which was examined after formalin fixation — presented a lymphocyte-like picture and failed to show much ergastoplasm. A more detailed evaluation has been omitted because the specimen was technically inadequate.

1. 1967.
2. Apitz (1940).

Leukemias

Figs. 116—140.

Lymph nodes are particularly involved in chronic lymphocytic leukemias and in poorly differentiated leukemias.

1. Chronic Lymphocytic Leukemia

Lymph nodes in chronic lymphocytic leukemias differ light microscopically from those in lymphosarcomas and the so-called lymphoblastic leukemias in that the majority of the cells present are typical small lymphocytes. In addition to these mature lymphocytes, there are always a few, sometimes many, lymphocyte precursors (lymphoblasts etc.). Occasionally, a distinct nodular pattern prevails. The focal cellular aggregates, however, are not to be mistaken for germinal centers.

The high degree of differentiation of lymphocytes in chronic lymphocytic leukemias is also evident electron microscopically. However, the morphology of each case is quite variable.

We have one case (Case 13), in which the patient simultanously suffered from a hemolytic anemia. The leukemic cells measure about 5—7 μ in diameter and exhibit a smooth surface. They show a high nuclear-cytoplasmic ratio. The nucleus is large and round and possesses irregular chromatin clumps, especially near the nuclear membrane. The cytoplasm is rich in free ribosomes and polyribosomes and is only meagerly supplied with vesicular or finely vacuolar endoplasmic reticulum, usually of the rough variety. Mitochondria are generally round and about 0.5—0.6 μ in diameter. Their matrices are slightly darker than is the cytoplasmic background. The Golgi apparatus is moderately well developed and composed of vesicular, vacuolar and lamellar elements. The fine structure described above compares fairly well with that of lymphocytes.

In another case (Case 9) with a clinically aleukemic course the neoplastic cells bear a certain resemblance to cells seen in macroglobulinemia, but there are no clinical or laboratory data to substantiate that diagnosis. The lymphocytes reveal many dilated vacuolar cisternae with a few ribosomes attached. In particular, the perinuclear cisternae are often greatly distended and furnished with almost electron-transparent amorphous substances. Usually many more free ribosomes are suspended in these cells than in the leukemic cells mentioned above.

2. Poorly Differentiated Leukemias and Erythremias[1]

Until recently the terminology of the poorly differentiated leukemias included many synonyms. Moreover, these leukemias were classified quite arbitrarily. Recent application of new histochemical methods allows a clear cut separation and distinction of most "blast-leukemias"[2].

We subdivide the "acute" leukemias into the following four types:

a) Lymphoblastic ("paraleukoblastic") leukemia, usually occurring in children, which responds relatively well to therapeutic measures (steroids):

---

1. FORTEZA BOVER, FORTEZA VILA, BAGUENA CANDELA and PRIETO GARCIA (1967).
2. HAYHOE, QUAGLINO and DOLL (1964); LÖFFLER (1966, 1969); LEDER, to be published.

PAS (granular) positive, other reactions negative.

b) Paramyeloblastic (poorly differentiated granulocytic) leukemia, occurring in children and adults, which is poorly responsive to therapy:

Peroxidase +, naphthol AS-D chloroacetate esterase +, Sudan black +, PAS finely granular +, acid phosphatase (+) (generally minute positive granules).

c) (Acute) monocytic leukemia, occurring particularly in adults, which, thus far, is poorly responsive to therapy, but may yield good therapeutic results with methotrexate[1]:

Naphthol AS acetate esterase +, alpha-naphthol acetate esterase +, acid phosphatase +, other reactions mentioned under paragraph a) and b) negative except for mixed type (granulocytic-mono-cytic leukemia).

d) Undifferentiated leukemia, occurring at any age, which is poorly responsive to therapy:

All histochemical reactions negative.

The *acute erythremias* are separate from these poorly differentiated leukemias. Acute erythremias usually occur in children and are resistant to therapy. According to LEDER[2] they show a strongly positive paranuclear unipolar acid phosphatase reaction. By means of this enzyme-histochemical method they can be distinguished from poorly differentiated leukoses. In addition, the erythremic cells in these cases which are in the process of maturation show a positive diffuse or granular PAS reaction. This is diagnostic of erythremias[3].

We have studied one or more cases of each of these four types of poorly differentiated leukemias both with the electron microscope and by histochemical techniques. This enables us to compare the ultrastructure with the histochemical pattern. We have also observed one case that, in all likelihood, may be diagnosed as erythremia. This case is reported at the conclusion of this chapter.

a) Poorly Differentiated Leukemia, Lymphoblastic Type

The lymph node of a girl, 5 years of age, was examined (Case 31). Histochemically, the PAS reaction is distinctly positive (coarse granules) both in imprints and cryostat sections. All enzyme-histochemical reactions are negative, disregarding tiny positive granules in the acid phosphatase reaction.

Electron microscopic preparations reveal cells that measure 7—10 µ in diameter and have an almost smooth cellular outline. Their nucleus is subject to great pleomorphism. It usually exhibits deep indentations and sometimes looks almost lobulated. Small coarse clumps of chromatin are focally concentrated at the nuclear margin. The nucleolus is large and round and is located in the center of the nucleus. It displays many amorphous areas and a few nucleolonemata. Irregularly shaped mitochondria are contained in the cytoplasm mainly at one side of the nucleus. They are smaller than 0.5 µ in d ameter and are provided with unevenly arranged mitochondrial cristae. The Golgi apparatus is moderately pronounced and is constituted of lamellar, vesicular and vacuolar objects. A few small granules accumulate especially around the Golgi zone. They are bound by a single membrane and furnished with a core of moderately electron-dense granular material surrounded by an electron-lucent halo. In addition, numerous ribosomes and polyribosomes are dispersed uniformly throughout the cytoplasm. Some vesicular smooth endoplasmic reticulum and long threads of rough endoplasmic reticulum are randomly distributed. Electron-dense glycogen granules dot parts of the cytoplasm in some cells.

1. HAYHOE, QUAGLINO and DOLL (1964); LÖFFLER (1966, 1969).
2. LEDER (1967b).
3. HAYHOE, QUAGLINO and DOLL (1964); LEDER, to be published.

These leukemic cells differ from normal lymphocytes particularly in having larger nuclei with larger nucleoli, more abundant cytoplasm with a higher glycogen content and smaller mitochondria.

b) Poorly Differentiated Leukemia, Granulocytic (Paramyeloblastic) Type

Three cases of paramyeloblastic leukemia were studied. In the first case (70-year-old male, Case 60) many cells reveal a finely granular PAS-positive reaction. Also, a positive granular acid phosphatase reaction is evident. It is generally weak, however, and not equally prominent in different cells. In several large cells the naphthol AS-D chloroacetate esterase reaction is moderately positive. The nonspecific esterase reactions are negative.

Electron microscopically, the cells measure approximately 8 μ in diameter. The nuclei are relatively large and of distinctly round configuration. The nuclear membrane appears as a sievelike structure, due to a tremendous number of discontinuities termed nuclear pores. The nucleoli are extremely prominent. A widespread distribution of chromatin is noted. There are large quantities of ribosomes and polysomes, but only a small complement of rough and smooth endoplasmic reticulum is disclosed. The Golgi apparatus is rather inconspicuous and finely vesicular. The mitochondria are round to ovoid and less than 0.5 μ in diameter, only occasionally more than 0.5 μ if of elongated configuration. Some multivesicular bodies and a few granules, about 0.5 μ in diameter, are contained in the cytoplasm. They consist of a homogeneous darkly stained core which is surrounded by a narrow clear halo. The outermost border is represented by a limiting membrane. Finally, there are a few additional small granules, approximately 0.2 μ in diameter, which are as electron-dense as the granules just mentioned, but which do not possess the clear halo. Also, a membrane cannot be made out clearly. It may be present, but it cannot be visualized.

The second case (Case 54) concerns a 24-year-old male. In imprints, numerous vacuoles are demonstrated. They apparently correspond to the granules seen in the acid phosphatase reaction. In Sudan black stains a positive "nucleus" may be demonstrated in the unstained halo. The naphthol AS-D chloroacetate and unspecific esterase reactions are negative. In addition to many "blasts" several chloroacetate-positive myelocytes and metamyelocytes are identified in smears of the peripheral blood. The myelocytic nature of the "blast-cell leukemia" therefore seems to be established.

The following findings set this case apart from the previous one: 1) the nucleolus is somewhat smaller; 2) the cytoplasm often contains large inclusions of neutral fat (however, only scarce granules with a clearly visible limiting membrane are noted); 3) the mitochondria are slightly larger (up to 0.6 μ).

In the third case (Case 57, 63-year-old male) the PAS, nonspecific esterase and naphthol AS-D chloroacetate esterase reactions are negative. The acid phosphatase reaction is slightly positive, the positive granules being unevenly distributed throughout the cytoplasm (same pattern as in Case 54). A Sudan stain was not done.

Electron microscopically the following findings are different from those in the first case: 1) the nucleoli are slightly smaller; 2) the cytoplasm of some cells contains a few or a moderate amount of granules (0.2—0.3 μ in diameter), delimited by a distinct membrane (occasionally neutral fat is included); 3) besides these "blasts" there are a few cells which, on the basis of their granular structure, take on the appearance of promyelocytes.

The three cases described exhibit different degrees of maturation. The first case is the most immature, the third one being the most mature. The presence of granules and deposition of neutral fat are typical, but not specific electron microscopic features, whereas glycogen is never found.

c) Poorly Differentiated Leukemia, Monocytic Type

We have studied two cases of monocytic leukemia. One showed only a slight tendency toward differentiation while the other one exhibited relatively well differentiated monocytes with a co-existent increase of granulocyte precursors in the tissue. However, these were not found in the peripheral blood (myelomonocytic leukemia).

The first patient (64-year-old male, Case 56) had lymphadenopathy, hepatosplenomegaly and leukemic infiltration of the skin. The illness lasted several months. His peripheral blood showed many esterase-positive monocytes and esterase-negative immature cells. In tissue sections occasional groups of promyelocytes were identified.

With the electron microscope the monocytes are of irregular shape, measuring about 10 μ in diameter. The nucleus is deeply indented and lobulated. The nucleoli are of medium size. The chromatin is diffusely suspended throughout with only an occasional suggestion of aggregation. Ribosomes and polysomes are common in the cytoplasm. Predominantly rough endoplasmic reticulum is abundant. The mitochondria are round, occasionally elongated. The Golgi apparatus is fairly prominent. Lysosomes, phagosomes and granules are completely lacking.

The second case (Case 55) is that of a 45-year-old woman with a rapidly progressive course. The patient died only few weeks after her illness was diagnosed. In the peripheral blood a considerable increase of monocytes and "monoblasts" are demonstrated. Numerous monocytes with a positive naphthol AS esterase reaction and a positive alpha-naphthol acetate esterase reaction are found in lymph node imprints. Also, the acid phosphatase reaction is positive. In addition, many granulocyte precursors show a positive naphthol AS-D chloroacetate esterase reaction and are also PAS-positive.

With the electron microscope the monocytic cells appear quite similar to those of the first case, except for the following differences: 1) large quantities of predominantly rough endoplasmic reticulum are present, but there are fewer ribosomes; 2) numerous round or elongated granules measuring 0.2—0.3 μ in diameter are also seen. A limiting membrane can be identified in only a few of them.

In addition to monocytes there are also significant numbers of promyelocytes, myelocytes and metamyelocytes. Their characteristic structure has been described by many authors[1]. The granules of the promyelocytes measure up to 0.5 μ in diameter. Their limiting membrane is prominently displayed. Their contents are not homogeneously black but somewhat lighter and mottled. These granules correspond to the azurophilic so-called promyelocyte granules. In contrast, the mature specific neutrophilic granules of myelocytes and metamyelocytes stain homogeneously black or grayish-black. They do not reveal a visible limiting membrane. They are considerably smaller (0.2—0.3 μ) than the azurophilic granules of promyelocytes, which also appear in more mature stages of granulocytopoiesis. They probably correspond to the so-called toxic granulation of polymorphonuclear leukocytes. In eosinophilic granulocytes and their precursors the well-known crystalloid structure of the granules is seen.

d) Undifferentiated Type

Our material did not include a case which we were unable to classify. We cannot, therefore, furnish electron microscopic findings matching this cytochemical type of leukemia.

---

1. e.g. KAIHOTSU (1967).

e) Acute Erythremia (Di Guglielmo's Disease)

A 13-year-old boy (Case 59) was suffering from an "immature acute leukosis" with generalized swelling of lymph nodes, liver and spleen, completely unresponsive to therapeutic measures. The patient died five weeks after admission to the hospital. A fairly strong acid phosphatase reaction was demonstrated in "leukemic" cells of imprints, particularly in the perinuclear area with concentration at nuclear indentations. Also, a few esterase- and PAS-positive normoblasts were encountered among the undifferentiated cells. These undifferentiated cells did not give any indication of enzymatic activity and were PAS-negative. The autopsy revealed "blast"-infiltration of bone marrow, liver, spleen, lymph nodes and thymus. There was also an extreme increase in typical, but PAS-positive, erythropoiesis in the bone marrow. The myelopoiesis was preserved in areas, being of the common-place myelocytic type. Considering the overall appearance, in our opinion only two interpretations are possible, namely that of an erythroleukemia or of an erythremia. However, the strong and characteristic acid phosphatase reaction in the blasts is more in keeping with an erythremia.

Electron microscopically, the proliferated cells measure approximately 8—10 μ in diameter and are delineated by an almost smooth cytoplasmic membrane. Their nucleus is ovoid or often quite irregular with deep indentations of the nuclear membrane. There are chromatin clumps in close proximity to the nuclear edge. A few moderately prominent or sometimes large nucleoli may be displayed in the nucleus which also contains some mitochondria concentrated at one side. They are ovoid or elongated and equipped with fairly irregular cristae. Usually, the cells have a well-developed Golgi apparatus with lamellar and vesicular elements. Typically, several electron-dense granules can be identified mainly around the Golgi apparatus. They are 0.2—0.5 μ in diameter and are bound by a distinct limiting membrane. They contain amorphous material of much more pronounced electron density than the granules in the lymphoblastic leukemic cells already mentioned. Considerable quantities of free ribosomes and polyribosomes are spread uniformly throughout the cytoplasm. An inconspicuous complement of long rough and smooth vesicular endoplasmic reticulum is haphazardly arranged in the cytoplasm.

Histiocytosis X

Figs. 141—147.

The term "Histiocytosis X"[1] includes three syndromes characterized by proliferation of reticulo-endothelial cells: Letterer-Siwe's disease, eosinophilic granulomatosis and lipoid granulomatosis (HAND-SCHÜLLER-CHRISTIAN). BASSET and his associates[2] have described virus particles in the proliferated cells and discussed a viral etiology. The three syndromes have in common a striking proliferation of large cells which may correspond to reticulum cells. Giant cells comparable to osteoclasts are also usually demonstrable. We had the opportunity to study two cases of histiocytosis X.

1. Letterer-Siwe's Reticulosis of Infants

In the first case ($1^1/_2$-year-old male, Case 17) the proliferating cells prove to be of varying size and shape and exhibit a diversity of ultrastructures of the cytoplasm and nucleus. In some areas of the lesions, where the cells are loosely arranged, they measure 10—15 μ in diameter and have irregular cytoplasm with many short cytoplasmic processes. Their nucleus is usually deeply indented. Delicate chromatin particles are diffusely suspended with only slight clumping near the nuclear membrane. The nucleolus is fairly pronounced. Variable amounts of finely vesicular smooth endoplasmic reticulum and small tubules of rough endoplasmic reticulum are scattered throughout the cytoplasm. Some cells contain prominent endoplasmic cisternae with many adhering ribosomes. Mitochondria measure 0.3—0.5 μ in diameter and are usually ovoid or somewhat elongated. A variable number of lysosomes or phagolysosomes is often noted. The ribosome and polyribosome content varies from one cell to another. Some cells have a large number of ribosomes and appear fairly electron dense, while others have clear cytoplasm with only a few ribosomes. In some areas close interaction between the surfaces of proliferating cells is accomplished by distinct interdigitation of the cytoplasmic processes. However, the ultrastructure of the cells correlates well with that of cells lacking such interdigitating processes. Collagen fibrils are often disposed in intercellular spaces and are closely attached to the cellular membrane. The fine structure of the giant cells is almost identical to that of the proliferating cells except for the multiple nuclei of giant cells.

2. Eosinophilic Granulomatosis of the Adult
(Lipoid Granulomatosis of Hand-Schüller-Christian without Lipoid Storage)

The second case (Case 16) represents an instance of histiocytosis X in an adult. A 48-year-old male experienced swelling of cervical and inguinal lymph nodes. No further abnormalities were noted clinically. An inguinal node was removed. The classical findings of histiocytosis X with some characteristic osteoclast-like giant cells were evident on examination with the light microscope.

1. BASSET, NEZELOF, MALLET and TURIAF (1965); BASSET and TURIAF (1965); RITTER (1966); CANCILLA, LAHEY and CARNES (1967); GEORGSSON and WESSEL (1967); TURIAF and BASSET (1967).
2. BASSET, NEZELOF, MALLET and TURIAF (1965).

The electron microscopic cytology resembles that of the first case described above. However, for the most part the cells have more abundant cytoplasm and an eccentrically placed nucleus. Also, the cytoplasm is furnished with many extensions and bears a certain similarity to that of epithelioid cells, though vacuolar structures containing electron-opaque and/or electron-lucent material are not present in the amounts seen in tuberculosis or sarcoidosis.

Electron microscopic studies by RITTER[1] and GEORGSSON and WESSEL[2] disclosed lipid granules in the cytoplasm of cells in patients with histiocytosis X. In semithin sections of our first (pediatric) case numerous cells display osmiophilic granules of variable size. Unfortunately, we were not able to trace them in electron microscopic preparations.

The cellular structure of our two cases militates against the concept that histiocytosis X is a true neoplasia.

1. 1966.
2. 1967.

Tumor Metastases

Figs. 148—160.

Immense numbers of diverse malignant tumors, particularly carcinomas, are capable of metastasizing into lymph nodes. In this atlas we would like to restrict our description to three examples: the lymphoepithelial carcinoma (SCHMINCKE-RÉGAUD), the squamous cell carcinoma, which is to be clearly separated from the lymphoepithelial carcinoma, and the malignant melanoma.

1. Lymphoepithelial Carcinoma (Schmincke-Régaud)

It is the general consensus currently to regard lymphoepithelial carcinoma of the nasopharynx as a special type of transitional cell carcinoma[1]. However, to date no conclusive information has been provided that might indicate whether it represents a carcinoma or a reticulum cell sarcoma.

We have studied our lymphoepithelial carcinoma with the electron microscope. The tumor cells possess an irregular polygonal cytoplasm and have a high nuclearcytoplasmic ratio. The nucleus is round, ovoid or sometimes irregular. Medium-sized chromatin particles are diffusely suspended within the nucleus. One or two large nucleoli are present. The cytoplasmic membrane shows complex ramifications and is connected by distinct desmosomes with neighboring cells forming a reticular meshwork. Ovoid mitochondria tend to be focally collected in the cytoplasm. Scanty endoplasmic reticulum, some of it rough, may be found. Tonofilaments are demonstrated in the cytoplasm. Some of the cells are connected by a desmosome. Numerous ribosomes as well as polyribosomes are widespread throughout the cytoplasm. Such a fine-structure is compatible with that of the basal cells of squamous epithelium. Typical normal lymphocytes, plasma cells and phagocytic reticulum cells are interspersed between the tumor cells.

The cells of the Schmincke tumor bear important similarities to normal cells of tonsils and thymus, as disclosed by light and electron microscopy.

Firstly, the tumor cells are morphologically similar to the epithelial cells, which superficially resemble reticulum cells and extend from the basal layer of the squamous epithelium of the crypts into the surrounding lymphoid tissue. Electron microscopically, they are exactly like the basal cell except for a difference in cellular shape (processes!). More importantly, the lymphoepithelial cells are not attached to a basement membrane.

Secondly, the tumor cells of the lymphoepithelial carcinoma are morphologically identical with the epithelial reticulum cells of the thymic cortex (but not the medulla)[2]. Hence, their morphology is also similar to that of the typical thymus carcinoma cells, as documented by one of our cases (Fig. 155).

However, the lack of intercellular bridges and of keratinization of the cells of Schmincke's tumor indicates a fundamental difference from the transitional cells of the tonsillar epithelium.

---

1. OTSUKA (1964); SVOBODA, KIRCHNER and SHANMUGARATNAM (1967).
2. ENATSU (1960); HOSHINO (1962, 1963); TANAKA (1962); CLARK (1963); LUNDIN and SCHELIN (1965); CLAWSON, COOPER and GOOD (1967); TOKER (1968).

These and other findings are interpreted by us as strong evidence in support of the view that the lymphoepithelial carcinoma represents a special histologic entity and is not a transitional cell carcinoma. Furthermore, the presence of numerous desmosomes and tonofibrils does not allow for the interpretation of Schmincke's tumor as reticulum cell sarcoma.

2. Squamous Cell Carcinoma

The ultrastructure of epidermoid carcinoma is incidental to the degree of differentiation of the tumor. The cells most frequently found have the appearance of cells of the prickle cell layer of the epidermis. In contrast to the cells in lymphoepithelial carcinoma, they are furnished with prominent intercellular bridges that are firmly bound together by distinct desmosomes. Numerous fine tonofilaments are visualized in the cytoplasm. Collagen fibers are usually deposited around the tumor cells and/or within the intercellular spaces. In contrast, lymphoepithelial carcinoma is generally devoid of collagen fibers.

3. Malignant Melanoma[1]

The ultrastructure of malignant melanoma cells is subject to variation, depending upon the degree of melanin production. The most undifferentiated cell displays an ovoid or somewhat irregularly shaped cytoplasm and an eccentric nucleus with prominent nucleoli. The characteristic feature of the cells is the presence of melanosomes. They vary in size and shape and include diverse substances. The typical melanosome contains electron-opaque aggregates of fine regularly cross-banded filaments. Electron microscopic data concerning melanin production have already been furnished by several investigators. The findings convey the impression that melanin is produced first by ribosomes, then concentrated in the Golgi apparatus, and finally stored in melanosomes.

---

1. BRAUNSTEINER, MLCZOCH and PAKESCH (1958); WELLINGS and SIEGEL (1959); STÄUBLI and LOUSTALOT (1962); RAPPAPORT, NAKAI and SWIFT (1963); OVERBECK and PHILIPP (1968); TOSHIMA, MOORE and SANDBERG (1968).

References

ACHONG, B. G., and M. A. EPSTEIN: Fine structure of the Burkitt's tumor. J. nat. Cancer Inst. **36**, 877—897 (1966).

ANDERSON, D. R.: Ultrastructure of normal and leukemic leucocytes in human peripheral blood. J. Ultrastruct. Res., Suppl. **9**, 1—42 (1966).

ANDRÉ, R., B. DREYFUS et M. BESSIS: La ponction ganglionnaire dans la maladie de Hodgkin, examinée au microscope electronique. Presse méd. **63**, 967—970 (1955).

APITZ, K.: Die Paraproteinosen. (Über die Störung des Eiweißstoffwechsels bei Plasmocytom.) Virchows Arch. path. Anat. **306**, 631—699 (1940).

BAINTON, D. F., and M. G. FARQUHAR: Origin of granules in polymorphnuclear leucocytes. Two types derived from opposite faces of the Golgi complex in developing granulocytes. J. Cell Biol. **28**, 277—301 (1966).

BAIRATI, A., and ST. AMANTE: Studies on the ultrastructure of the lymph nodes. 1. The reticular network. Z. Zellforsch. **63**, 644—672 (1964).

BASSET, F., C. NEZELOF, R. MALLET et J. TURIAF: Nouvelle mise en evidence, par la microscopie electronique de particules d'allure virale dans une seconde forme clinique de l'histiocytose X, le granulome eosinophile de l'os. C. R. Acad. Sci. (Paris) **261**, 5719—5720 (1965).

—, et M. J. TURIAF: Identification par la microscopie electronique de particules de nature probablement virale dans les liaisons granulomateuses d'une histiocytose X pulmonaire. C. R. Acad. Sci. (Paris) **261**, 3701—3703 (1965).

BERARD, C., G. T. O'CONOR, L. B. THOMAS, and H. TORLONI: Histopathological definition of Burkitt's tumour. Bull. World Hlth Organ. **40**, 601—607 (1969).

BERNHARD, W.: Elektronenmikroskopischer Beitrag zum Studium der Kanzerisierung und der malignen Zustände der Zelle. Verh. dtsch. Ges. Path. **45**, 8—37 (1966).

—, and N. GRANBOULAN: The fine structure of the cancer cell nucleus. Exp. Cell Res. Suppl. **9**, 19—53 (1963).

—, et D. LAMBERT: Ultrastructure des tumeurs de Burkitt de l'enfant African. In: The lymphoreticular tumours in Africa (F. C. ROULET, ed.), p. 270—284. Basel and New York: S. Karger 1964.

—, and R. LEPLUS: Fine structure of the normal and malignant human lymph node. Oxford: Pergamon; Paris: Gauthier-Villars; New York: MacMillan 1964.

BESSIS, M.: Ultrastructure of lymphoid and plasma cells in relation to globulin and antibody formation. Lab. Invest. **10**, 1040—1067 (1961).

—, and J. P. THIÉRY: Electron microscopy of human white blood cells and their stem cells. Int. Rev. Cytol. **12**, 199—241 (1961).

— — Études au microscope electronique sur la leucémie humaine. I. Les leucémies granulocytaires. Nouv. Rev. franç. Hémat. **1**, 703—728 (1961).

— — II. Les leucémies lymphocytaires. Comparison avec la leucémie de la souris de souche AK. Nouv. Rev. franç. Hémat. **2**, 387—414 (1962).

— — III. Leucémies à cellules-souches, erythrémies, réticulo-lympho-sarcomes, maladie de Hodgkin, plasmocytomes. Nouv. Rev. franç. Hémat. **2**, 577—601 (1962).

BINGGELI, M. F.: Abnormal intranuclear and cytoplasmic formations associated with a chemically induced, transplantable chicken sarcoma. J. biophys. biochem. Cytol. **5**, 143—152 (1959).

BLOOM, G. D.: Electron microscopy of neoplastic mast cell: A study of the mouse mastocytoma mast cell. Ann. N.Y. Acad. Sci. **103**, 53—86 (1963).

— A nucleus with cytoplasmic features. J. Cell Biol. **35**, 266—268 (1967).

BRANDES, D., B. SCHOFIELD, and E. ANTON: Nuclear mitochondria. Science **149**, 1373 (1965).

BRAUNSTEINER, H., K. FELLINGER, and F. PAKESCH: Electron microscopic investigations on sections from lymph node and bone marrow in malignant blood diseases. Blood **12**, 278—294 (1957).

— F. MLCZOCH u. F. PAKESCH: Elektronenmikroskopische Untersuchungen über die Struktur von intracellulärem Melanin beim Melanoblastom. Klin. Wschr. **36**, 262—263 (1958).

—, and F. PAKESCH: Electron microscopy and the functional significance of a new cellular structure in plasmocytes. A review. Blood **10**, 650—654 (1955).

— — u. O. THALHAMMER: Elektronenmikroskopische Untersuchungen über die Morphologie des Toxoplasma gondii und das Wesen des Farbtestes nach Sabin-Feldmann. Wien. Z. inn. Med. **38**, 10—27 (1957).

BRITTIN, G. M., Y. TANAKA, and G. BRECHER: Intranuclear· inclusions in multiple myeloma and macroglobulinemia. Blood **21**, 335—351 (1963).

BROOKS, R. E., and B. V. SIEGEL: Normal human lymph node cells. An electron-microscopic study. Blood **27**, 687—705 (1966).

— — Nuclear bodies of normal and pathological human lymph node cells: An electron microscopic study. Blood **29**, 269—275 (1967).

BUCCIARELLI, E.: Intranuclear cisternae resembling structures of the Golgi complex. J. Cell Biol. **30**, 664—665 (1966).

BÜTTNER, D. W., u. E. HORSTMANN: Haben die Sphaeridien in den Zellkernen kranker Gewebe eine pathognomonische Bedeutung? Virchows Arch. path. Anat. **343**, 142—163 (1967).

CANCILLA, P. A., M. E. LAHEY, and W. H. CARNES: Cutaneous lesions of Letterer-Siwe disease. Electron microscopic study. Cancer (Philad.) **20**, 1986—1991 (1967).

CLARK, S. L., JR.: The reticulum of lymph node in mice studied with the electron microscope. Amer. J. Anat. **110**, 217—257 (1962).

— The thymus in mice of strain 1291 J studied with the electron microscope. Amer. J. Anat. **112**, 1—34 (1963).

CLARK, W. H.: Electron microscope studies of nuclear extrusions in pancreatic acinar cells of the rat. J. biophys. biochem. Cytol. **7**, 345—352 (1960).

CLAWSON, C. C., M. D. COOPER, and R. A. GOOD: Lymphocytic fine structure in the Bursa of Fabricius, the thymus, and the germinal center. Lab. Invest. **16**, 407—421 (1967).

DE MAN, J. H. C.: Rod-like tabular structures in the cytoplasm of histiocytes in "Histiocytosis X". J. Path. Bact. **95**, 123—126 (1968).

DMOCHOWSKI, L., T. YUMOTO, C. E. GREY, E. DESIGNER, R. L. HALES, P. L. LANGFORD, H. G. TAYLOR, E. J. FREIREICH, C. C. SHULLENBERGER, J. A. SHIVELY, and C. D. HOWE: Electron microscopic studies of human leukemia and lymphoma. Cancer (Philad.) **20**, 760—777 (1967).

DOHI, S., S. HANAOKA, and S. AMANO: Electron microscopic studies on the plasma cell. Acta path. jap. **7**, 1—11 (1957).

DORFMAN, R. F.: The fine structure of a malignant lymphoma in a child from St. Louis, Missouri. J. nat. Cancer Inst. **38**, 491—504 (1967).

ENATSU, T.: Electron microscopic studies on the thymus, with special reference to epithelial cells. Igaku Kenkyu **30**, 309—325 (1960).

EPSTEIN, M. A., and B. G. ACHONG: Fine structural organization of human lymphoblast of a tissue culture strain (E B 1) from Burkitt's lymphoma. J. nat. Cancer Inst. **34**, 241—253 (1965).

— Y. M. BARR, and B. G. ACHONG: The behaviour and morphology of a second tissue culture strain (EB₂) of lymphoblasts from Burkitt's lymphoma. Brit. J. Cancer **19**, 108—115 (1965).

—, and P. B. HERDSON: Cellular degeneration associated with characteristic nuclear fine structural changes in the cells from two cases of Burkitt's malignant lymphoma syndrome. Brit. J. Cancer **17**, 56—58 (1963).

FEDORKO, M. E., and J. G. HIRSCH: Cytoplasmic granule formation in myelocytes. An electron microscope radioautographic study on the mechanism of formation of cytoplasmic granules in rabbit heterophilic myelocytes. J. Cell Biol. **29**, 307—316 (1966).

FERNANDO, N. V. P., and H. Z. MOVAT: The fine structure of connective tissue. III. The mast cell. Exp. molec. Path. **2**, 450—463 (1963).

FORTEZA BOVER, G., J. BAGUENA CANDELA, J. FORTEZA VILA, E. BARBERA GUILLEM y R. BAGUENA CANDELA: Estudio de la ultrastructura de la célula gigante de Sternberg y de sus relaciones con otras células immunocompetentes. Medicina esp. **59**, Num. 351 (1968).

— G. FORTEZA VILA, R. BAGUENA CANDELA, A. JUAN BORDON, F. PRIETO GARCIA y E. BARBERA GUILLEM: Ultrastructura de las células blastos de los cultivos de leucocitos humanos estimulados par la fitohemaglutinina. Med. esp. **57**, 346—354 (1967).

— — — y F. PRIETO GARCIA: Ultrastructura de las células de leucemias agudas humanas. Sangre **12**, 121—157 (1967).

— F. PRIETO GARCIA, E. BARBERA GUILLEM, R. BAGUENA CANDELA y J. FORTEZA VILA: Ultrastructura del paragranuloma de Hodgkin. Med. esp. **59**, 47—75 (1968).

FRAJOLA, W. J., M. H. GREIDER, and B. A. BOURONCLE: Cytology of the Sternberg-Reed cell as revealed by the electron microscope. Ann. N.Y. Acad. Sci **73** (I), 221—236 (1958).

FRASKA, J. M., O. AUERBACH, V. PARK, and W. STOECKENIUS: Electron microscope observation of nuclear evagination in bronchial epithelium. Exp. molec. Path. **4**, 340—355 (1965).

FREMAN, J. A., and M. S. SAMUELS: The ultrastructure of a "fibrillar formation" of leukemic human blood. Blood **13**, 725—731 (1958).

FRESEN, O., u. H. J. WELLENSIEK: Elektronenoptische Befunde am retikulumzelligen Gewebe. Zbl. allg. Path. path. Anat. **97**, 406—407 (1958).

FRESEN, O., u. H. J. WELLENSIEK: Zur elektronenmikroskopischen Struktur des Lymphknotens. Verh. dtsch. Ges. Path. 42, 353—363 (1959).

GALL, E. A.: The cytological identity and interrelation of mesenchymal cells of lymphoid tissue. Ann. N.Y. Acad. Sci. 73, 120—130 (1958).

—, and T. B. MALLORY: Malignant lymphoma. A clinico-pathologic survey of 618 cases. Amer. J. Path. 18, 381—412 (1942).

GAUDECKER, B. V., u. K. HINRICHSEN: Elektronenmikroskopische Untersuchungen zur Zytologie von Thymusrinde und Keimzentrum. Z. Zellforsch. 65, 139—162 (1965).

GEORGSSON, G., u. W. WESSEL: Elektronenmikroskopische und enzymatisch-analytische Untersuchung eines eosinophilen Granuloms. Virchows Arch. path. Anat. 343, 177—188 (1967).

GERNHAM, P. C. C., J. R. BAKER, and R. G. BIRD: The fine structure of cystic form of toxoplasma gondii. Brit. med. J. 1962I, 83—84.

GOWANS, J. L., and E. J. KNIGHT: The route of recirculation of lymphocytes in the rat. Proc. roy. Soc. Edinb. B 159, 257—282 (1964).

GROPP, A., u. R. FISCHER: Untersuchungen zur phytohämagglutininstimulierten Umwandlung von menschlichen Blutlymphozyten zu blastenartigen Zellen. Virchows Arch. path. Anat. 338, 64—77 (1964).

GUSEK, W.: Zur Ultrastruktur und Genese von Mastzellen und Mastzellen-Granula. Proc. Europ. Regio. Confer. Electr. Microsc. 2, 912—916 (1960).

— Submikroskopische Untersuchungen zur Feinstruktur aktiver Bindegewebszellen. Veröffentl. morph. Path., Heft 64. Stuttgart: Gustav Fischer 1962.

HAN, S. S.: The ultrastructure of the mesenteric lymph node of the rat. Amer. J. Anat. 109, 183—225 (1961).

HARVEN, E. DE, B. CLARKSON, and A. STRIFE: Electron microscopic study of human leukemic cells in tissue culture. Cancer (Philad.) 20, 911—925 (1967).

HAYHOE, F. G. J., D. QUAGLINO, and R. DOLL: The cytology and cytochemistry of acute leukaemias. London: Her Majesty's Stationery Office 1964.

HIBBS, R. G., G. E. BURCH, and J. H. PHILLIPS: Electron-microscopic observations in human mast cell. Amer. Heart J. 60, 121—127 (1960).

— J. H. PHILLIPS, and G. E. BURCH: Electron microscopy of human tissue mast cells. J. Amer. med. Ass. 174, 508—510 (1960).

HORIE, A.: An electron microscope observation of the lymphatic nodule. Fukuoka Acta med. 52, 135—141 (1961).

HOSHINO, T.: The fine structure of ciliated vesicle-containing reticular cell in the mouse thymus. Exp. Cell Res. 27, 615—617 (1962).

— Electron microscopic studies of the epithelial reticular cells of the mouse thymus. Z. Zellforsch. 59, 513—529 (1963).

HOSOKAWA, K.: Electron microscopic study on lymphatic tissue. Bull. Gikei Med. Acad. 76, 1274—1284 (1960).

HÜBNER, G., u. F. PAULUSSEN: Die Feinstruktur des Gewebes nach protrahierter Formolfixierung. Virchows Arch. Abt. B Zellpath. 1, 107—119 (1968).

HULDT, G.: Experimental toxoplasmosis. Studies of the multiplication and spread of toxoplasma in experimentally infected rabbits. Acta path. microbiol. scand. 67, 401—423 (1966).

INMAN, D. R., and E. H. COOPER: Electron microscopy of human lymphocytes stimulated by phytohemagglutinin. J. Cell Biol. 19, 441—445 (1963).

IZARD, J.: A class of dense reticular cells with long processes in the mouse thymus. Anat. Rec. 157, 264 (1967).

—, and E. DE HARVEN: Increased numbers of a characteristic type of reticular cell in the thymus and lymph nodes of leukemic mice: An electron microscope study. Cancer Res. 28, 421—433 (1968).

KAIHOTSU, NAOKI: Electron microscopic studies on the maturation process of neutrophilic leukocytes. Kôbe J. med. Sci. 13, 47—66 (1967).

KAUTZ, J., Q. B. DE MARSCH, and W. THORNBURG: A polarizing and electron microscope study of plasma cell. Exp. Cell Res. 13, 596—599 (1957).

KIKKAWA, Y., and B. GUEFT: Toxoplasma cyst in the human heart. An electron microscopic study. J. Parasit. 50, 217—225 (1964).

KUHN, C.: Nuclear bodies and intranuclear globulin inclusions in Waldenström's macroglobulinemia. Lab. Invest. 17, 404—415 (1967).

LEDER, L. D.: The origin of blood monocytes and macrophages. Blut 16, 86—98 (1967a).

— Die fermentcytochemische Erkennung normaler und neoplastischer Erythropoesezellen in Schnitt und Ausstrich. Blut 15, 289—293 (1967b).

LEDUC, E. H., and J. W. WILSON: An electron microscope study of intranuclear inclusions in mouse liver and hepatoma. J. biophys. biochem. Cytol. 6, 427—430 (1959).

LENNERT, K.: Diagnose und Ätiologie der Piringerschen Lymphadenitis. Verh. dtsch. Ges. Path. 42, 203—208 (1959).

— Lymphknoten. Cytologie und Lymphadenitis. In: Handbuch der speziellen pathologischen Anatomie und Histologie, Bd. I, 3 A. Berlin-Göttingen-Heidelberg: Springer 1961.

References

LENNERT, K., Pathologie und Klassifikation der malignen Lymphome. Kongr. Japan. Soc. Path. 1969.
— R. CAESAR, and H. K. MÜLLER: Electron microscopic studies of germinal centers in man. In: Germinal centers in immune responses, p. 49—59. Berlin-Heidelberg-New York: Springer 1966.
—, u. J. MESTDAGH: Lymphogranulomatosen mit konstant hohem Epitheloidzellgehalt. Virchows Arch. Abt. A Path. Anat. **344**, 1—20 (1968).
—, u. H. R. NIEDORF: Nachweis von desmosomal verknüpften Retikulumzellen im follikulären Lymphom (BRILL SYMMERS). Virchows Arch. Abt. B Zellpath. **4**, 148—150 (1969).
LÖFFLER, H.: Cytochemische Untersuchungen bei unreifzelligen Leukosen. Ihre Bedeutung als Grundlage für die Klassifizierung und Therapie. Habil.-Schr. Gießen 1966.
— Eine Klassifizierung als Grundlage der Behandlung unreifzelliger Leukosen. Hämatologie u. Bluttransfusion **8**, 105—107 (1969).
LOW, F. N., and J. A. FREEMAN: Electron microscopic atlas of normal and leukemic human blood. New York: McGraw Hill 1958.
LUDVIK, I.: Vergleichende elektronenoptische Untersuchungen an Toxoplasma gondii und Sarcocystitenella. Zbl. Bakt., I. Abt. Orig. **166**, 60—65 (1956).
— Neue elektronenmikroskopische Befunde am Toxoplasma gondii. Zbl. allg. Path. path. Anat. **101**, 540 (1960).
LUKES, R. J.: Relationship of histologic features of clinical states in Hodgkin's disease. Amer. J. Roentgenol. **90**, 944—955 (1963).
— L. F. CRAVER, T. C. HALL, H. RAPPAPORT, and P. RUBEN: Report of the nomenclature committee. Cancer Res. **26**, 1311 (1966).
LUNDIN, P. M., and U. SCHELIN: Ultrastructure of the rat thymus. Acta path. microbiol. scand. **65**, 379—394 (1965).
MALDONALDO, J. E., A. L. BROWN JR., E. D. BAYRD, and G. L. PEASE: Ultrastructure of the myeloma cell. Cancer (Philad.) **19**, 1613—1627 (1966a).
— — — — Cytoplasmic and intranuclear electron-dense bodies in the myeloma cell. Light and electron microscopy observations. Arch. Path. **81**, 484—500 (1966b).
MARUYAMA, K., and M. HANAOKA: Relationship between the fine structure and antigen distribution in the germinal center of the lymph node. Proc. Jap. Soc. R.E.S. **6**, 146 (1966).
—, and T. MASUDA: Electron microscopic observation on the germinal center of the lymph node in guinea pigs sensitized with sheep erythrocytes. Ann. Rep. Inst. Virus Res. Kyoto Univ. **7**, 149—152 (1964).

MATSUURA, S.: Electron microscopic studies on the histogenesis of tubercles in the lymph node. Bull. Kobe Med. Coll. **13**, 549—560 (1958).
MILANESI, S.: Sulla presenza di dispositivi di giunzione tra le cellule dendrifiche dei follicoli linfatici del linfonodo. Boll. Soc. ital. Biol. sper. **41**, 1223—1225 (1965).
MILLER, J. J., and G. J. V. NOSSAL: Antigens in immunity. VI. The phagocytic reticulum of lymph node follicle. J. exp. Med. **120**, 1075—1086 (1964).
MIURA, A.: Electron microscopic observation of reticuloendothelial cells in the lymph node. Proc. Jap. Soc. R.E.S. **5**, 246—259 (1965).
MOE, R. E.: Fine structure of the reticulum and sinuses of lymph nodes. Amer. J. Anat. **112**, 311—336 (1963).
— Electron microscopic appearance of the parenchyma of lymph nodes. Amer. J. Anat. **114**, 341—369 (1964).
MOLLO, F., and A. STRAMIGNONI: Nuclear projections in blood and lymph node cells of human leukemia and Hodgkin's disease and in lymphocytes cultured with phytohemagglutinin. Brit. J. Cancer **21**, 519—523 (1967).
MORI, Y.: Electron microscopic observation of cells in the lymph node. J. Kyushu Hem. Soc. **10**, 221—237 (1960).
MOVAT, H. Z., and N. V. P. FERNANDO: The fine structure of lymphoid tissue. Exp. mol. Path. **3**, 546—568 (1965).
— — The fine structure of the lymphoid tissue during antibody formation. Exp. molec. Path. **4**, 155—188 (1966).
MÜLLER-HERMELINK, H. K., u. R. CAESAR: Elektronenmikroskopische Untersuchung der Keimzentren in menschlichen Tonsillen. Z. Zellforsch. **96**, 521—547 (1969).
NOSSAL, G. J. V., A. ABBOT, and J. MITCHELL: Antigen in immunity. XIV. Electron microscopic radioautographic studies of antigen capture in the lymph node medulla. J. exp. Med. **127**, 263—276 (1968).
— — —, and Z. LUMMUS: Antigen in immunity. XV. Ultrastructural features of antigen capture in primary and secondary lymphoid follicles. J. exp. Med. **127**, 277—290 (1968).
OETTGEN, H. F.: Electron microscopic observations in biopsy material of Burkitt's tumour. In: Treatment of Burkitt's tumour, ed. by J. H. BURCHENAL and D. P. BURKITT. UICC Monogr. Series, Vol. 8. Berlin-Heidelberg-New York: Springer 1967.
OGAWA, K.: Elektronenoptische Untersuchungen bei einem Fall von sog. Hodgkin-Sarkom. Frankfurt. Z. Path. **71**, 677—693 (1962).

OGINO, N., and C. YONEDA: The fine structure and mode of division of toxoplasma gondii. Arch. Ophthal. **75**, 218—227 (1966).

OLISA, E. G.: The fine structure of reproducing toxoplasma gondii. Parasitology **53**, 643—649 (1963).

OTSUKA, H.: A so-called lymphoepithelioma of the nasopharynx. Mod. Med. (Saishin Igaku) **19**, 1708—1718 (1964).

OVERBECK, L., u. E. PHILIPP: Elektronenmikroskopische Untersuchungen am Melanomalignom der Vulva. Zugleich ein Beitrag zur Melaninbildung in Tumorzellen. Z. Geburtsh. Gynäk. **168**, 167—177 (1968).

— — Zur Melanogenesis in menschlichen Tumorzellen. Naturwissenschaften **55**, 232—233 (1968).

PARKER, J. W., H. WAKASA, and R. J. LUKES: The morphologic and cytochemical demonstration of lysosomes in lymphocytes incubated with phytohemagglutinin by electron microscope. Lab. Invest. **14**, 1736—1743 (1965).

PETRIS, S. DE, G. KARLSBAD, and B. PERNIS: Filamentous structures in the cytoplasm of normal mononuclear phagocytes. J. Ultrastruct. Res. **7**, 39—55 (1962).

PIRINGER-KUCHINKA, A.: Eigenartiger mikroskopischer Befund an exzidierten Lymphknoten. Verh. dtsch. Ges. Path. **36**, 352—362 (1953).

POLICARD, A., A. COLLET et S. PREGERMAIN: Étude au microscope electronique des mastocytes des tissues chez les mammifères. Rev. Hémat. **15**, 374—384 (1960).

RAPPAPORT, H.: Tumors of the hematopoietic system. Atlas of tumor path., sect. III, fasc. 8. Washington: Armed Forces Institute of Pathology 1966.

— T. NAKAI, and H. SWIFT: The fine structure of the normal and neoplastic melanocytes in the Syrian hamster with particular reference to carcinogen induced melanotic tumors. J. Cell Biol. **16**, 171—186 (1963).

REBUCK, J. W., and G. A. LOGRIPPO: Characteristics and interrelationships of the various cells in the reticuloendothelial cell, macrophage, lymphocyte and plasma cell series in man. Lab. Invest. **10**, 1068—1093 (1961).

RITTER, R. A., JR.: Histiocytosis X. A case report with electron microscopic observations. Cancer (Philad.) **19**, 1155—1164 (1966).

ROBERTSON, D. M., and J. D. MACLEAN: Nuclear inclusions in malignant gliomas. Arch. Neurol., **13**, 287—296 (1965).

ROGER, G. E.: Electron microscopy of mast cells in the skin of young mice. Exp. Cell Res. **11**, 393—402 (1956).

ROULET, F. C.: The lymphoreticular tumors in Africa. Basel and New York: S. Karger 1964.

SEMAN, G., C. ROSENFELD, C. JASMIN et R. CAMAIN: Remarques à propos de deux cultures et d'une biopsie de tumeur de BURKITT vues au microscope électronique. Rev. franç. Étud. clin. biol. **13**, 83—87 (1968).

SMITH, D. E., and Y. S. LEWIS: Electron microscopy of the tissue mast cell. J. biophys. biochem. Cytol. **3**, 9—14 (1957).

SMITH, G. F., and P. T. O'HARA: Structure of nuclear pockets in human leukocytes. J. Ultrastruct. Res. **21**, 415—423 (1968).

SORENSEN, G. D.: An electron microscopic study of popliteal lymph nodes from rabbits. Amer. J. Anat. **107**, 73—96 (1960).

— Electron microscopic observations on the fate of colloidal gold in popliteal lymph nodes of rabbits. Anat. Rec. **139**, 276 (1961).

STÄUBLI, N., and P. LOUSTALOT: Electron microscopy of transplantable melanotic and amelanotic hamster melanomas. Cancer Res. **22**, 84—88 (1962).

STOECKENIUS, W.: Zur Feinstruktur der Granula menschlicher Gewebsmastzellen. Exp. Cell Res. **11**, 656—658 (1956).

— Weitere Untersuchungen am lymphatischen Gewebe. Verh. dtsch. Ges. Path. **41**, 304—312 (1957).

— Elektronenmikroskopische Untersuchungen am lymphatischen Gewebe. Habil.-Schr. Hamburg 1958.

STOECKENIUS jr., W., u. P. NAUMANN: Elektronenmikroskopische Untersuchungen zur Antikörperbildung in der Milz. Verh. des 6. Kongr. der Europ. Ges. für Haematol., S. 4—9. Basel u. New York: S. Karger 1958.

SVOBODA, D. J., F. R. KIRCHNER, and K. SHANMUGARATNAM: The fine structure of nasopharyngeal carcinomas. In: Cancer of the nasopharynx, ed. C. S. MUIR and K. SHANMUGARATNAM, p. 163—172. Copenhagen: Scandinavian University Books 1967.

SWARTZENDRUBER, D. C.: Desmosomes in germinal centers of mouse spleen. Exp. Cell Res. **40**, 429—432 (1965).

— The fine structure of lymphatic tissue germinal center. Amer. J. Path. **48**, 613—626 (1966).

—, and M. G. HANNA: Electron microscopic autoradiography of germinal center cells in mouse spleen. J. Cell Biol. **25**, 109—119 (1965).

SWIFT, H.: Cytochemical studies on nuclear fine structure. Exp. Cell Res. Suppl. **9**, 54—67 (1963).

TANAKA, H.: Electron microscopic observation of lymphatic cells in the lymph node and thymus, especially on lymphogonia. Acta haematol. jap. **20**, 237—254 (1957).

TANAKA, H.: Comparative cytologic studies by means of an electron microscope on monocytes, subcutaneous histiocytes, reticulum cells in the lymph nodes and peritoneal macrophages. Ann. Report Inst. Virus Res. Kyoto Univ., Ser. A **1**, 87—149 (1958).

— Mesenchymal and epithelial reticulum in lymph nodes and thymus of mice as revealed in the electron microscopy. Ann. Report, Inst. Virus Res. Kyoto Univ. **5**, 146—169 (1962).

— Electron microscopic observation of reticulo-endothelial cells. Jap. Handbook Hematol. **1**, 364—396 (1963).

TOKER, C.: Thymoma. An ultrastructural study. Cancer (Philad.) **21**, 1157—1163 (1968).

TOSHIMA, S., G. E. MOORE, and A. A. SANDBERG: Ultrastructure of human melanoma in cell culture. Cancer (Philad.) **21**, 202—216 (1968).

— N. TAKAGI, J. MINOWADA, G. E. MOORE, and A. A. SANDBERG: Electron microscopic and cytogenetic studies of cells derived from Burkitt's lymphoma. Cancer Res. **27**, 753—771 (1967).

TURIAF, J., et F. BASSET: Un cas d'histiocytose X pulmonaire avec presence de particules de nature probablement virale dans les lesions granulomateuses pulmonaires examinées au microscope electronique. Bull. Soc. méd. Hôp. Paris **116**, 1197—1208 (1965).

WEBER, A., S. WHIPP, E. USENIK, and S. FROMMES: Structural changes in the nuclear body in the adrenal zona fasciculata of the calf following the administration of ACTH. J. Ultrastruct. Res. **11**, 564—576 (1964).

WELLENSIEK, H. J.: Zur submikroskopischen Morphologie von Plasmazellen mit Russelschen Körperchen und Eiweißkristallen. Beitr. path. Anat. allg. Path. **118**, 173—202 (1957).

WELLINGS, S. R., and B. V. SIEGEL: Role of Golgi apparatus in the formation of melanin granules in human malignant melanoma. J. Ultrastruct. Res. **3**, 147—154 (1959).

— — Electron microscopy of human malignant melanoma. J. nat. Cancer Inst. **24**, 437—460 (1960).

WELSCH, R. A.: Electron microscopic localization of Russell bodies in the human plasma cell. Blood **16**, 1307—1312 (1960).

WILDFÜHR, W.: Elektronenmikroskopische Untersuchungen an Toxoplasma gondii. Z. ges. Hyg. **10**, 541—546 (1964).

YAMORI, T.: On phagocytes: Their structures and participation in inflammation. Acta path. jap. **14**, 1—43 (1964).

— Exhibition on Congr. Japan. Soc. Path. 1969.

ZUCKER-FRANKLIN, D.: The ultrastructure of cells in human thoracic duct lymph. J. Ultrastruct. Res. **9**, 325—339 (1963).

The Cell Types

Lymphocyte

Characteristic Cells of the Germinal Center
    Germinocyte
    Germinoblast

Basophilic Stem Cell

Plasma Cell Series
    Plasma Cell
    Proplasmacyte
    Plasmablast

Reticulum Cell
    Nonphagocytic Reticulum Cell
    Phagocytic Reticulum Cell

Retothelial Cells

Mast Cell

Other Cell Types

Fig. 1a   Lymphocyte
The nucleus has distinct chromatin clumps. A few rounded mitochondria and free ribosomes are evident in the cytoplasm.
Case 15. ×17,500

Fig. 1b   Germinocyte
Chromatin clumping is somewhat less pronounced.
Case 49. ×20,000

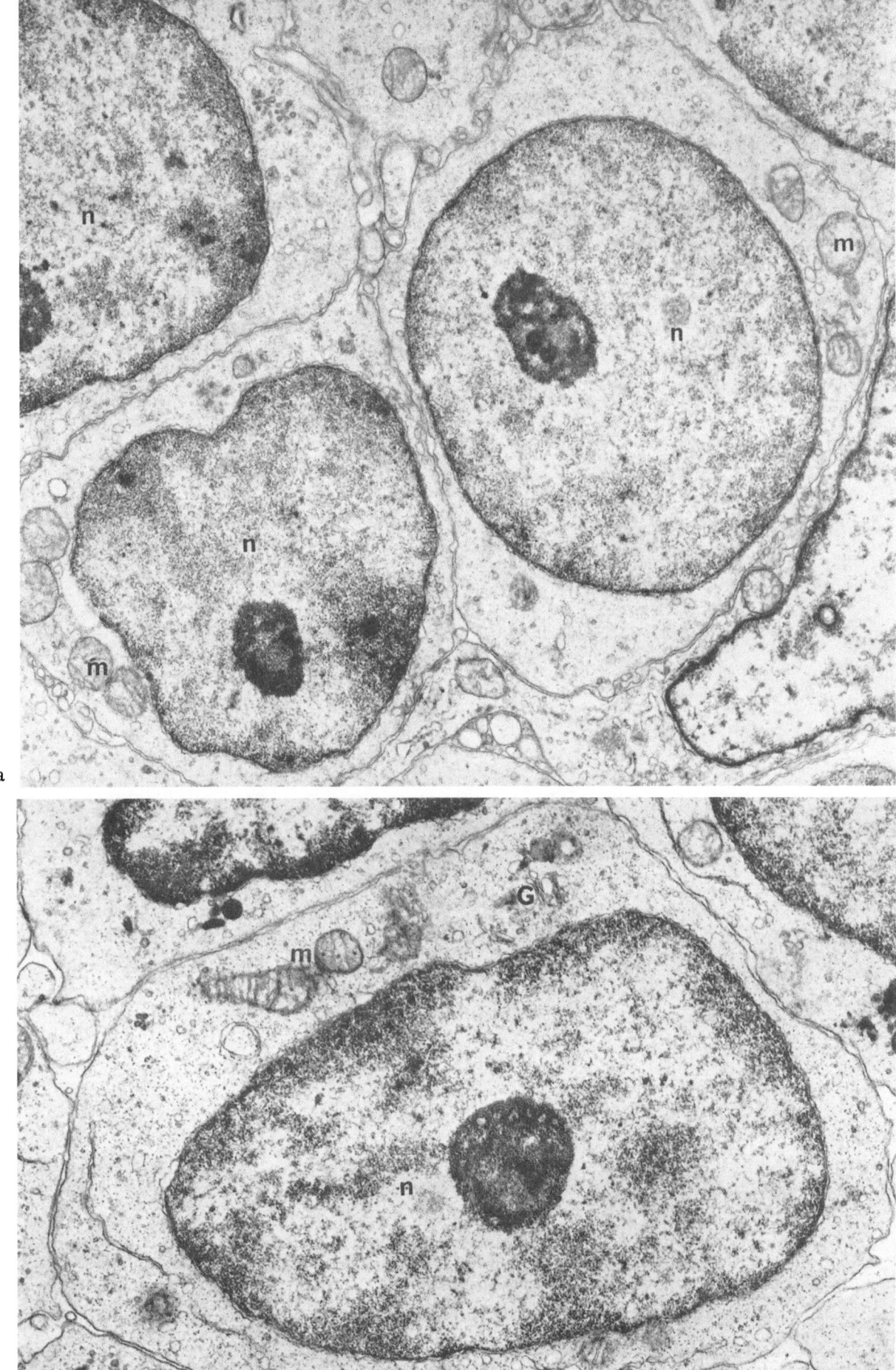

a

b

Fig. 2   Germinocytes
Almost uniformly dispersed slightly marginated chromatin, rod-shaped mitochondria and a moderately prominent Golgi apparatus can be seen. Located near the center of the picture is a germinocyte with two nuclear pockets, one of them with vacuolar and vesicular elements resembling a Golgi apparatus.

Hyperplastic tonsil. Contributed by H.-K. MÜLLER-HERMELINK. ×17,500

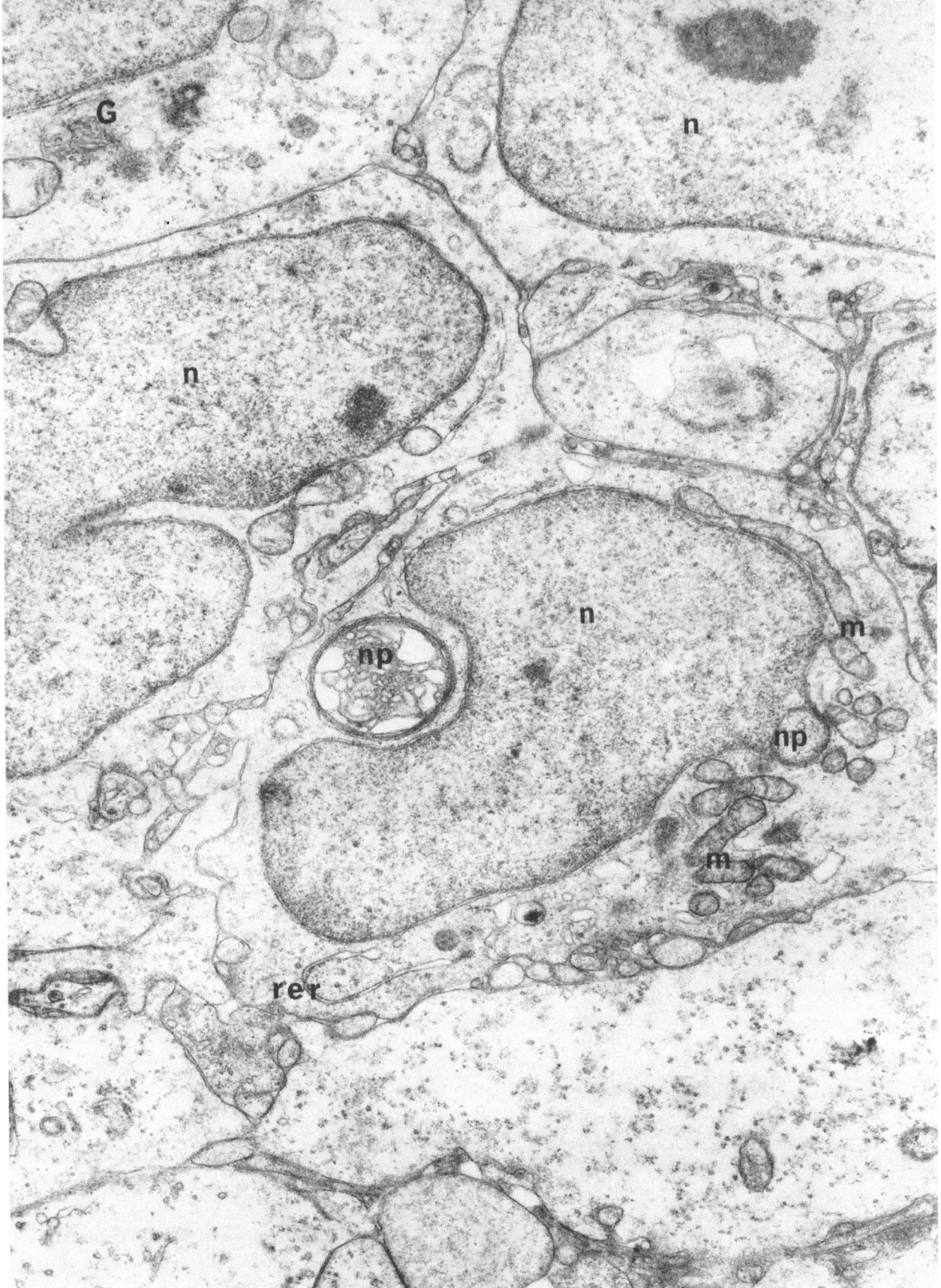

Fig. 3   Germinoblast
A large nucleolus is present in the nucleus. Numerous ribosomes and polyribosomes are disseminated throughout the cytoplasm.
Case 31. ×17,500

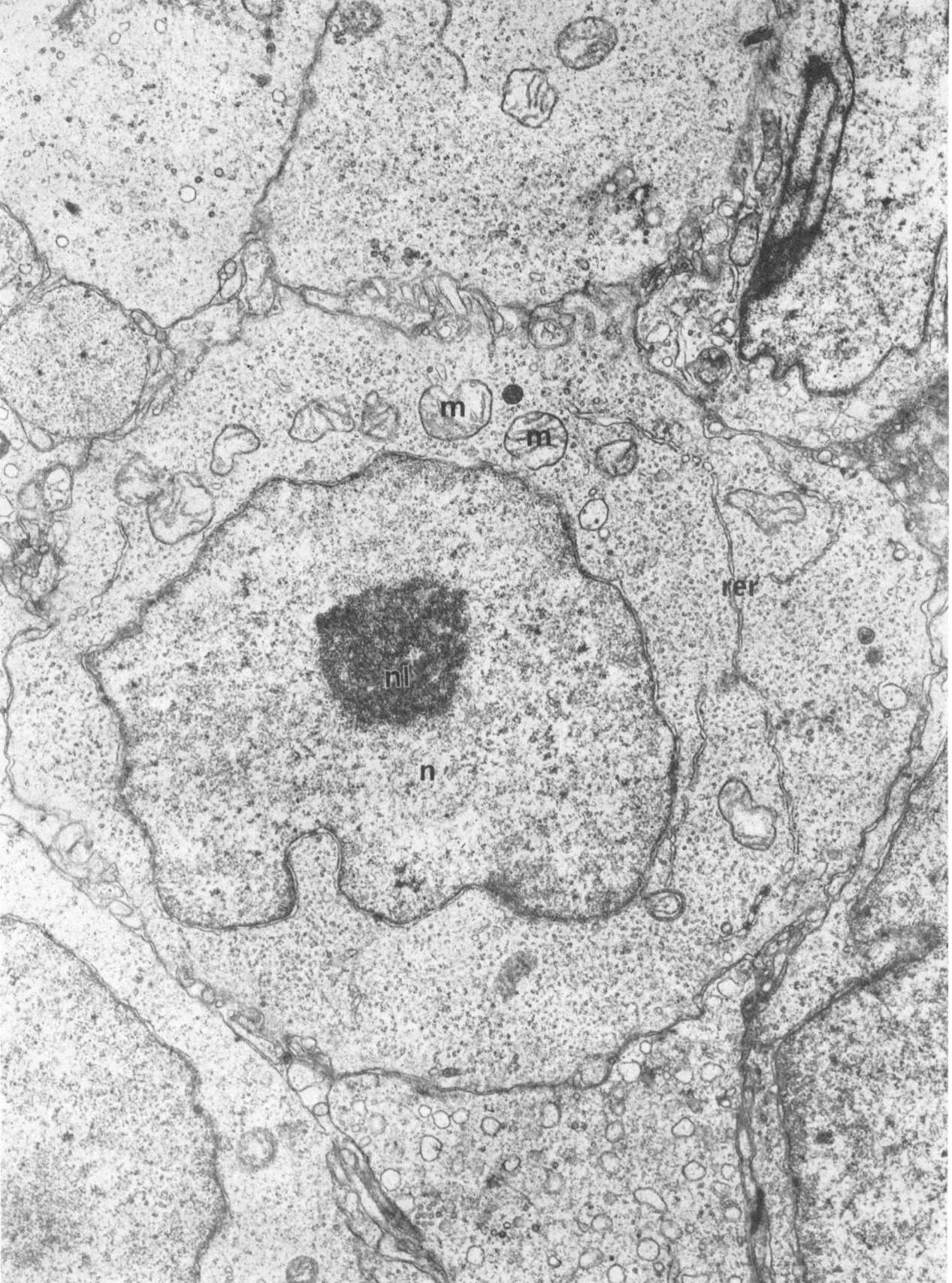

Fig. 4   Basophilic stem cell of germinal center

A giant nucleolus is apposed to the nuclear membrane in the region of a nuclear pocket. Numerous ribosomes, polysomes and little serpentine rough endoplasmic reticulum are recognized in the cytoplasm.

Case 31. ×17,500

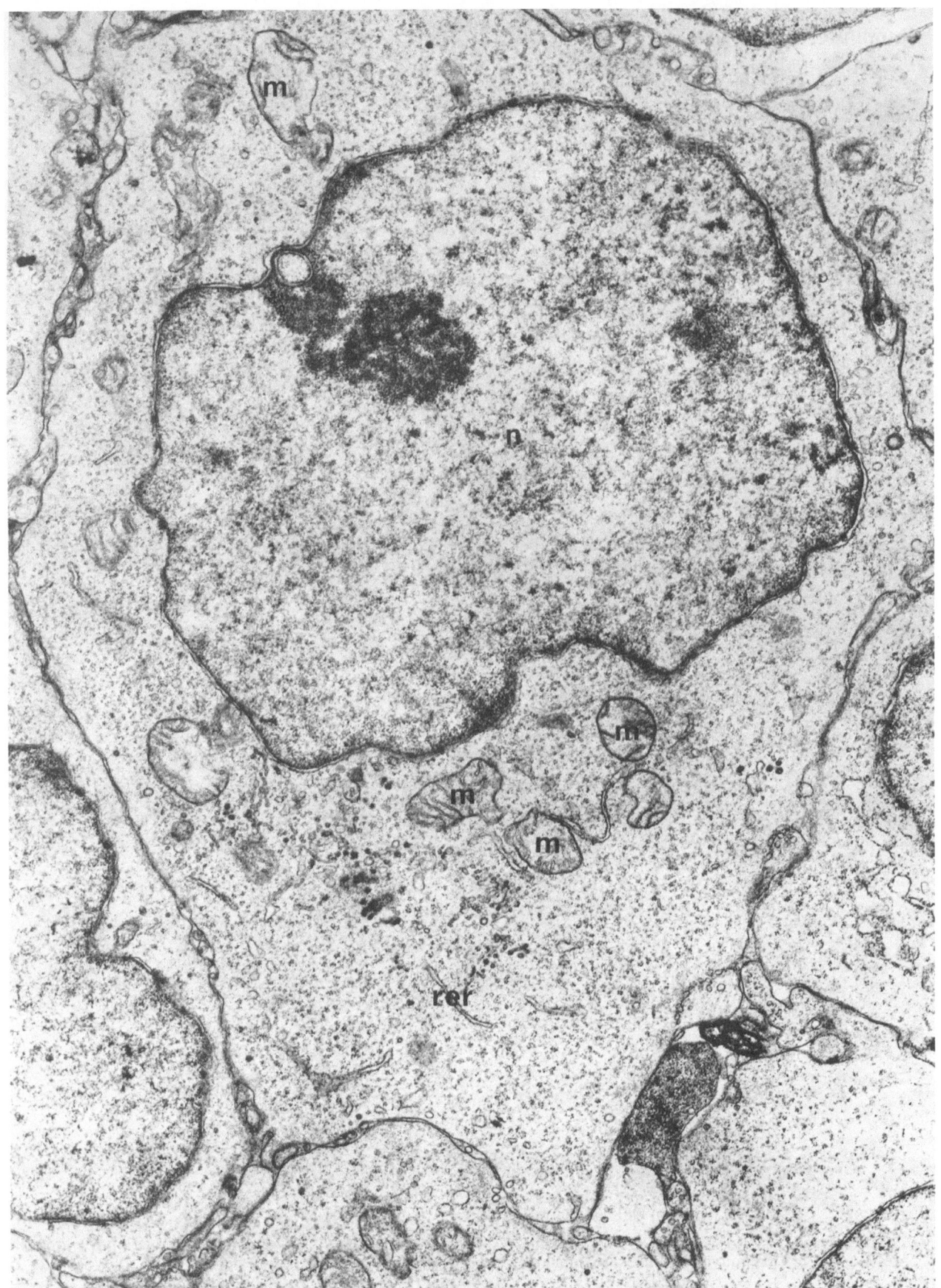

Fig. 5   Basophilic stem cells in the pulp
Note large prominent nucleoli (*nl*), large ovoid mitochondria and scarcity of rough endo-
plasmic reticulum. Two nuclear pockets are shown in the upper right cell.
Case 15. ×17,500

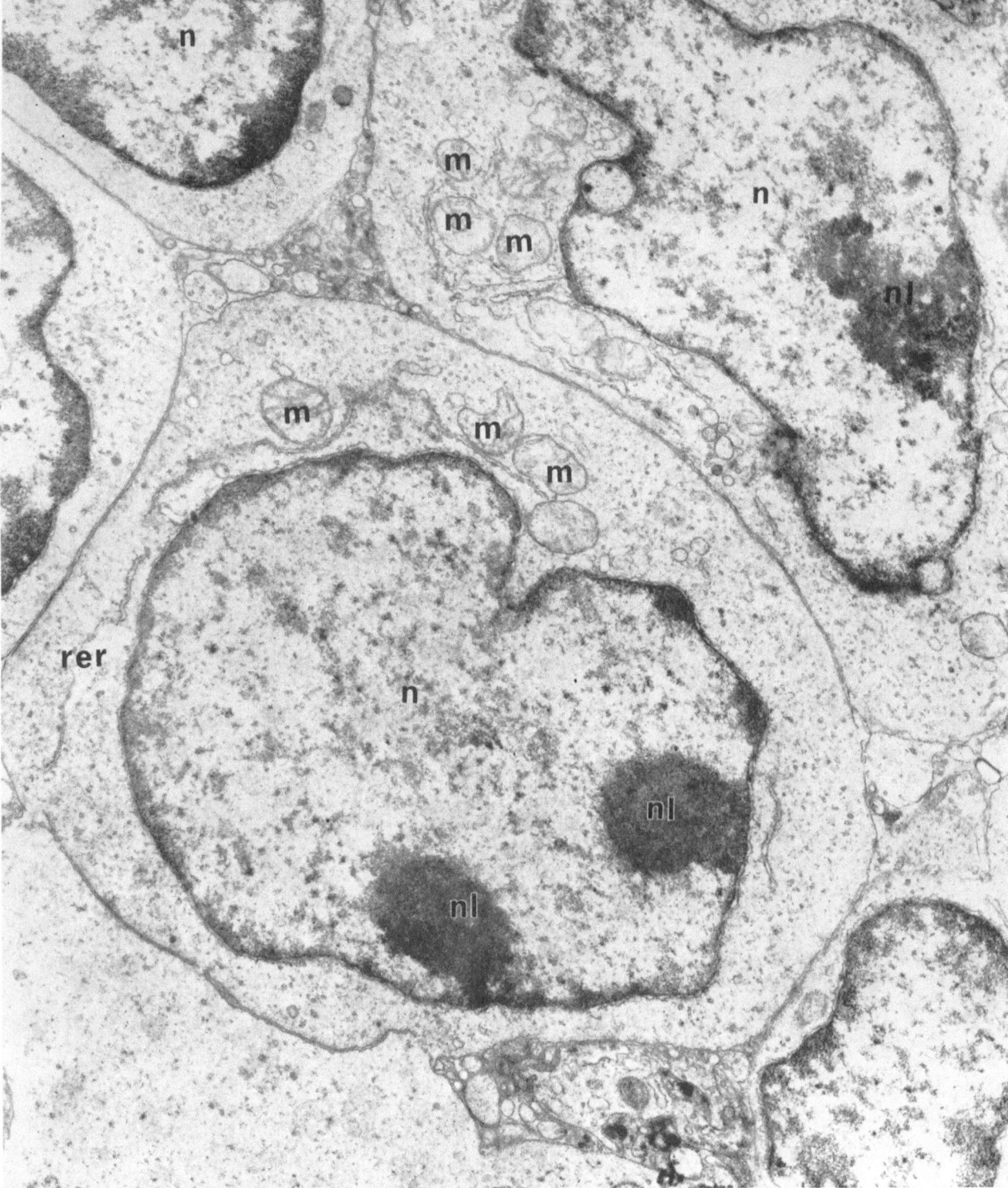

Fig. 6   Plasma cell
Extremely prominent Golgi apparatus and enormous accumulation of rough lamellar endoplasmic reticulum in parallel arrangement characterize the cytoplasm of the plasma cell.

Case 63. ×20,000

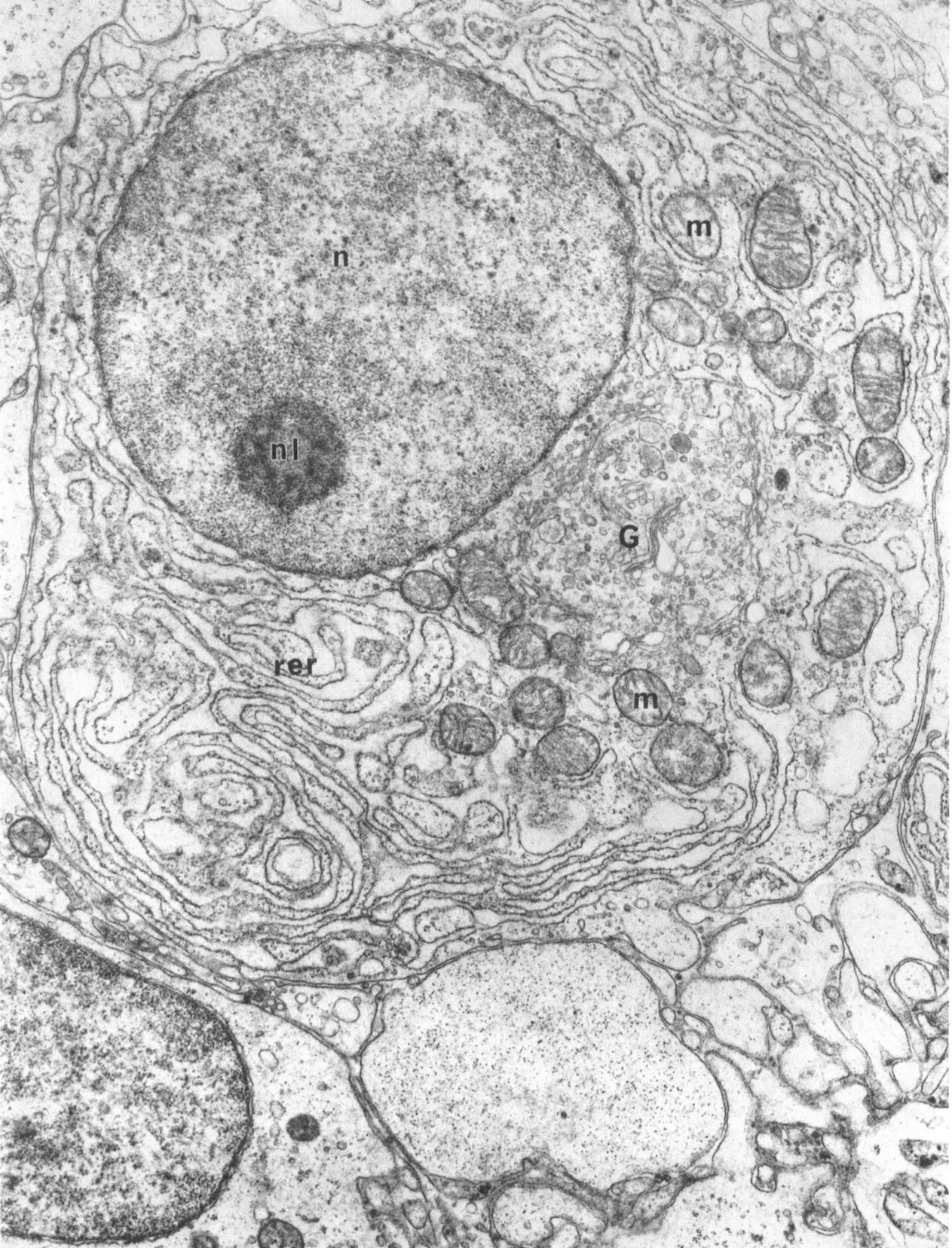

Fig. 7   Plasma cells with so-called lipochondrion
              The lipid droplet in the upper right cell corresponds to the so-called lipochondrion seen
              with the light microscope.
              Case 33. ×17,500

Inset   So-called lipochondrion with concentric runs of rough endoplasmic reticulum
              Case 40. ×23,500

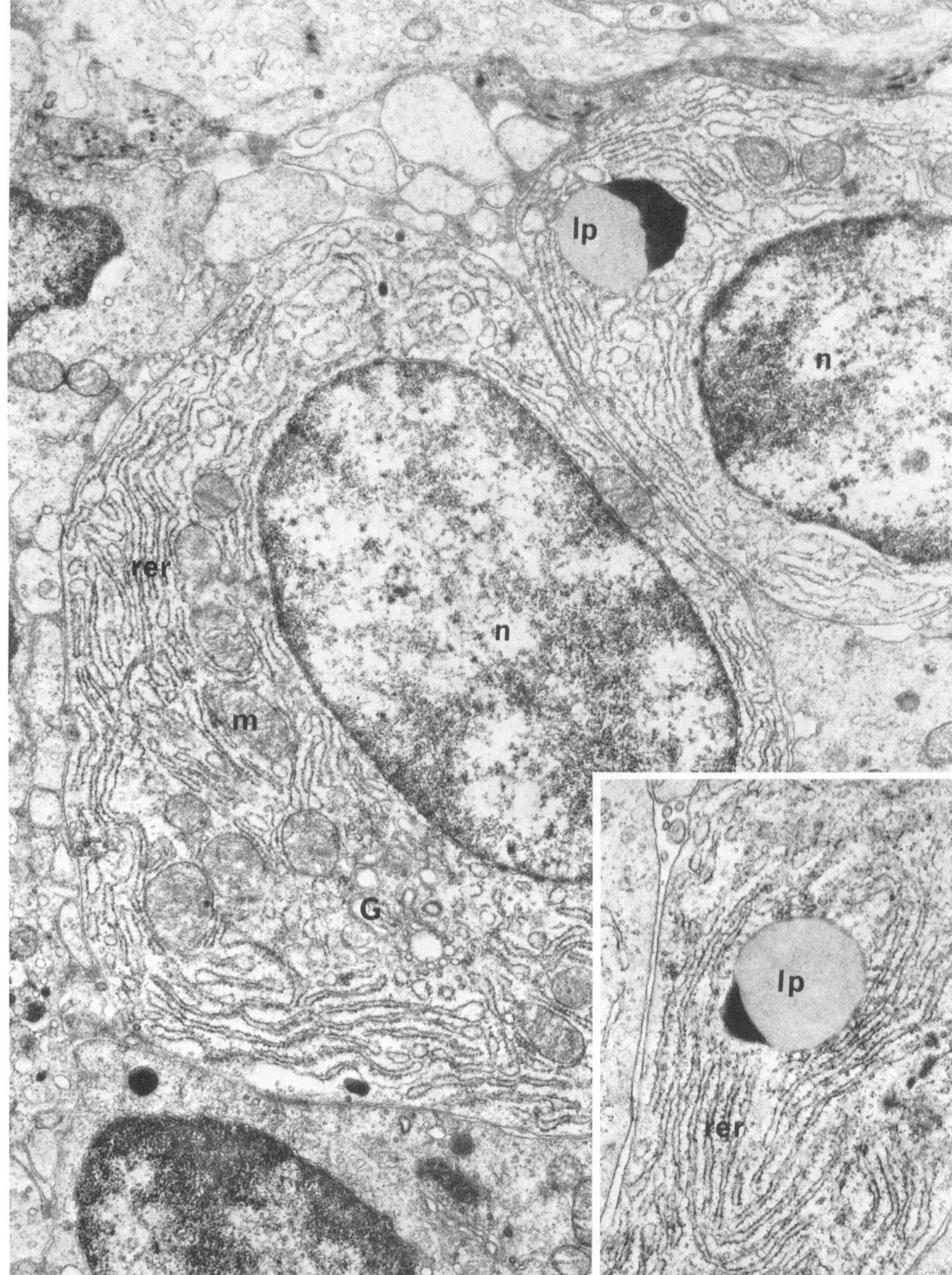

Fig. 8   Proplasmacyte
Lamellar rough endoplasmic reticulum is reduced in amount as compared with mature plasma cell.
Case 63. ×17,500

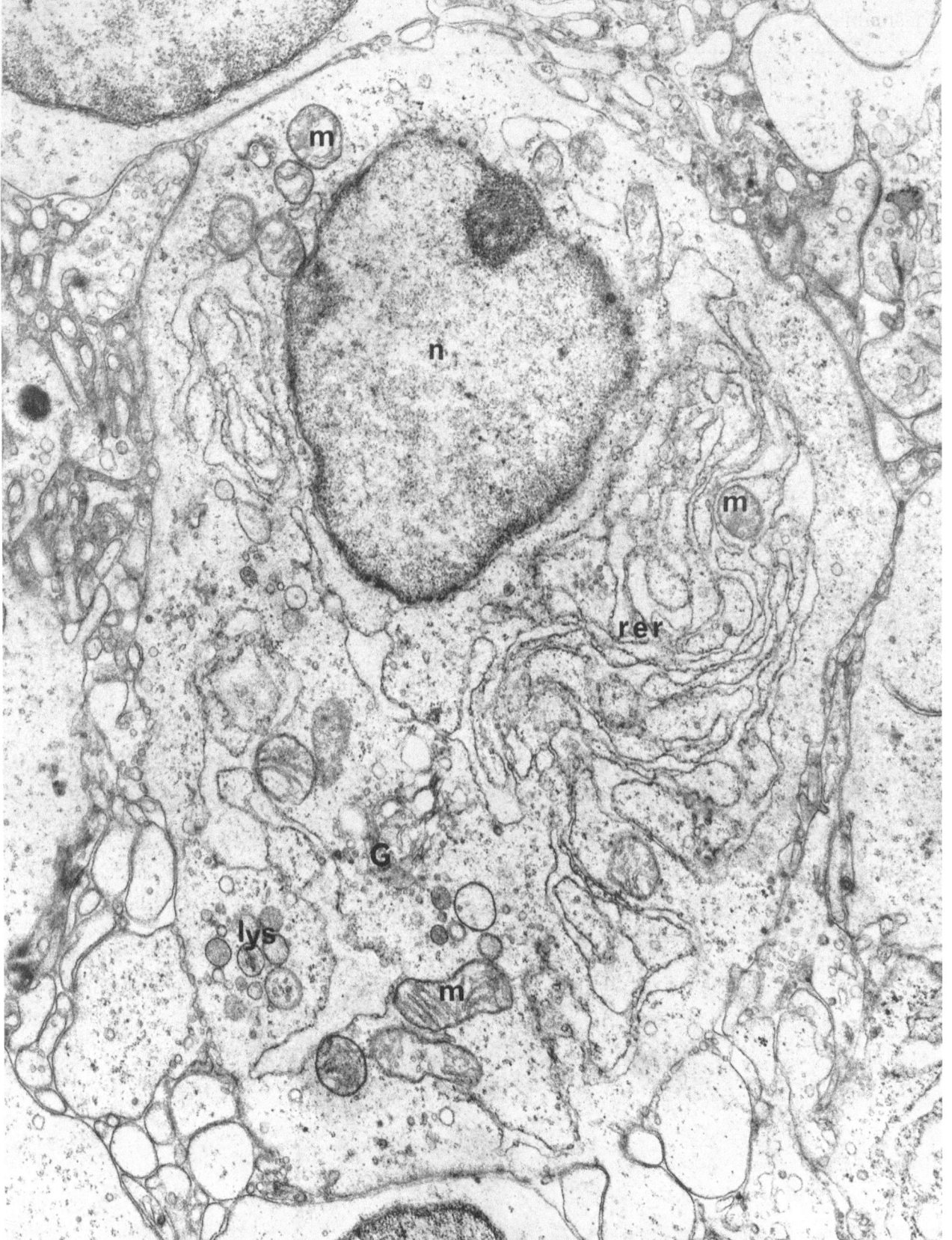

Fig. 9   Plasmablast
Note large nucleoli and moderate amount of rough endoplasmic reticulum.
Hyperplastic tonsil. Contributed by H.-K. MÜLLER-HERMELINK. ×17,500

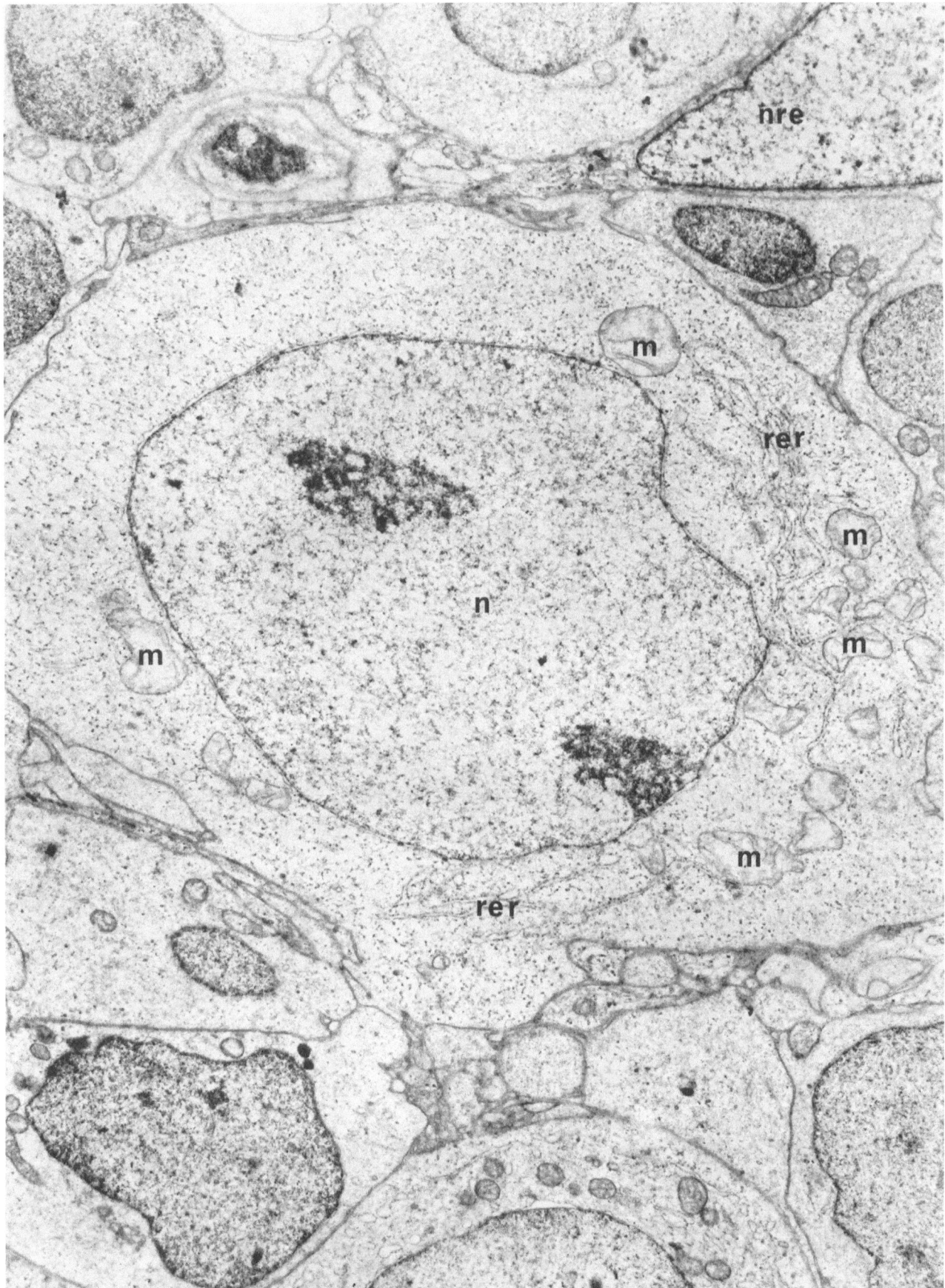

Fig. 10   Nonphagocytic long-branching reticulum cell

The cytoplasm is provided with irregular mitochondria, somewhat swollen rough endoplasmic reticulum and filaments. A reticulum fiber composed of a few collagen fibrils
and embedded in ground substance is demonstrated.

Case 47. ×17,500

Inset   Higher magnification of the filaments in the reticulum cell

×32,500

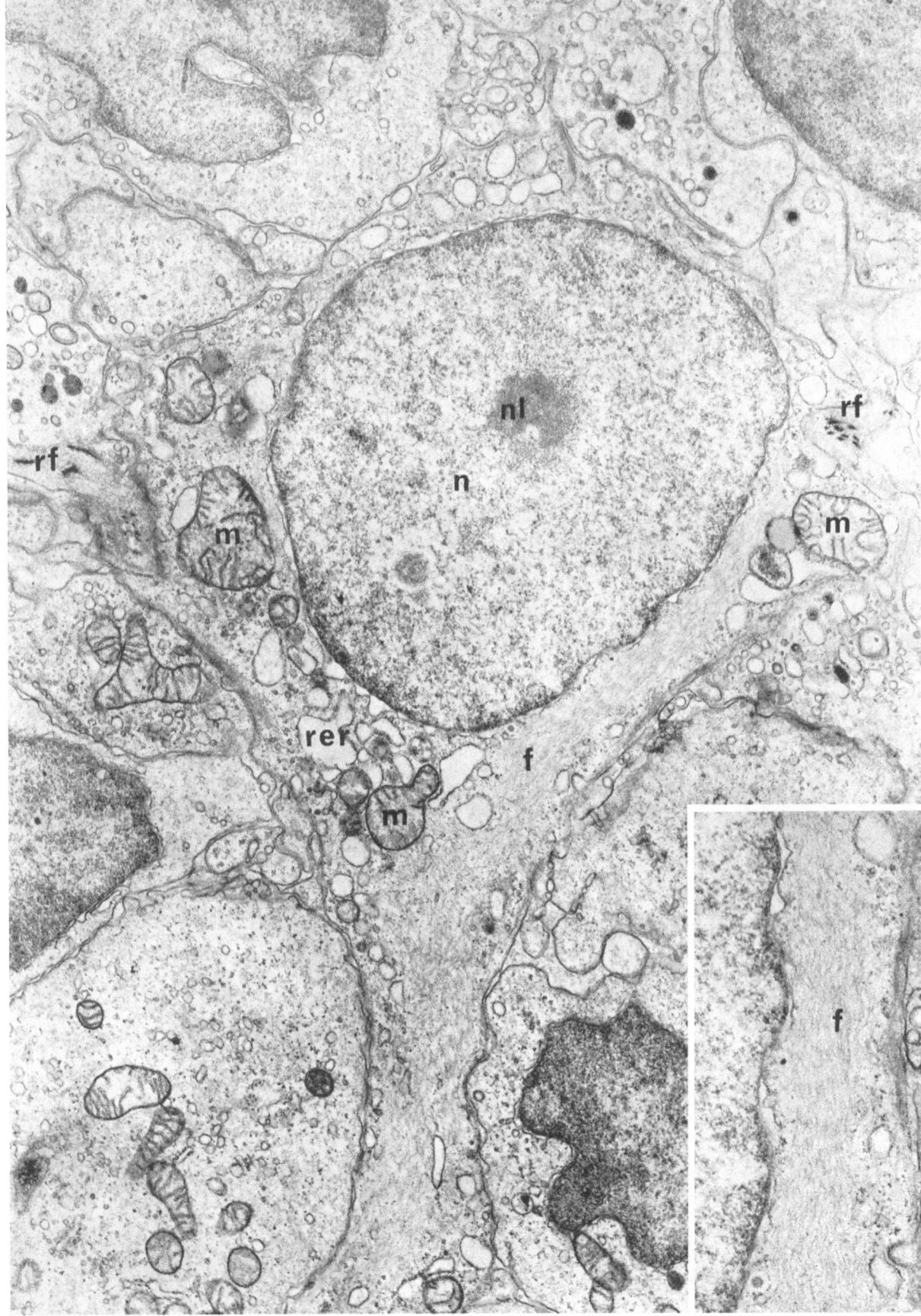

Fig. 11   Long-branching reticulum cell with desmosomal connection in germinal center adjacent to a capillary.

Case 63. ×17,500

Inset   Higher magnification of the desmosome.

×35,000

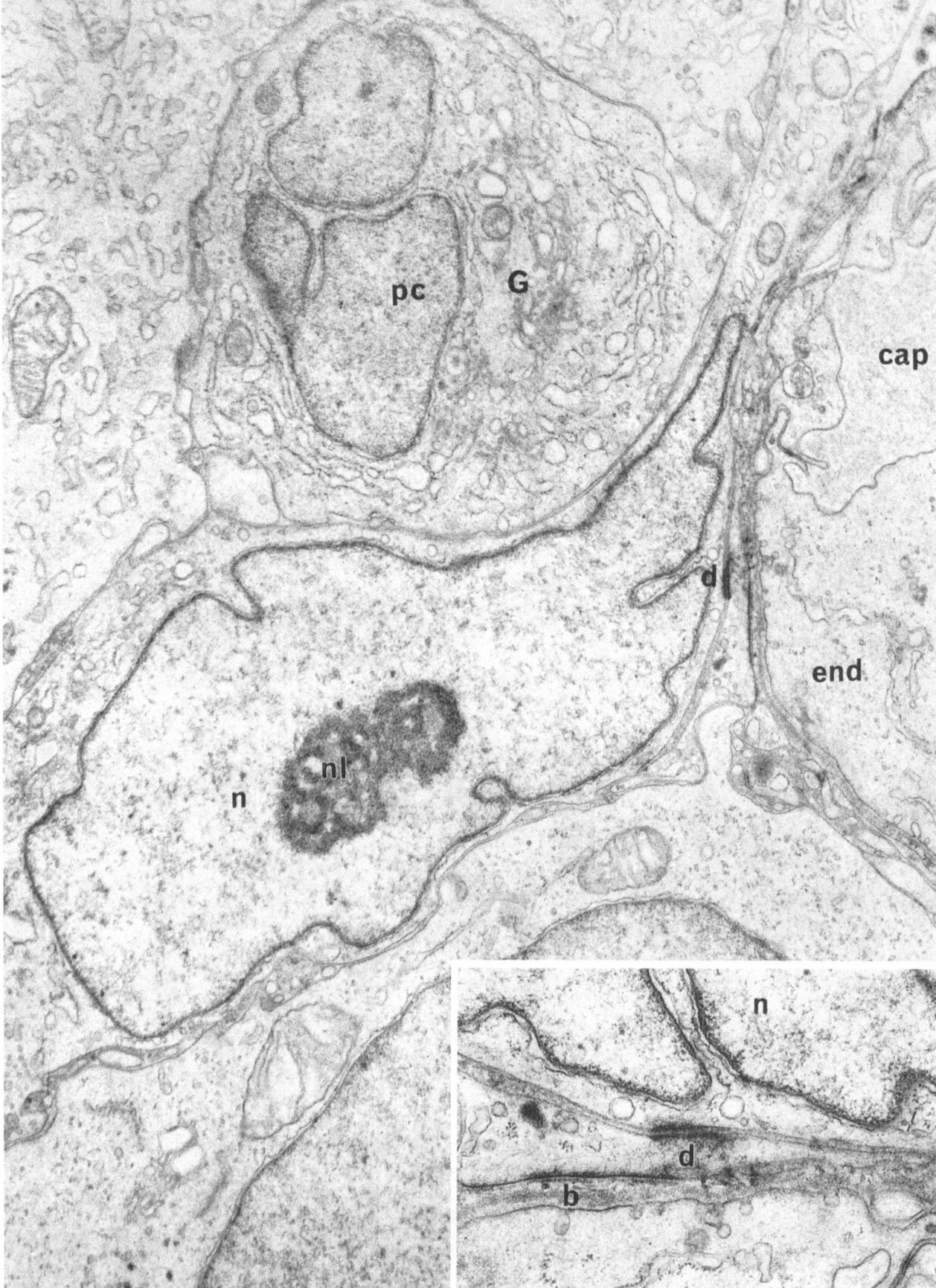

Fig. 12   Nonphagocytic reticulum cell in germinal center without demonstrable desmosomes
The cell in the center exhibits poorly developed Golgi complexes and fine cytoplasmic extensions. Above the cell in the center a thick reticulum fiber is displayed consisting of bundles of collagen fibers embedded in a ground substance. The fine structure of the cell is identical with that of a long-branching reticulum cell with desmosomal attachments, but desmosomes are not visible in this section.

Tonsil. ×17,500

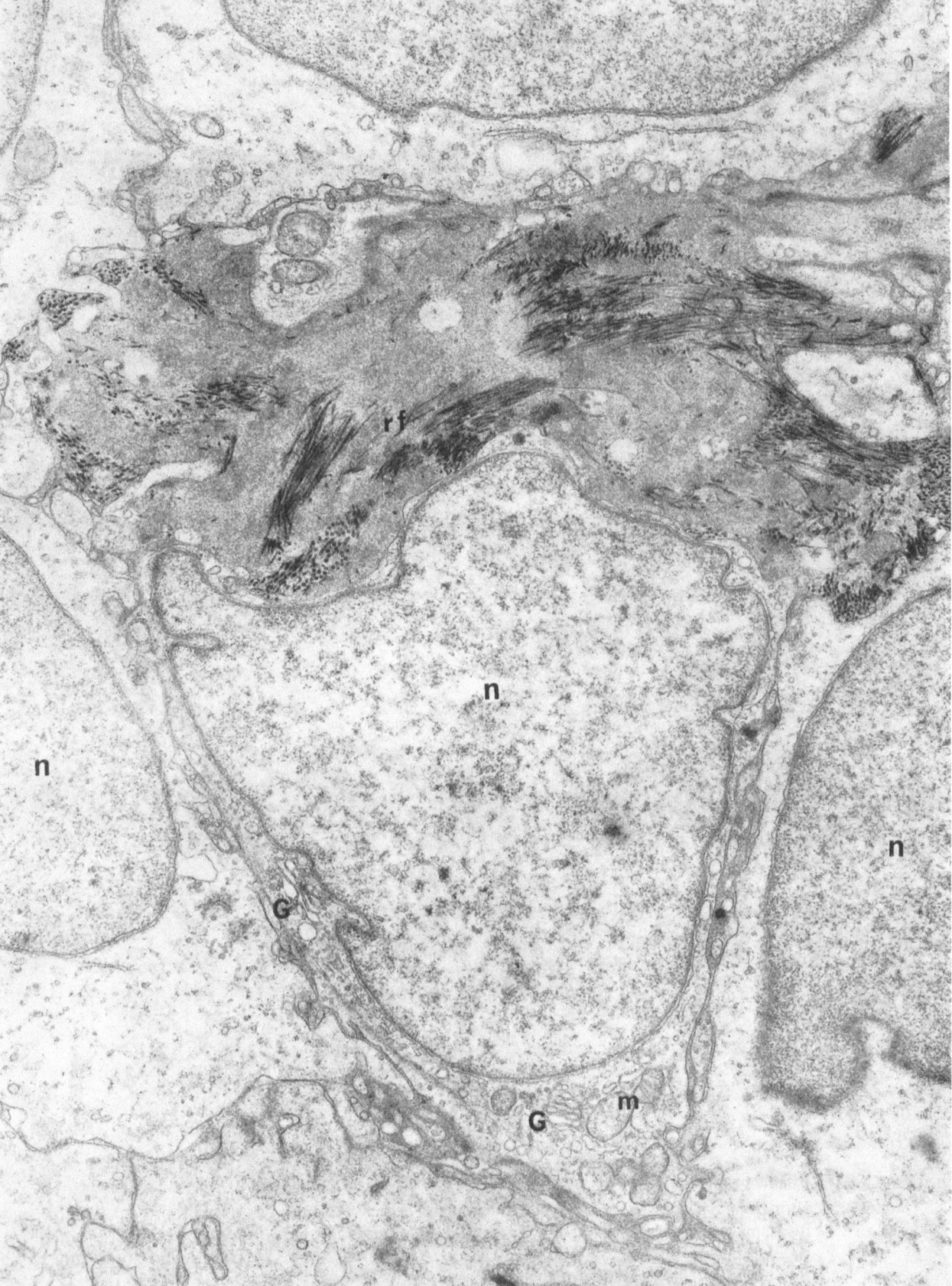

Fig. 13   Phagocytic reticulum cell

Numerous lysosomes are seen in the cytoplasm. Some of them presumably represent phagolysosomes (probably those near the upper right nuclear border). Many processes arise from the cytoplasm, but they are not as delicate as those of the nonphagocytic long-branching reticulum cell. Some of the mitochondria are rod-shaped.

Case 49. ×17,500

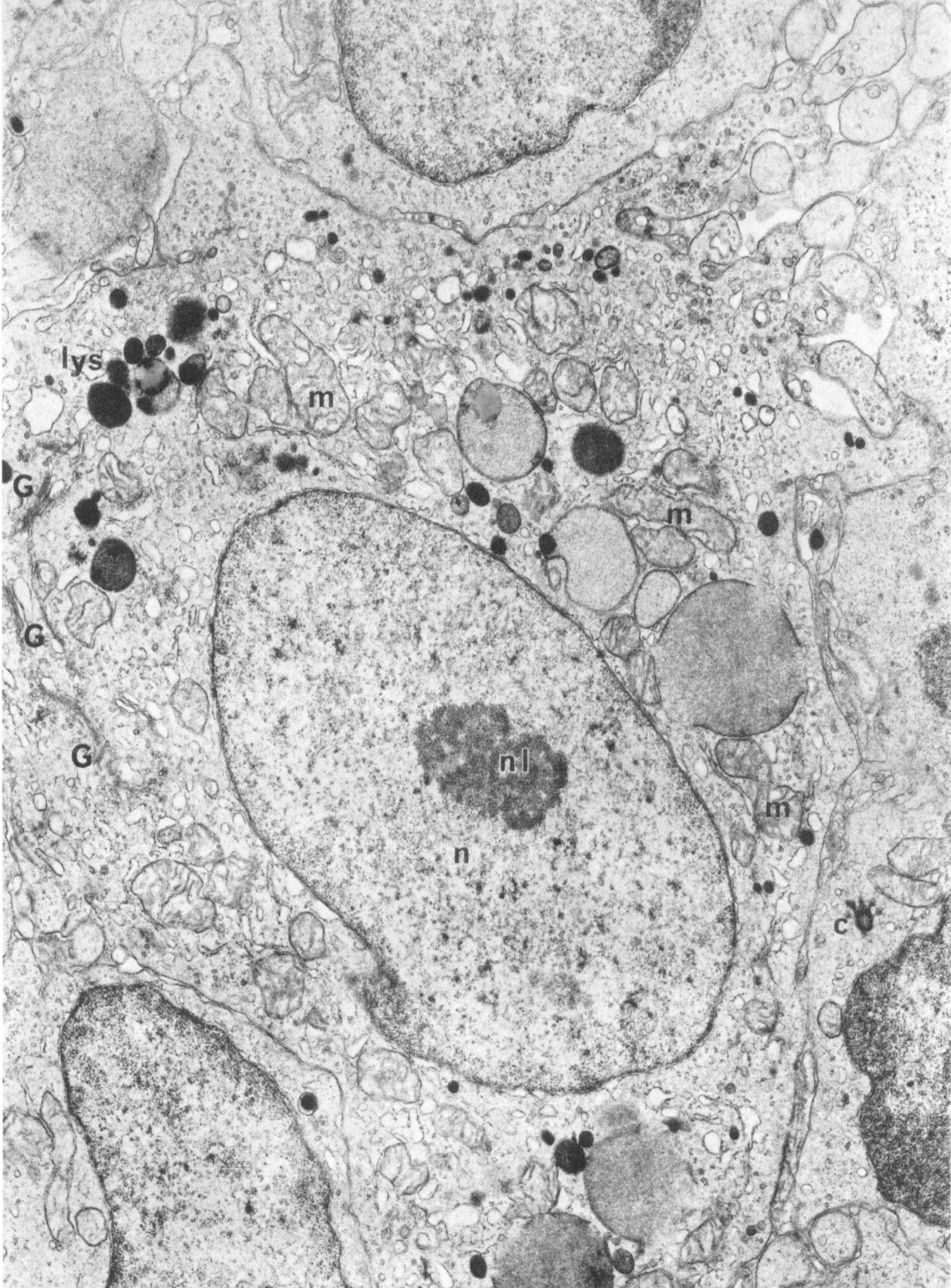

Fig. 14   Phagocytic reticulum cell (*pre*) and nonphagocytic reticulum cell (*nre*) in germinal center

The nonphagocytic reticulum cell possesses scanty cytoplasm with fine processes, while the cytoplasm of the phagocytic reticulum cell is abundant and contains many lysosomes and phagolysosomes.

Tonsil. ×17,500

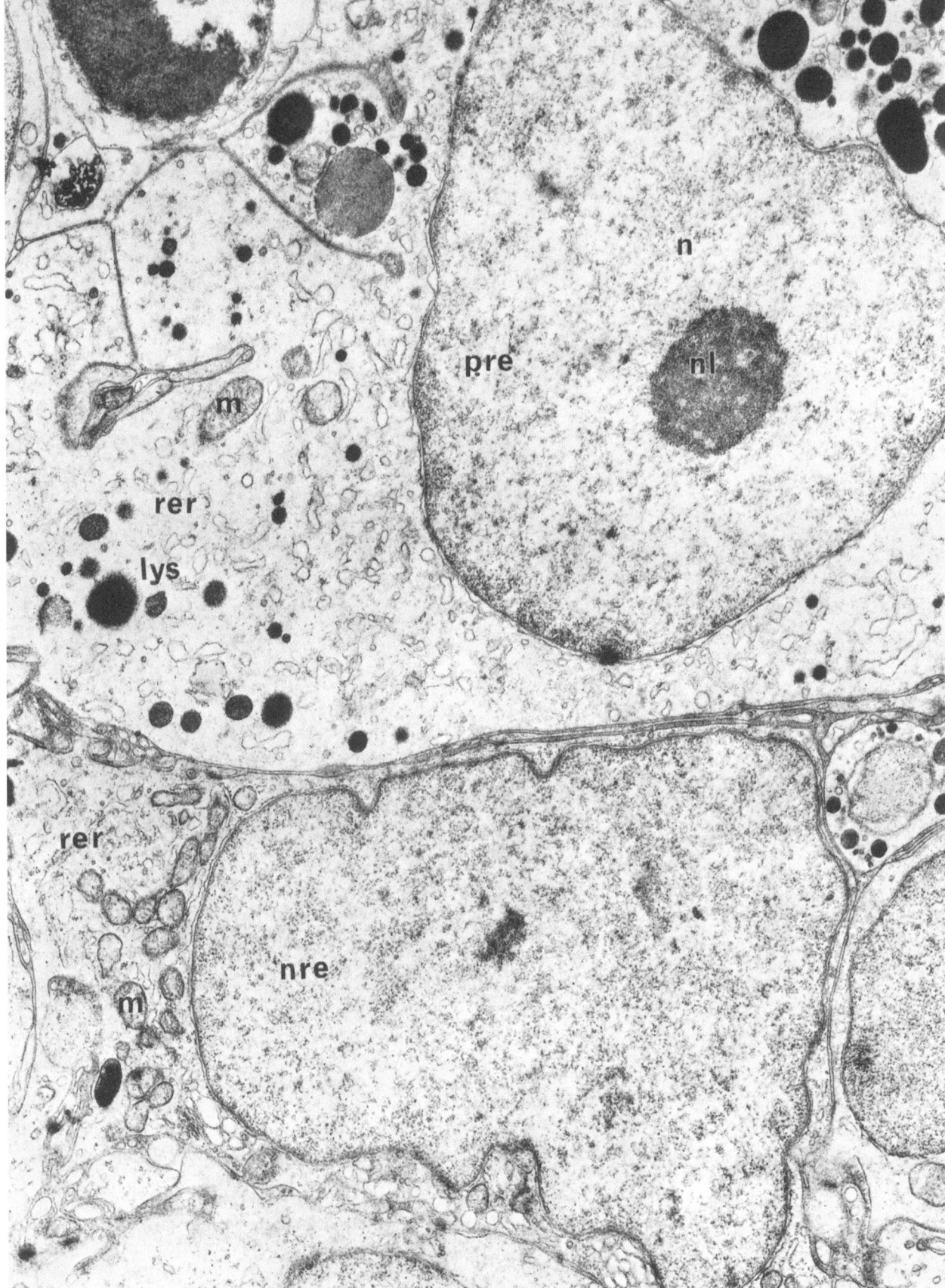

Fig. 15a and b Cytoplasmic processes of both nonphagocytic and phagocytic reticulum cells in germinal centers

The nonphagocytic reticulum cell (a) displays finer processes than the phagocytic reticulum cell (b). Two nonphagocytic reticulum cells are united by desmosomes (arrows, a). The intercellular spaces between the processes of the nonphagocytic cells are filled with electron-dense material.

Tonsil.   a) ×8,750   b) ×11,250

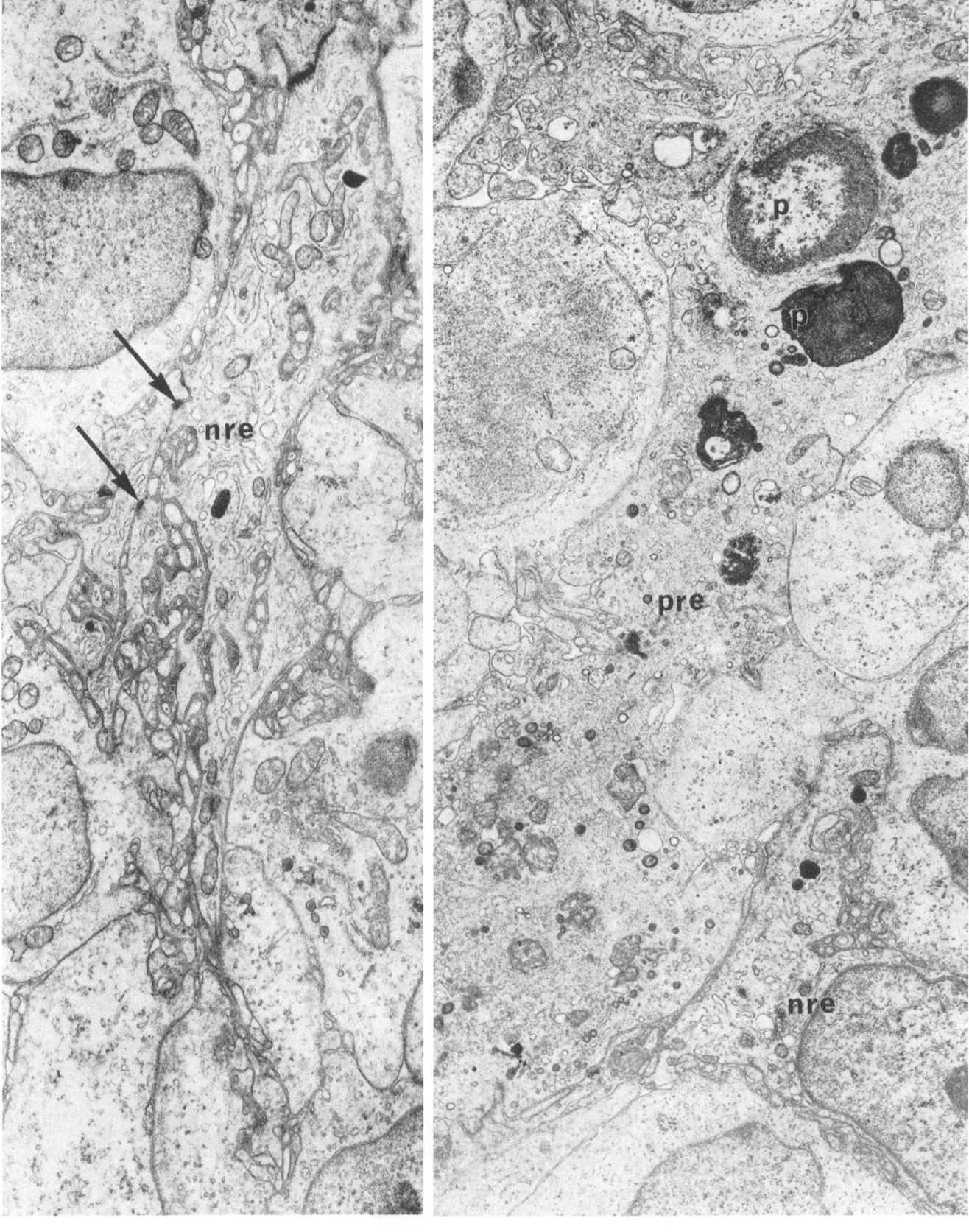

a

b

Fig. 16   Retothelium on the capsular side of the marginal sinus

The basement membrane of the retothelial cells is in contact with the lymph node capsule. The cytoplasm of the retothelial cells encloses rod-shaped mitochondria, filaments, multi-vesicular bodies and small vesicles. Some of the latter perhaps are pinocytotic vesicles. The capsule shows numerous bundles of collagen fibrils. A neutrophilic granulocyte and an immature histiocyte occupy most of the sinus lumen.

Case 35. ×17,500

Basement membrane ↑

Fig. 17   Retothelium on the inner side of the marginal sinus

The retothelial cell does not seem to rest upon a basement membrane but appears to be closely attached to the ground substance of reticulum fibers. Immature histiocytes are located in the sinus lumen.

Case 35. ×17,500

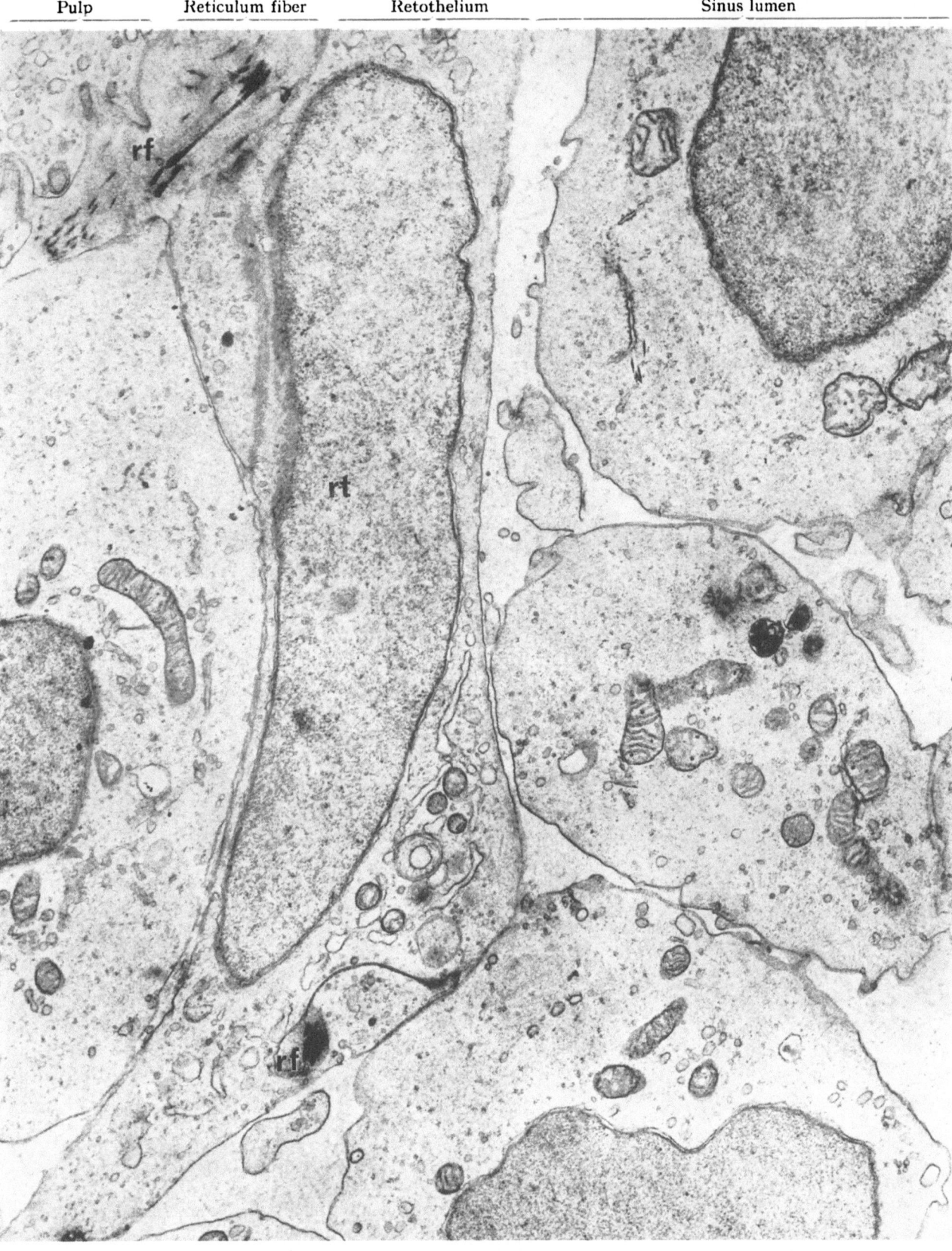

Fig. 18a and b   Retothelial cells in the sinus lumen

    a) The cytoplasm reveals small rod-shaped mitochondria, small vesicles and filaments. The reticulum fibers are surrounded by the cytoplasm of retothelial cells.

    b) Desmosomes unite processes of the retothelial cells. The desmosomes (arrows) are only poorly demonstrated due to osmium fixation.

Case 35.   a) ×17,500   b) ×12,500

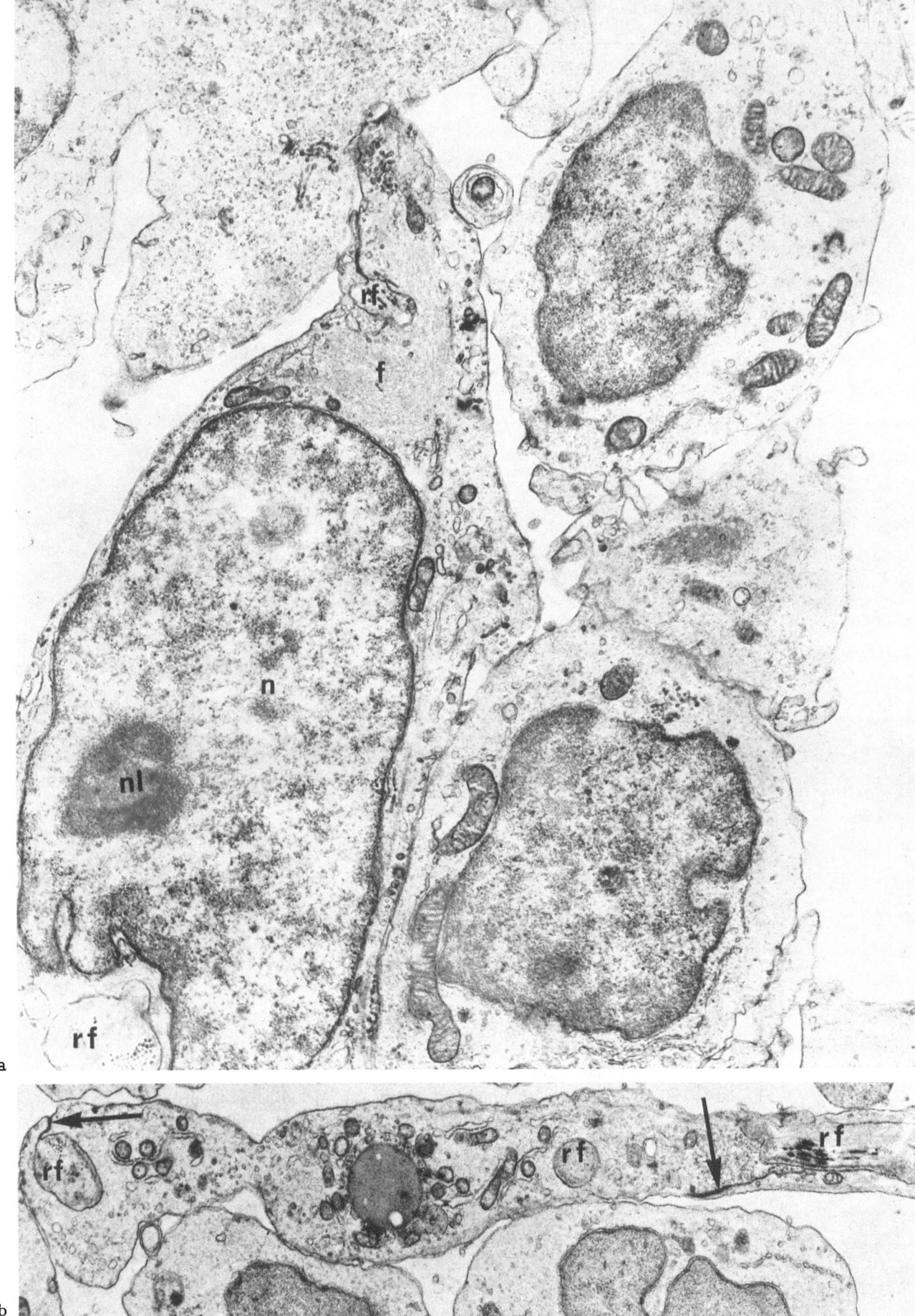

a

b

Fig. 19   Retothelium in medullary sinus
The large cell is supplied with abundant cytoplasm including prominent cytoplasmic organelles. A small number of phagolysosomes and a large amount of endoplasmic reticulum, both smooth and rough, are encountered. Cytoplasmic filaments are also present. The arrow indicates a cellular connection (desmosome). The reticulum fiber at the lower right is surrounded by another retothelial cell. At this magnification the reticulum fiber reveals fine filamentous elements in addition to collagen fibers.

Case 47. ×17,500

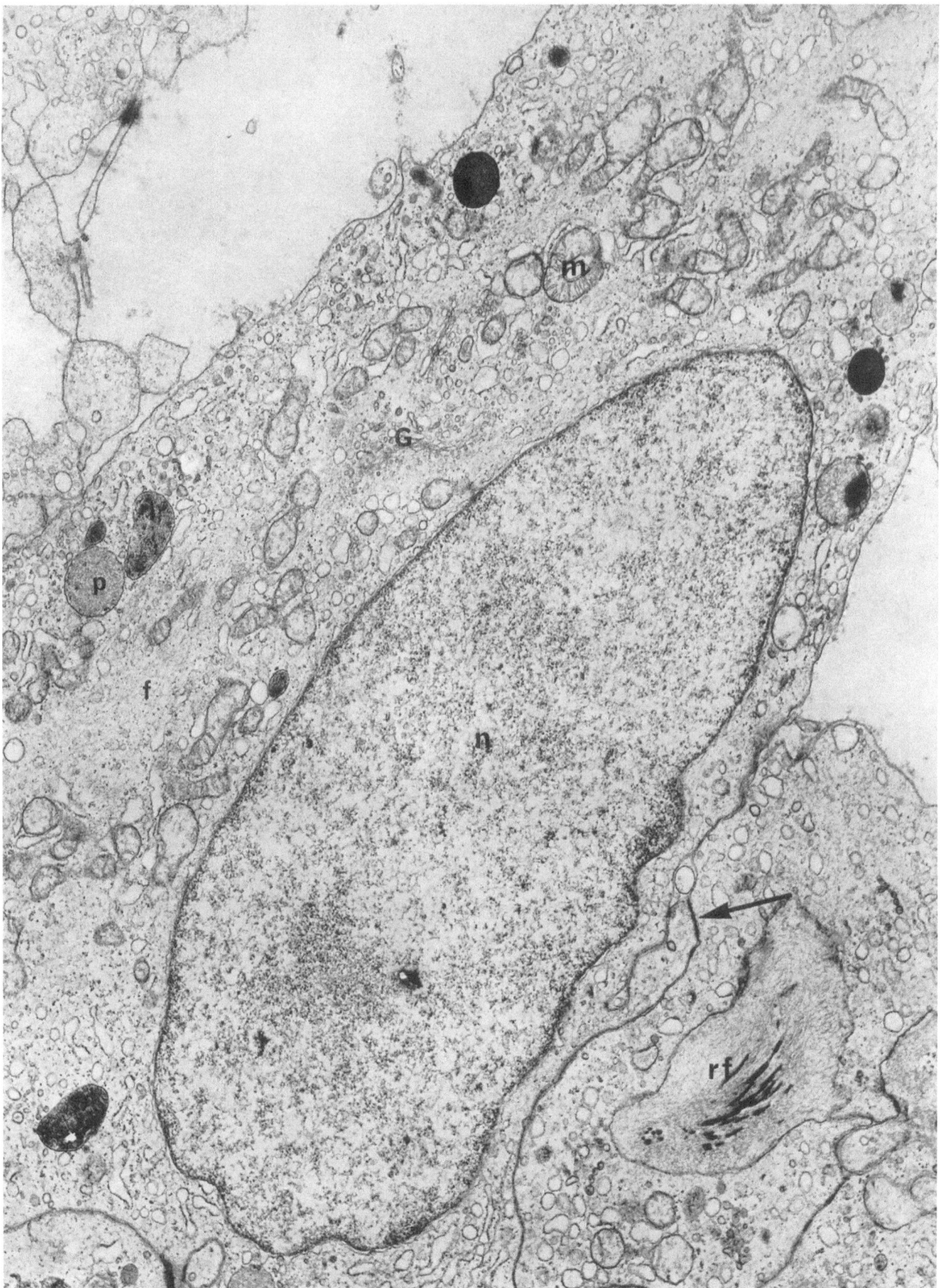

Fig. 20   Mast cell

This cell is distinguished by a conspicuous accumulation of specific granules made up of granular appearing substances of variable electron density. Some lipid droplets are illustrated in the cytoplasm. A centriole is present (upper left).

Case 47. ×21,000

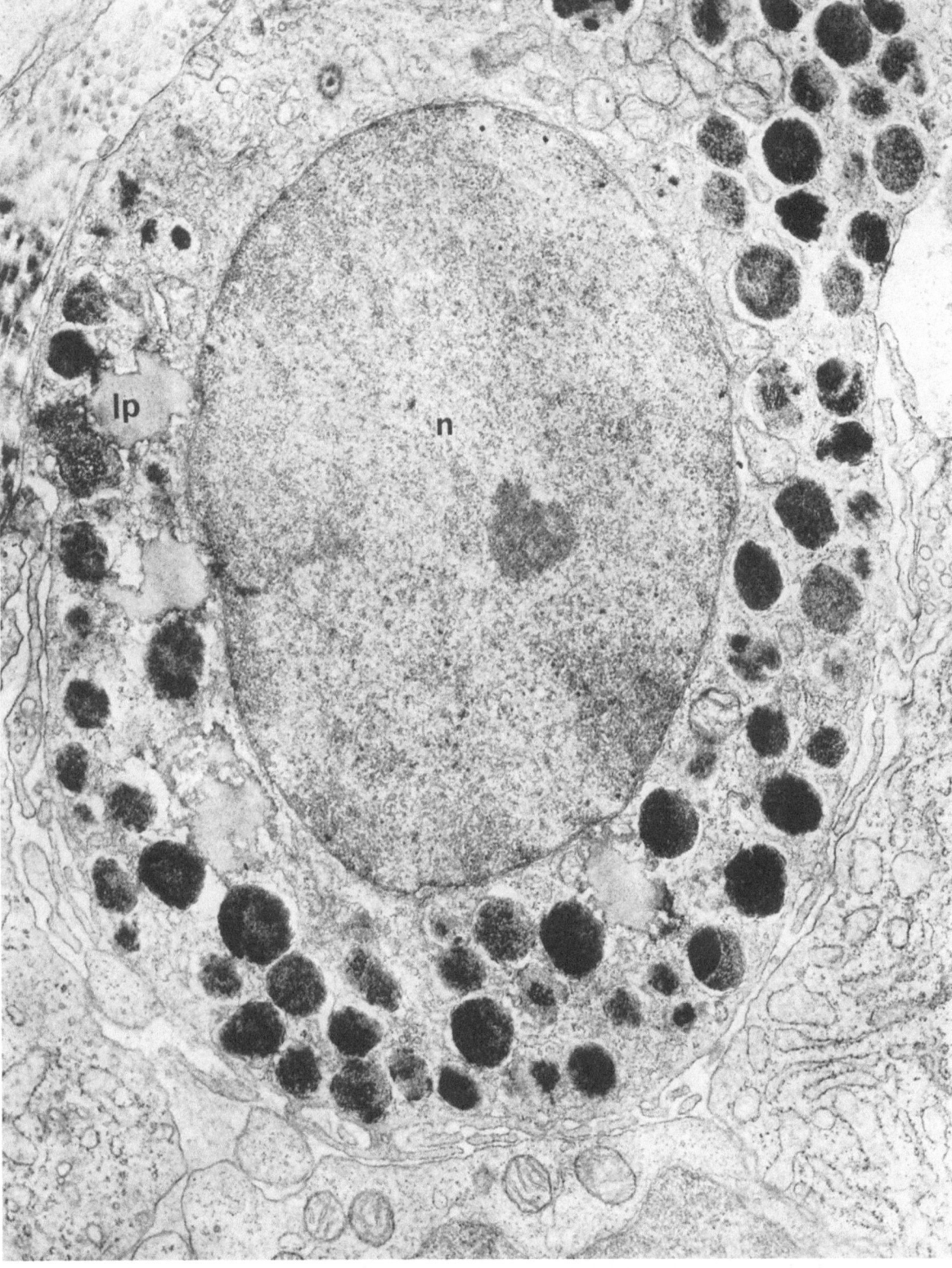

Fig. 21a   Neutrophilic polymorphonuclear leukocyte
This cell contains two types of specific granules: one is completely electron-opaque whereas the other appears less electron-dense. A limiting membrane of the granules is prominent in the latter type while it is somewhat indistinct in the former because of its high electron density. Three lipid vacuoles are also seen in the picture.

Fig. 21b   Eosinophilic leukocyte
Note characteristic granules containing crystal-like particles.

Case 50.   a) ×17,500   b) ×17,500

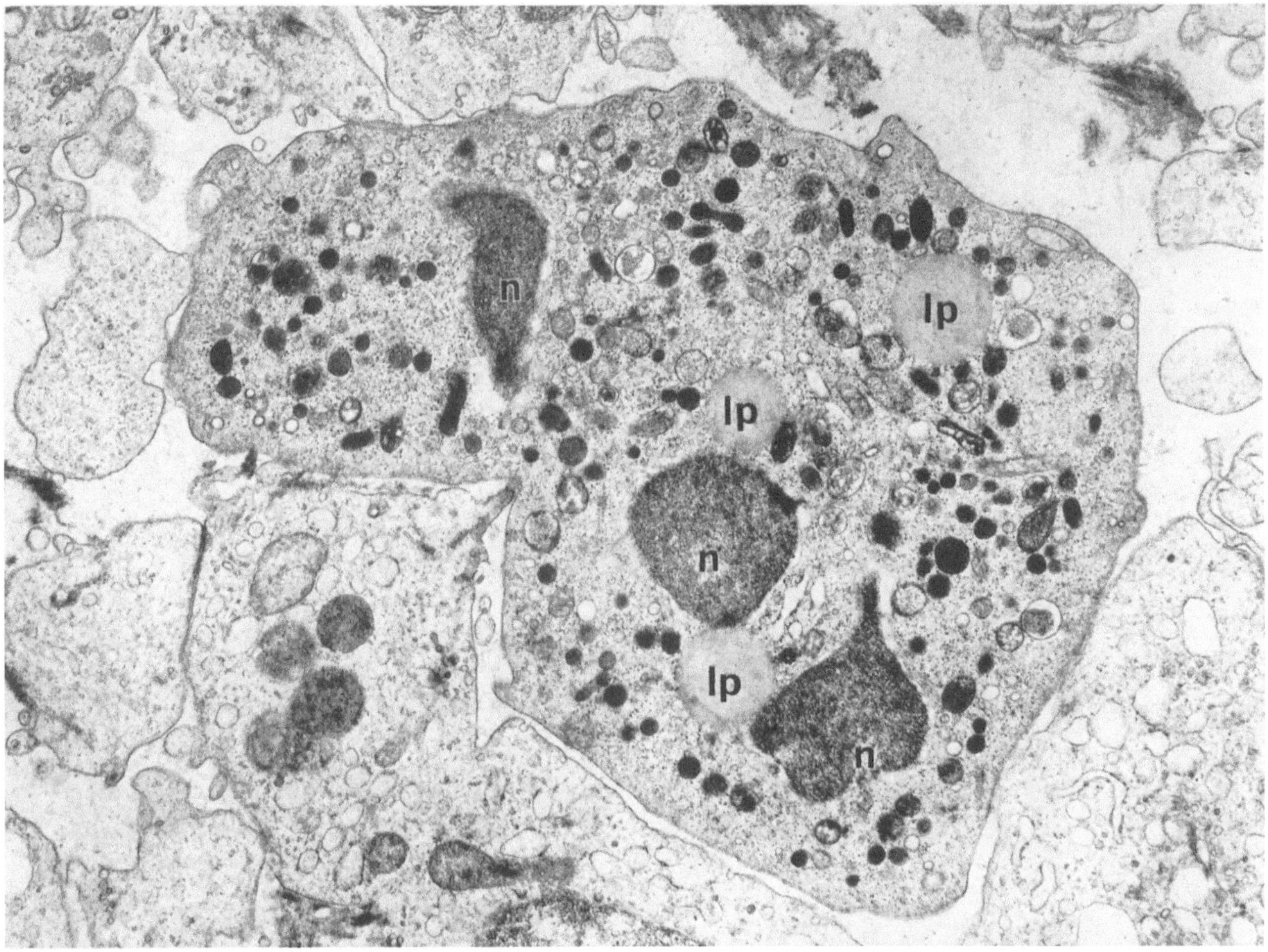

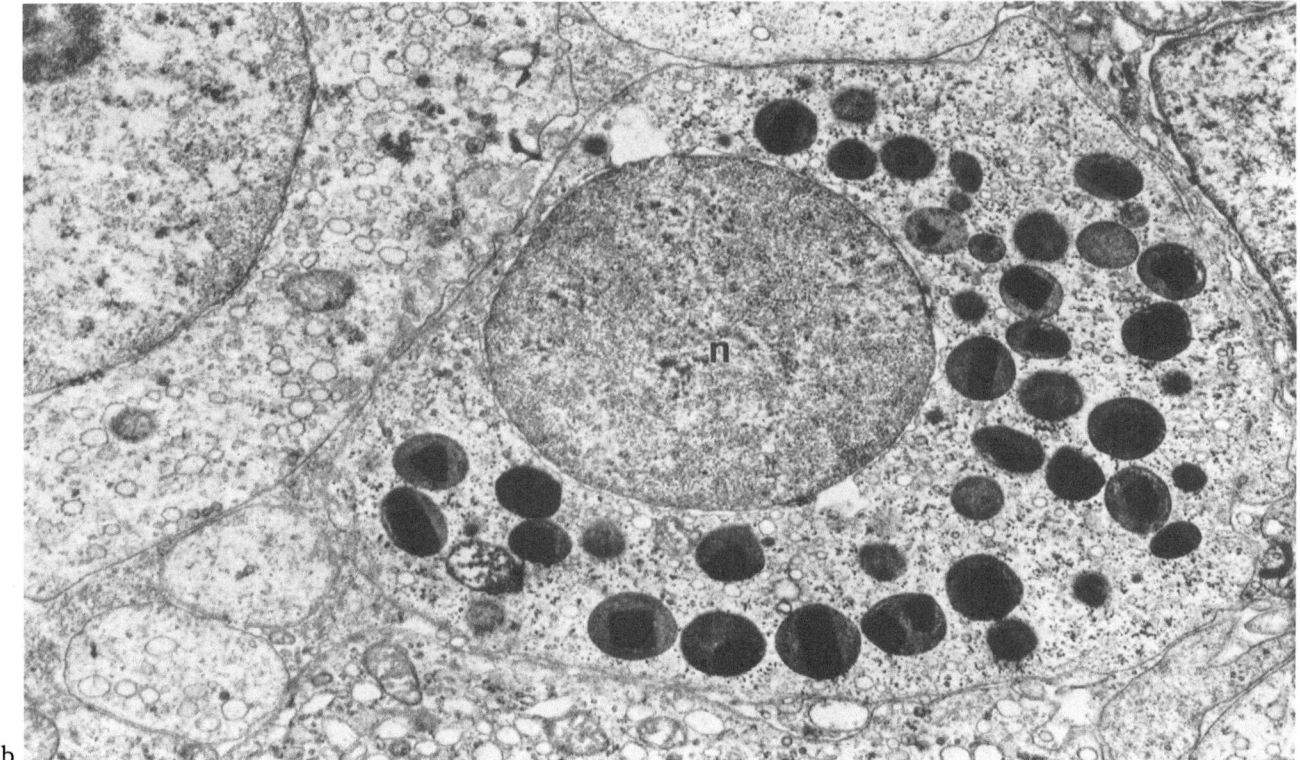

Main Structural Units of the Lymph Node
Germinal Center
Blood Vessels

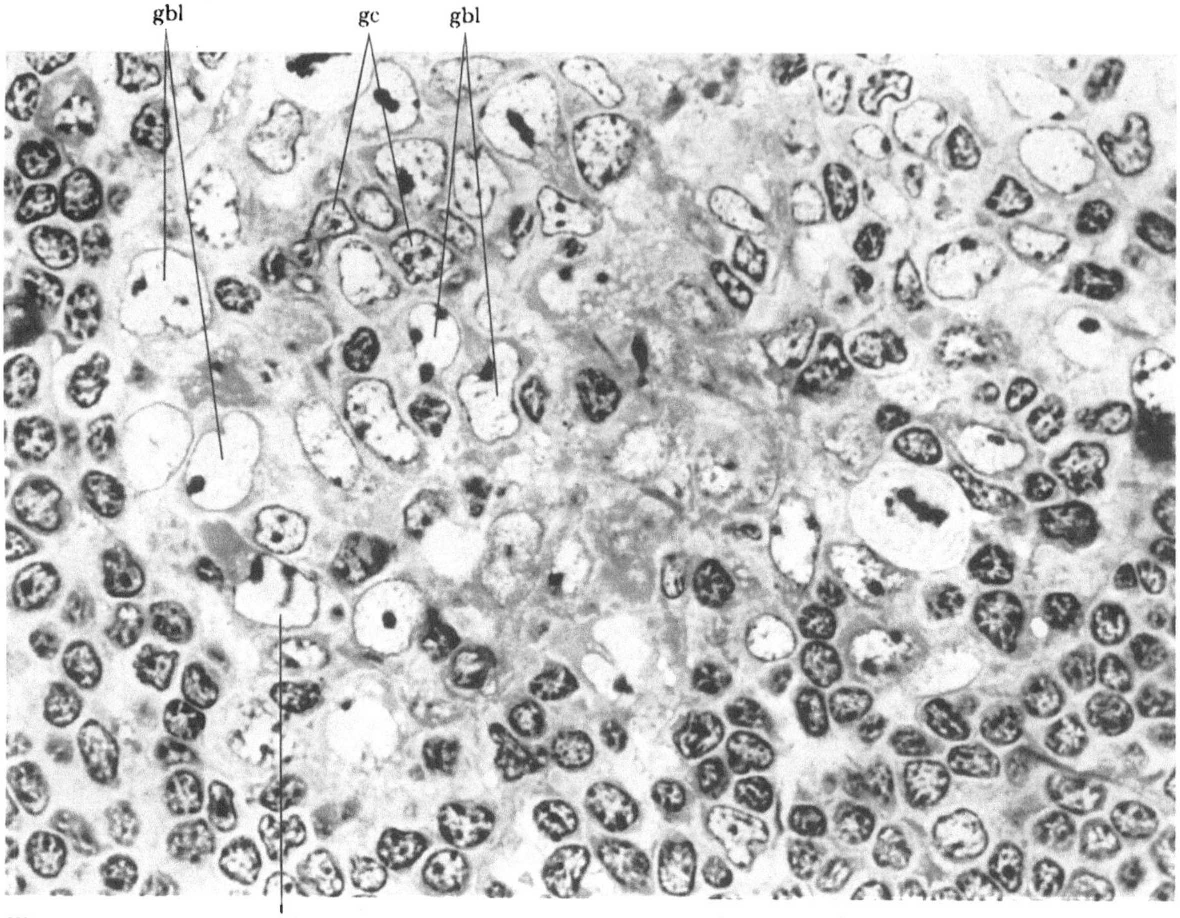

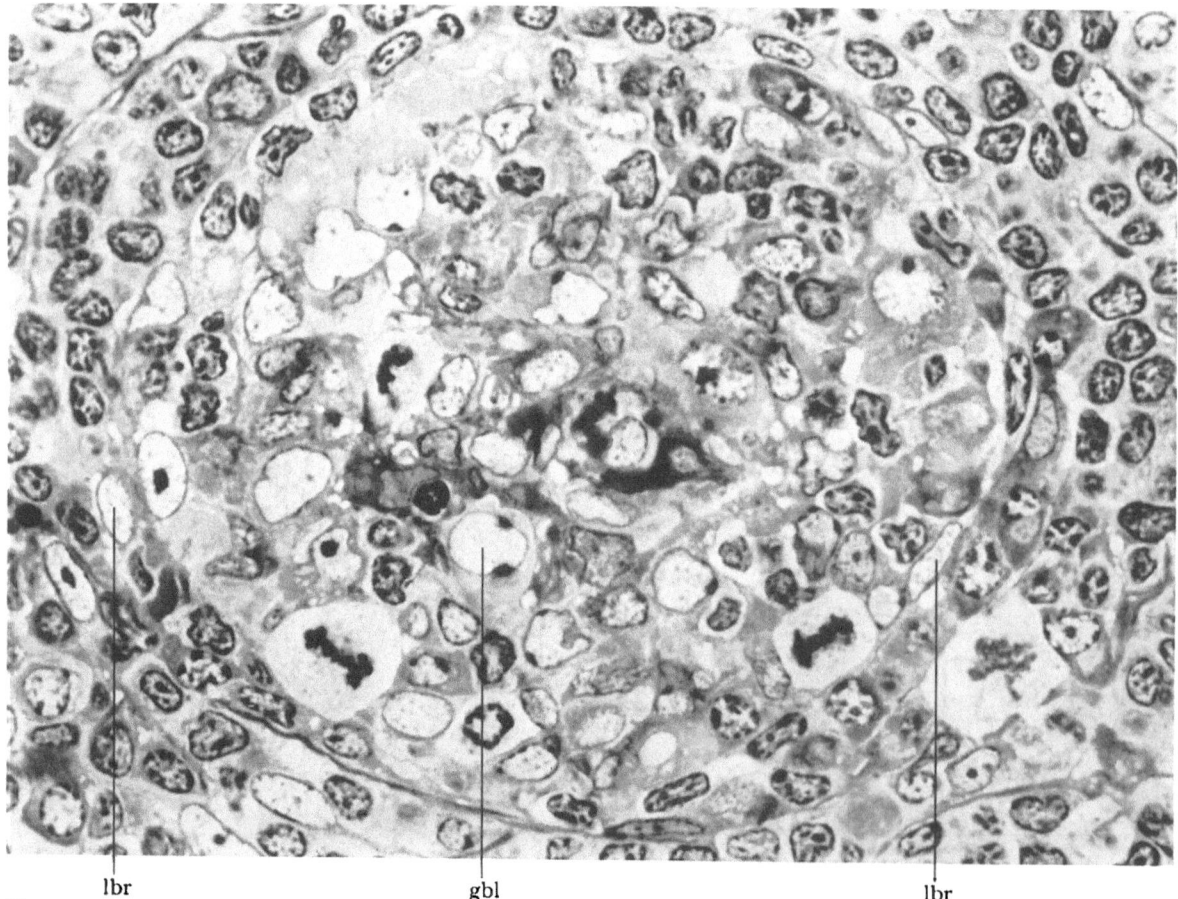

gbl   gc   gbl

gbl

Fig. 22

lbr   gbl   lbr

Fig. 23

Fig. 22   Germinal center in a tonsil

Germinoblasts (*gbl*) and germinocytes (*gc*) in germinal center. Lymphocytes encircling periphery of germinal center are evident at the left and lower margin.

Photomicrograph provided by H.-K. Müller-Hermelink

Osmium tetroxide fixation, preceded by fixation in buffered neutral formalin. Semithin section. Methylene blue-Azure II. ×1,230

Fig. 23   Another germinal center of the same tonsil illustrated in Fig. 22 (overall view)

In addition to the germinal center cells such as germinoblasts (*gbl*), several long-branching reticulum cells (*lbr*) are seen in the periphery of the germinal center in apposition to the lymphocytic cuff. Few plasma cells are also present.

Photomicrograph provided by H.-K. Müller-Hermelink

Osmium tetroxide fixation, preceded by fixation in buffered neutral formalin. Semithin section. Methylene blue-Azure II. ×1,230

Fig. 24   Germinal center of tonsil (mounted figure, see next two pages)

Nonphagocytic long-branching reticulum cells with long processes are demonstrated. Some of the processes seem to constitute a form of boundary (arrows) between the germinal center and the surrounding lymphocytic cuff. Basophilic stem cells, germinoblasts, germinocytes, plasma cells, lymphocytes and phagocytic reticulum cells are also present.

In collaboration with H.-K. Müller-Hermelink

×5,000

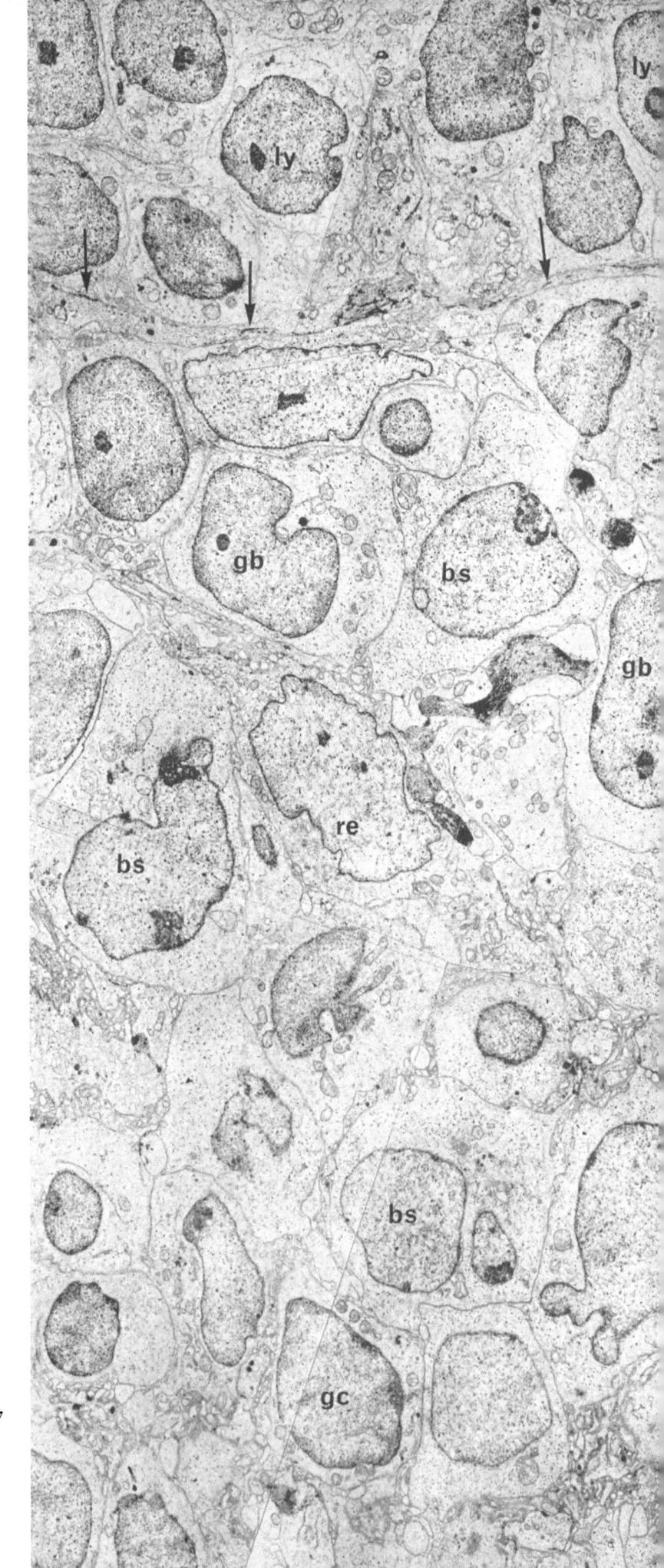

Fig. 24. For legend see page 97

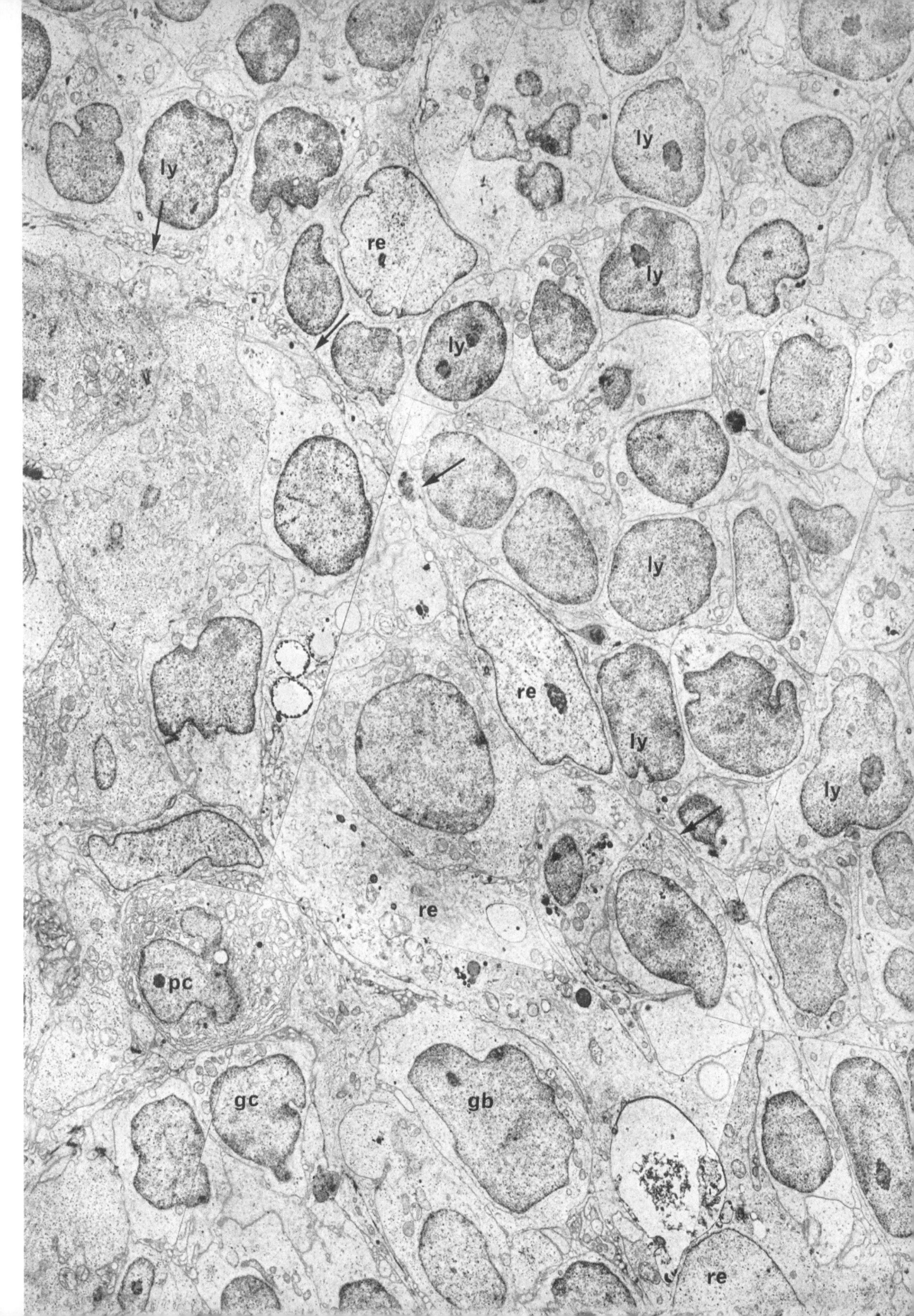

Fig. 25   Border between germinal center and the surrounding lymphocytic cuff in lymph node

The slender cytoplasmic processes (arrows) of the nonphagocytic long-branching reticulum cells extend from the upper left to the lower right between the germinal center (upper right) and the surrounding lymphocytic cuff (lower left). A portion of the cytoplasm of a phagocytic reticulum cell, a plasma cell, a germinoblast and lymphocytes are also present.

Case 37. ×8,750

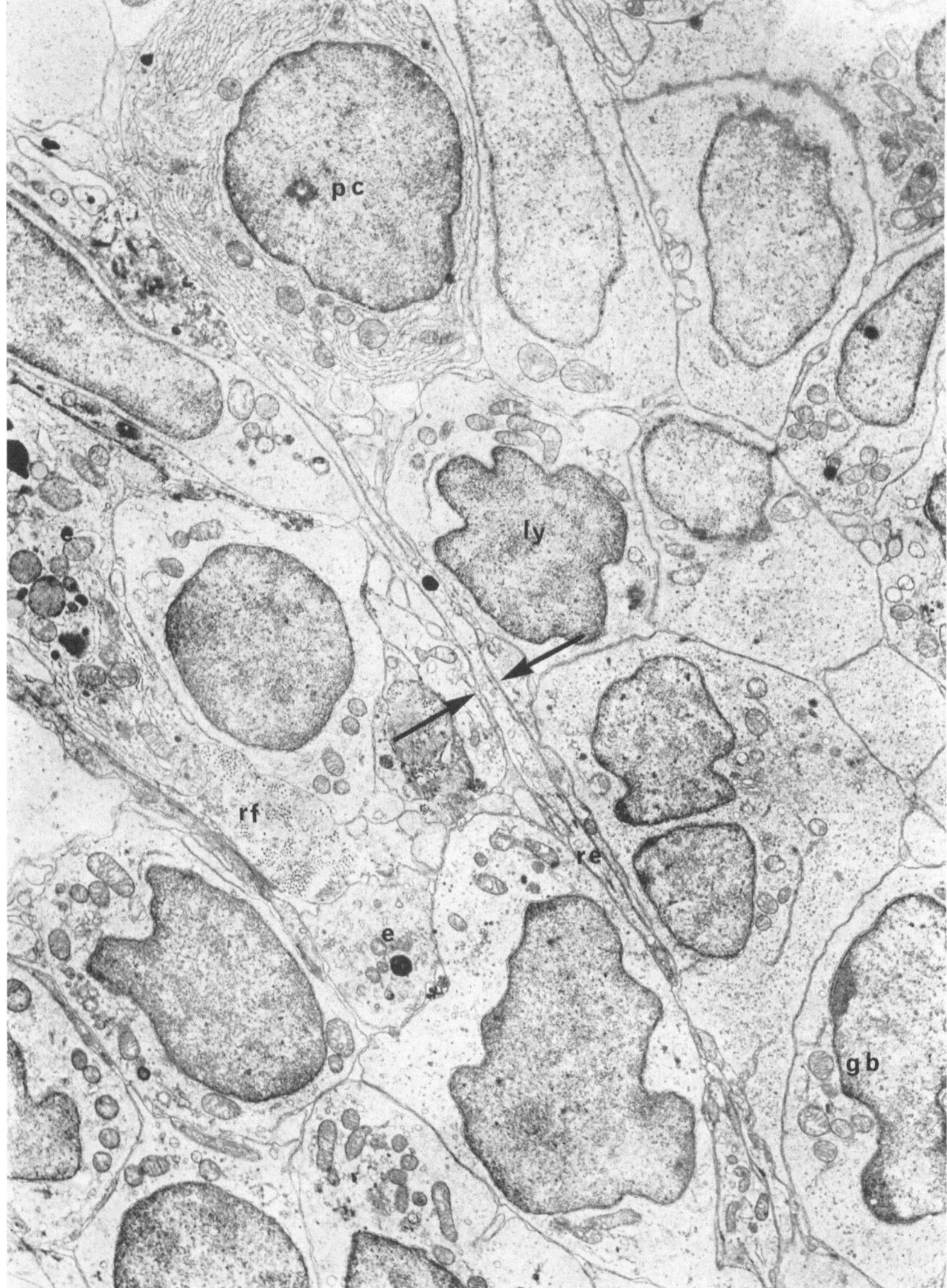

Fig. 26   5'-nucleotidase reaction in germinocyte (pH 7,2)
Distinct granules gather for the most part in immediate vicinity of cellular membrane.
Some isolated scattered granules are probably artifacts.
Contributed by E. KAISERLING (method see KAISERLING and CAESAR, 1969)
×22,500

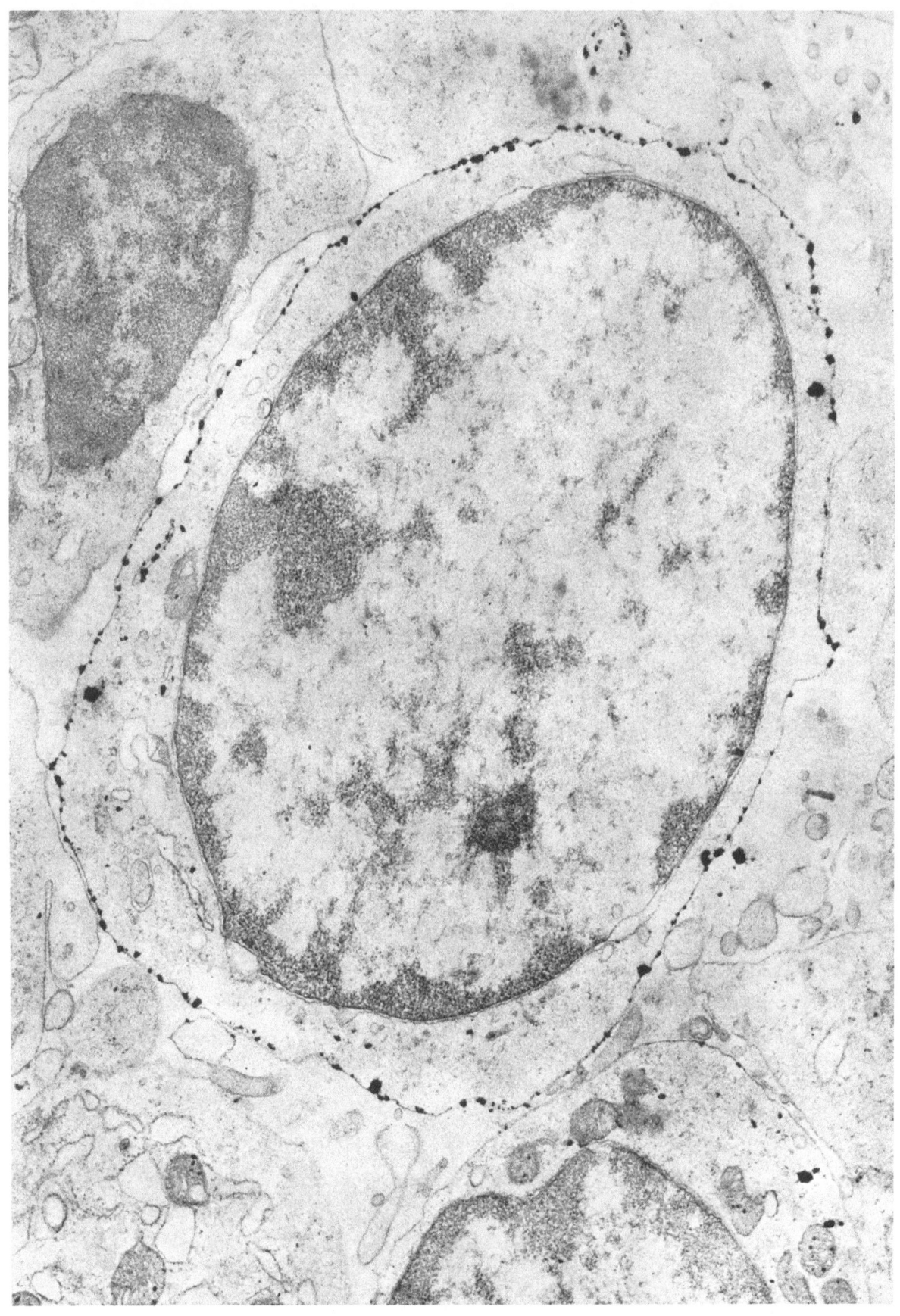

Fig. 27   Capillary with pericyte
The cytoplasm of the capillary endothelium exhibits variable electron density. A pericyte
invested in a distinct basement membrane is attached to the capillary.

Case 33. ×20,000

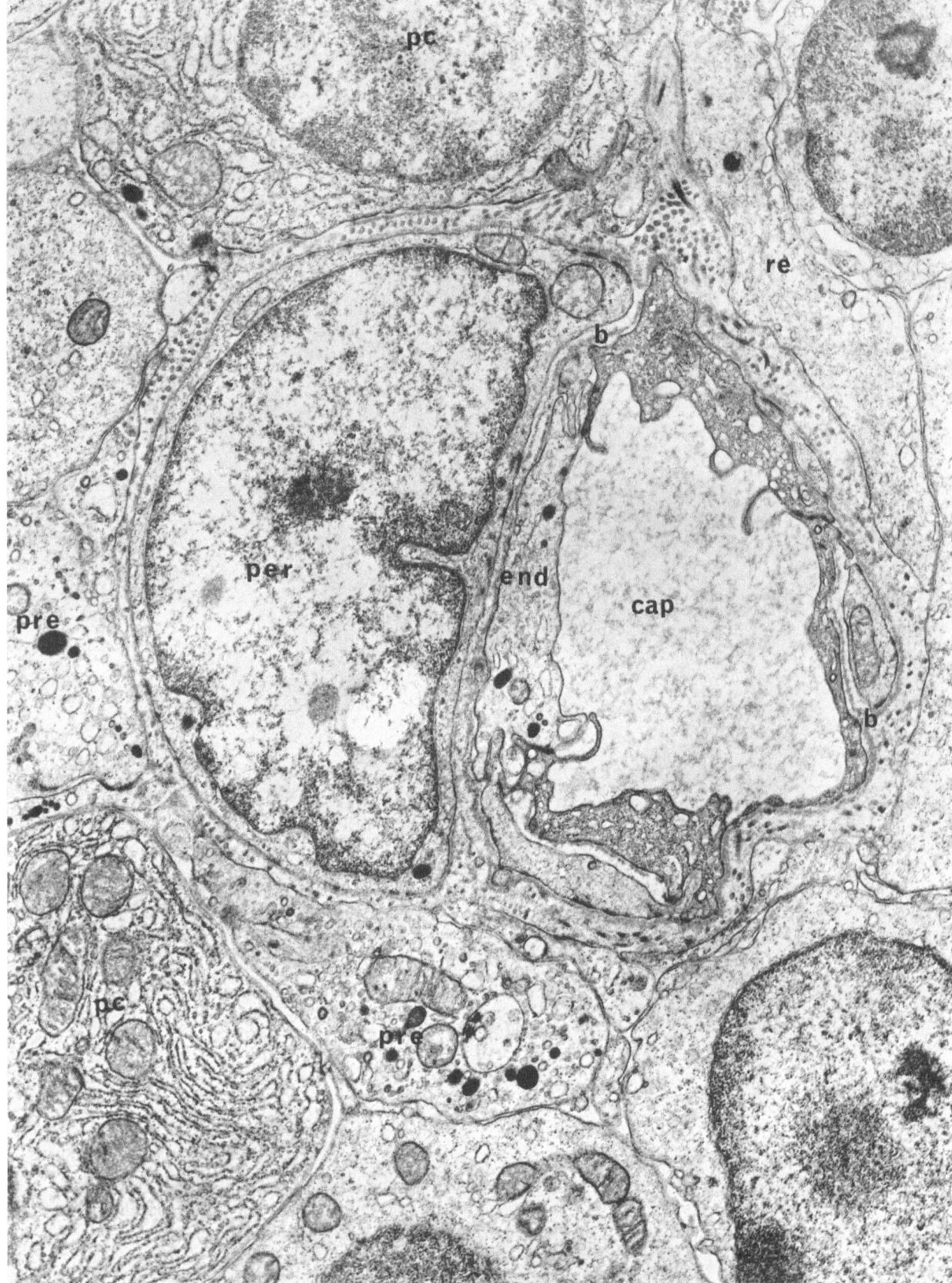

Fig. 28   Fenestrated capillary

Arrows indicate fenestrations. Small process of smooth muscle cell is seen on top of the capillary. Slender cytoplasmic extensions of reticulum cells encircle the capillary. One so-called lipochondrion is evident in an endothelial cell.

Case 44. ×17,500

Inset Higher magnification of fenestrations (arrows)

×32,500

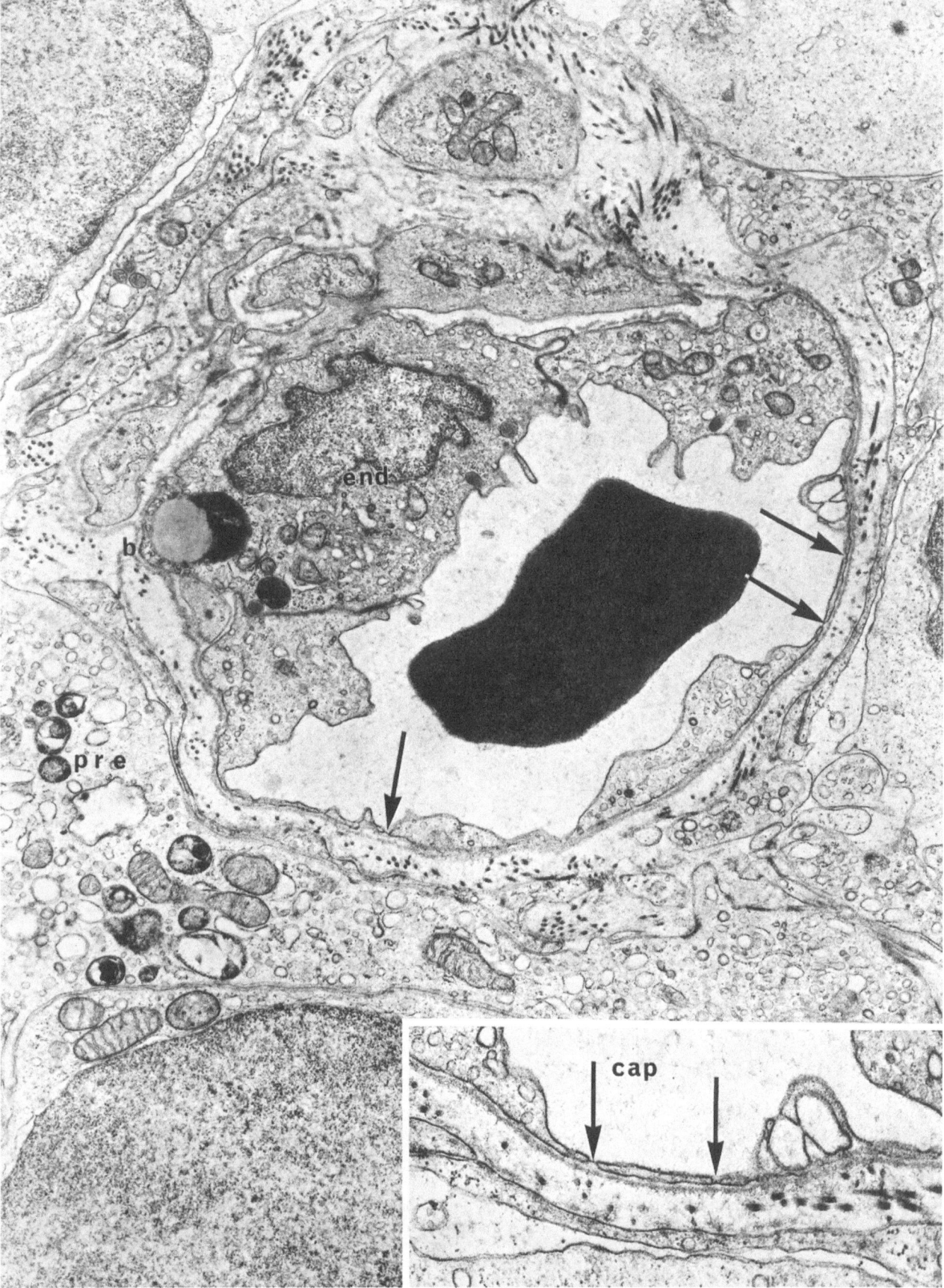

Fig. 29   Precapillary vessel (terminal arteriole)
            The cytoplasm of the endothelium is somewhat thicker than that of the capillary endo-
            thelial cells and is surrounded by many bundles of collagen fibrils. A macrophage (pre)
            surrounds the vessel.
            Case 40. ×17,500

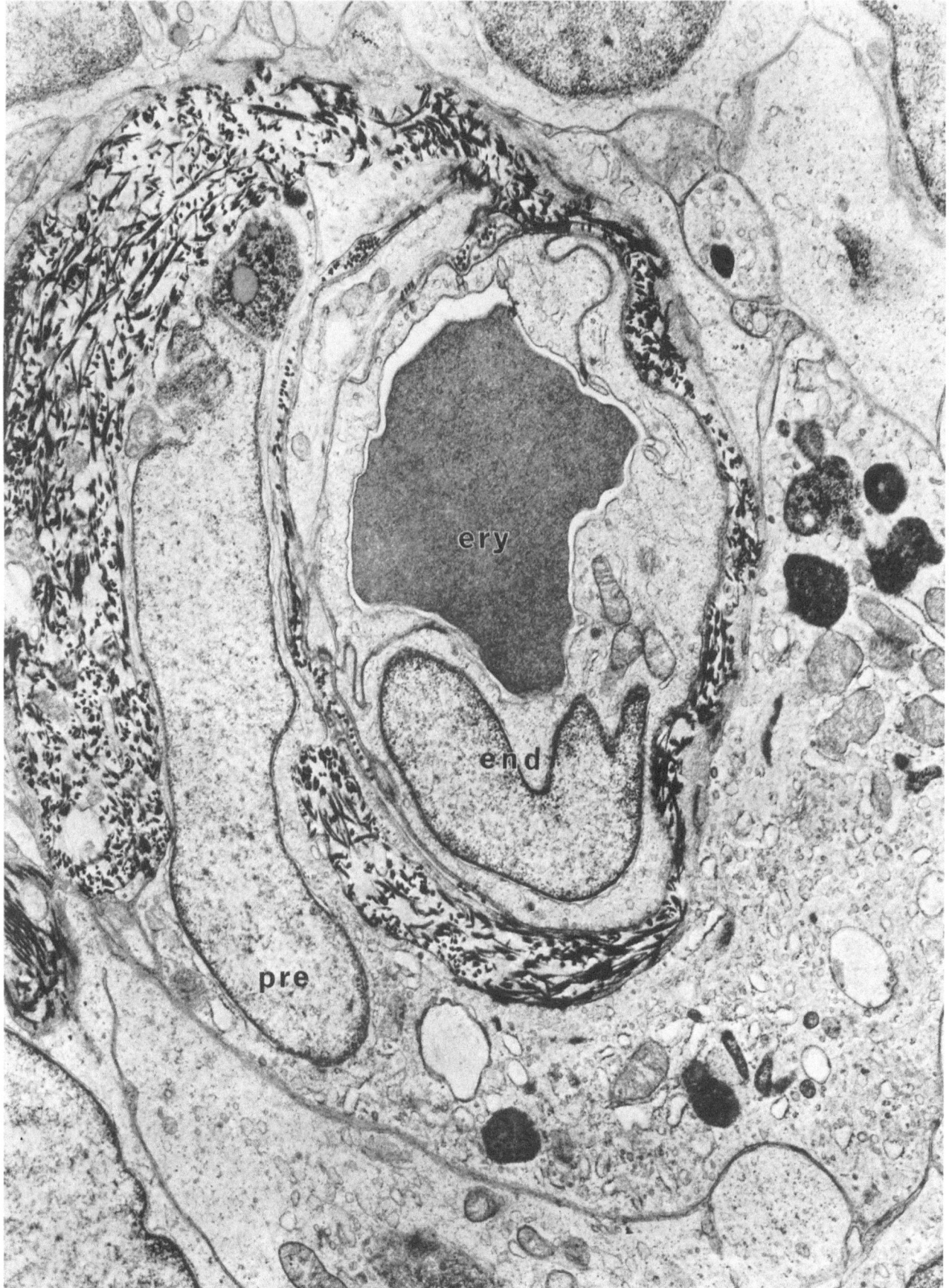

Fig. 30   Postcapillary venule
The cytoplasm of the endothelium is much wider than in the other blood vessels.
Case 40. ×8,750

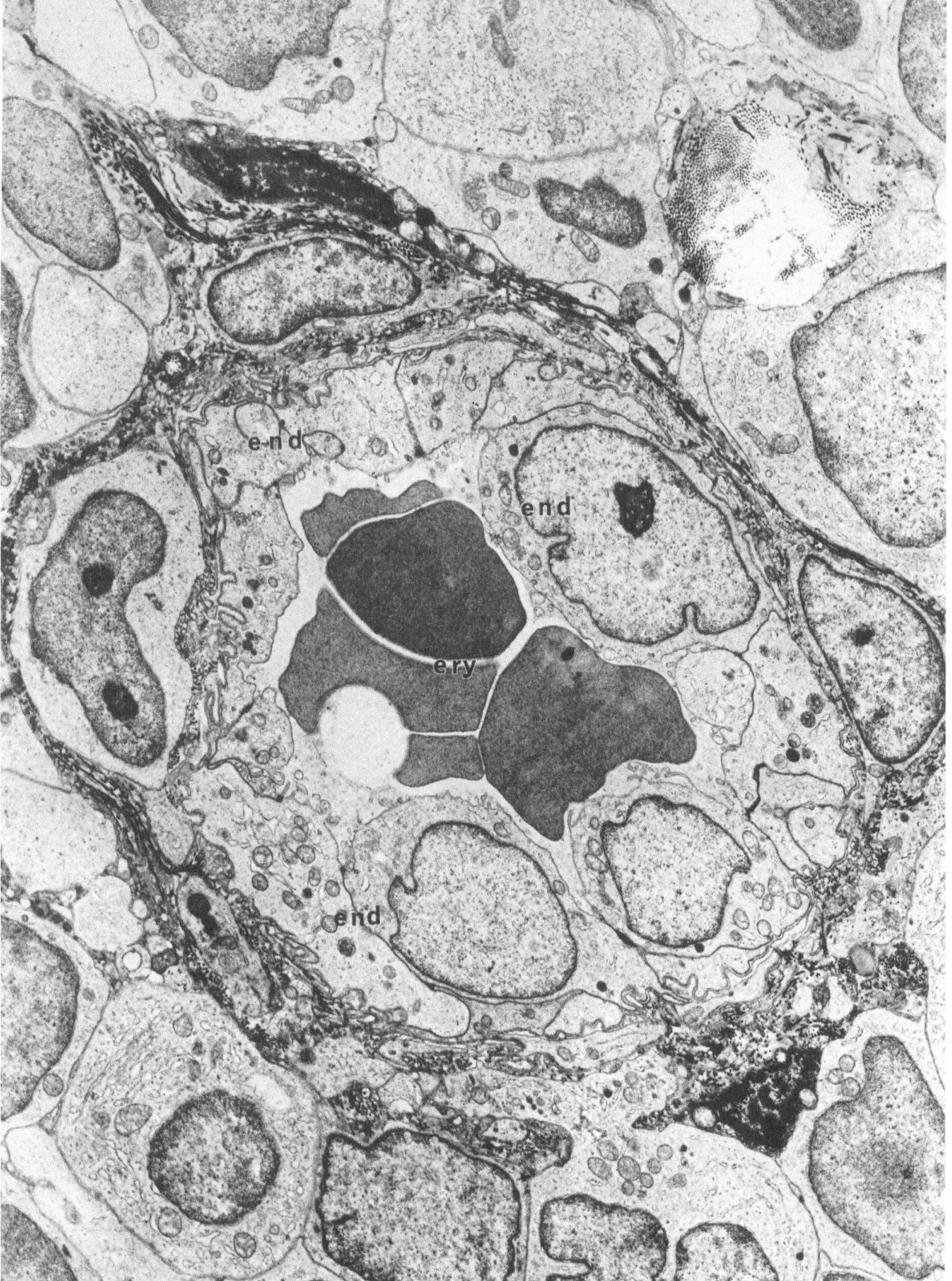

Fig. 31 a and b   Endothelium of postcapillary venule

    a) Portions of endothelial cells, one with a prominent Golgi apparatus. Note lysosomes.

    b) Many lysosomes and phagolysosomes. Filamentous structures in the perinuclear area of endothelial cells.

    Case 8. $\times 17,500$

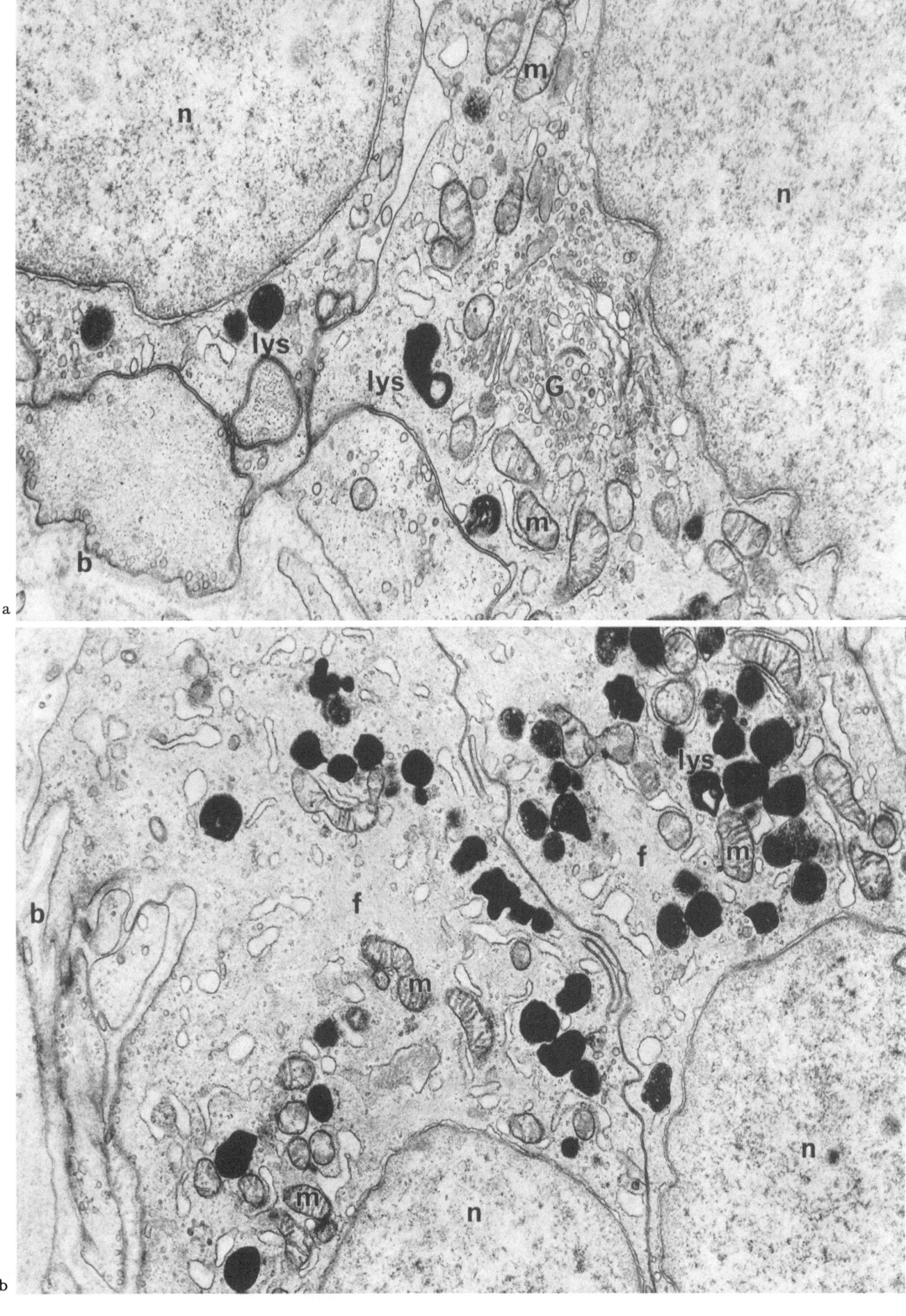

Plasma Cell Hyperplasia (Plasmacytosis)
Diverse Hyperplasia of the Pulp
Sinus Catarrh
Immature Sinus Histiocytosis

Fig. 32   Plasma cell proliferation in germinal center
Seven plasma cells, one of which is immature, one nonphagocytic longbranching reticulum cell, one phagocytic reticulum cell and one germinoblast are seen.
Case 37. ×10,000

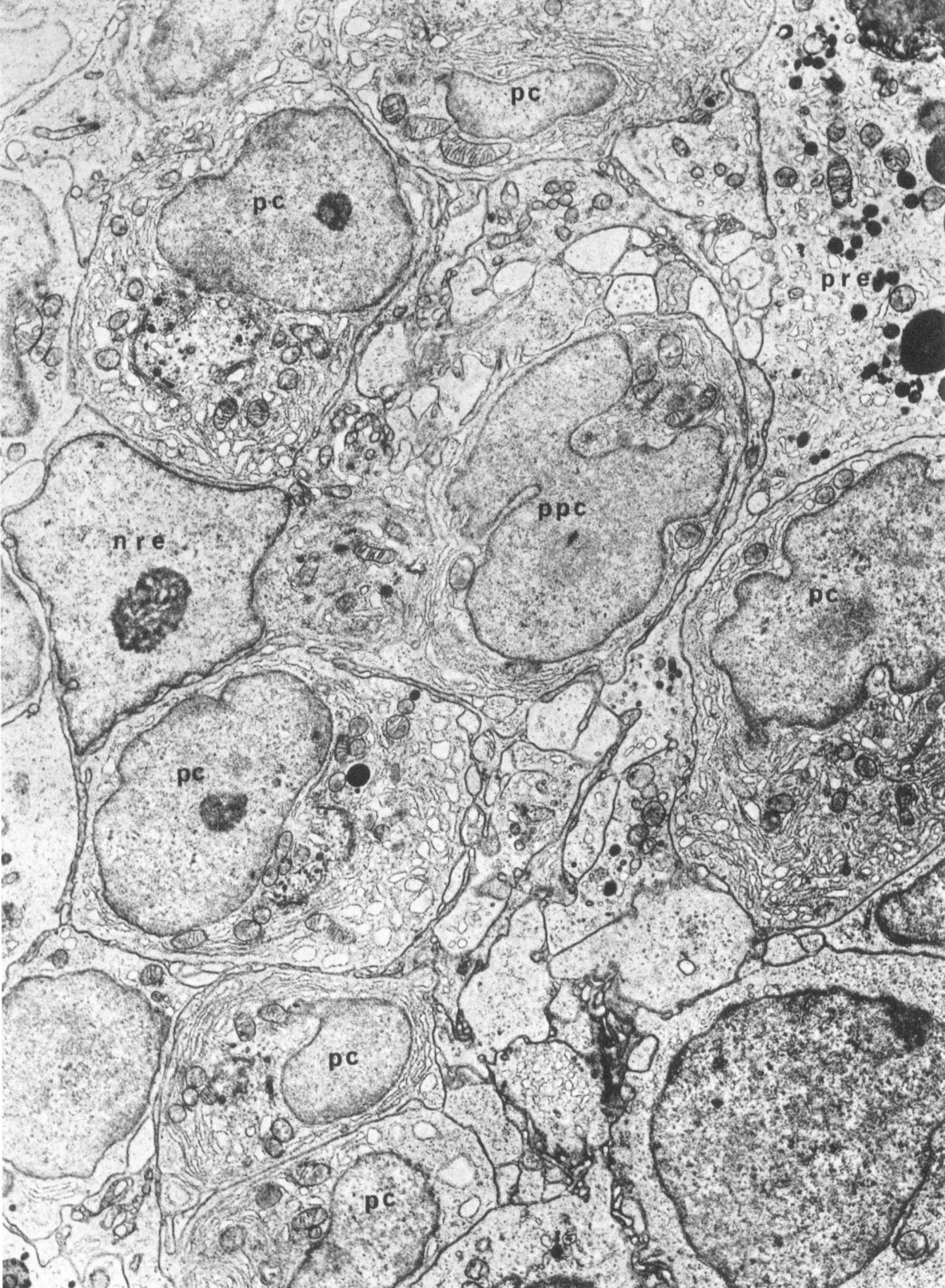

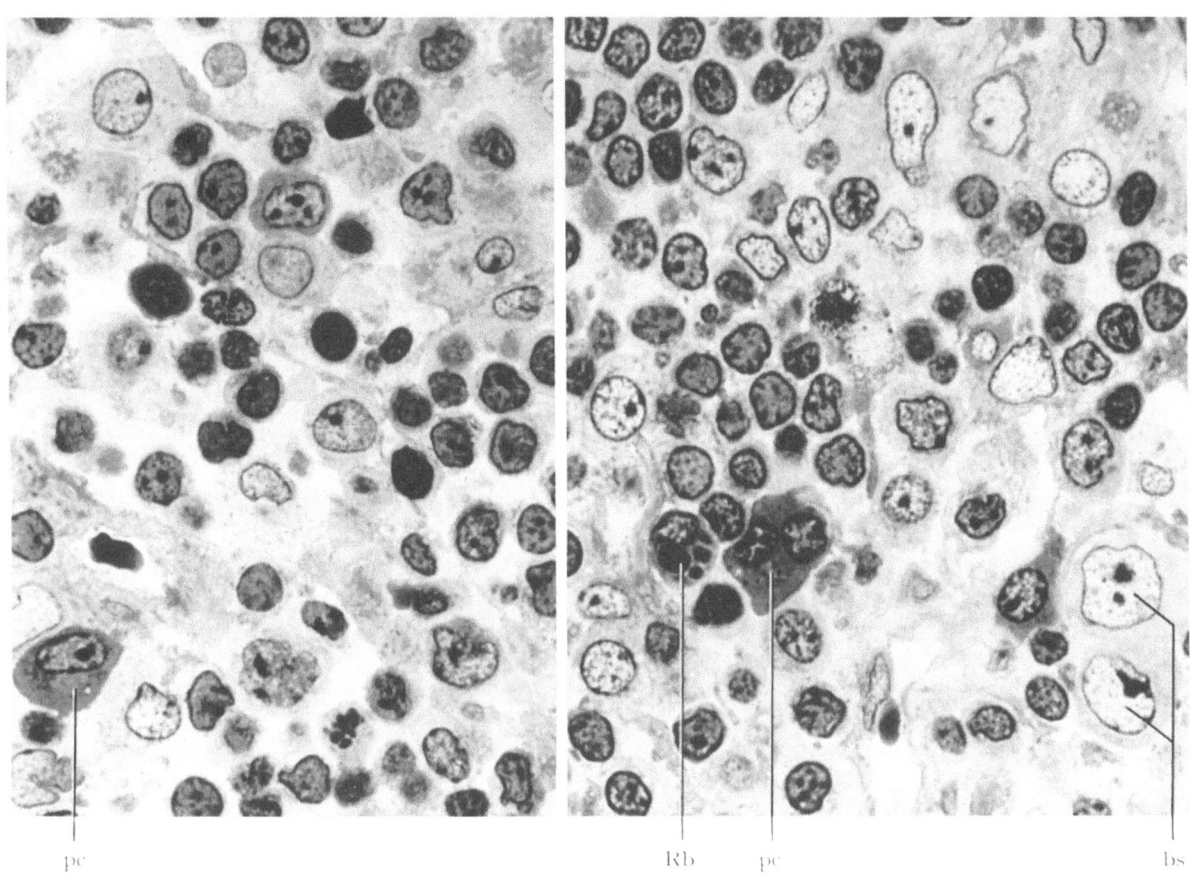

pc   Rb pc   bs

Fig. 33.   Diverse hyperplasia of the pulp

Basophilic stem cells (*bs*), plasma cells (*pc*) with Russell bodies (*Rb*), reticulum cells, lymphoid cells of variable size and configuration ("lymphatische Reizformen", "activated lymphocytes") and mitoses are seen. PIRINGER's lymphadenitis.

Osmium tetroxide fixation. Semithin section. Methylene blue-Azure II
Case 15. ×1,230

Fig. 34   Diverse hyperplasia of the pulp (mounted figure, see next two pages)
Lymphocytes of varying size and basophilic stem cells are seen. One reticulum cell contains
a large distinct nucleolus. Another reticulum cell displays phagocytosis.

Case 35. ×8,500

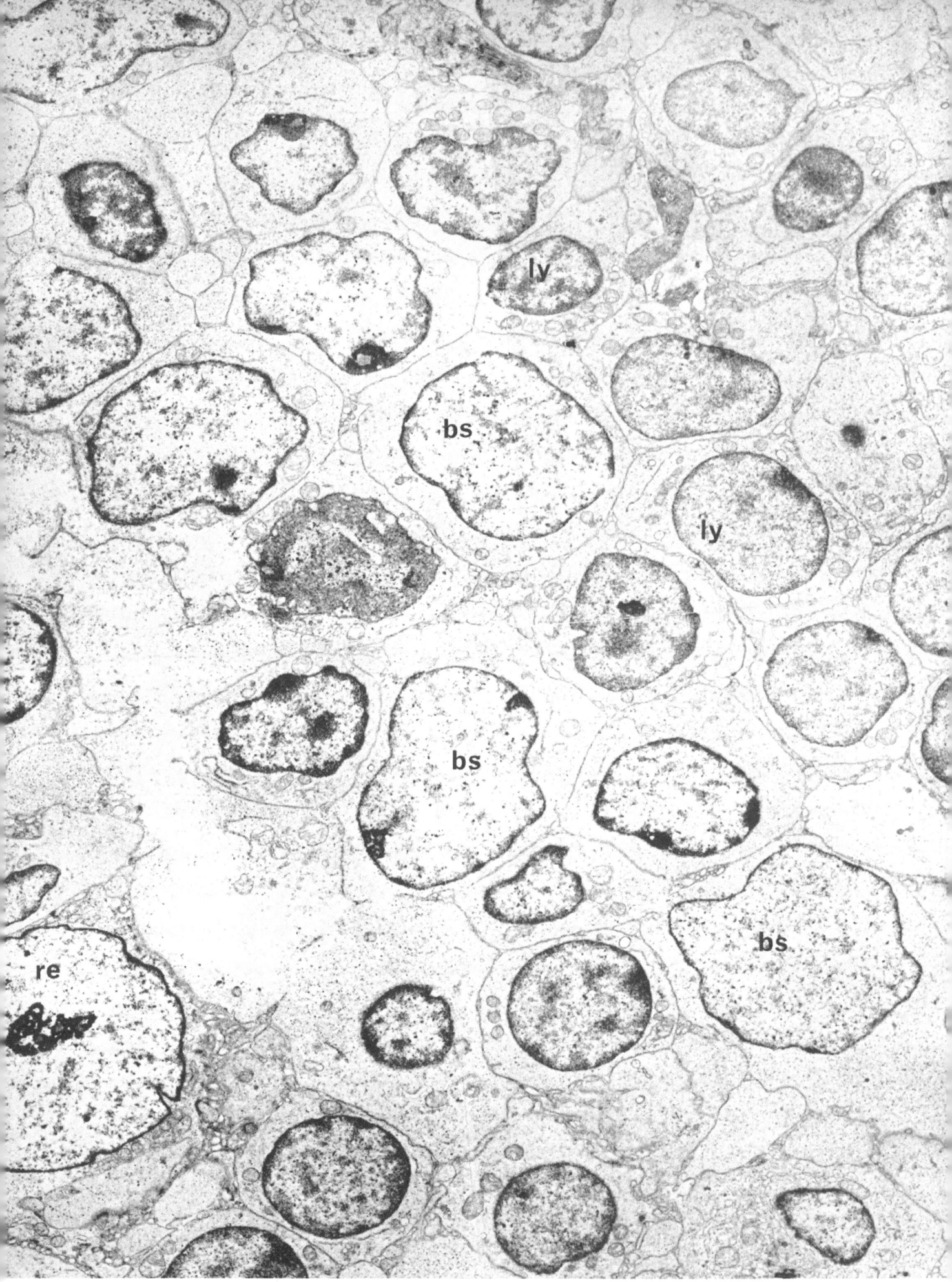

Fig. 34. For legend see p. 119

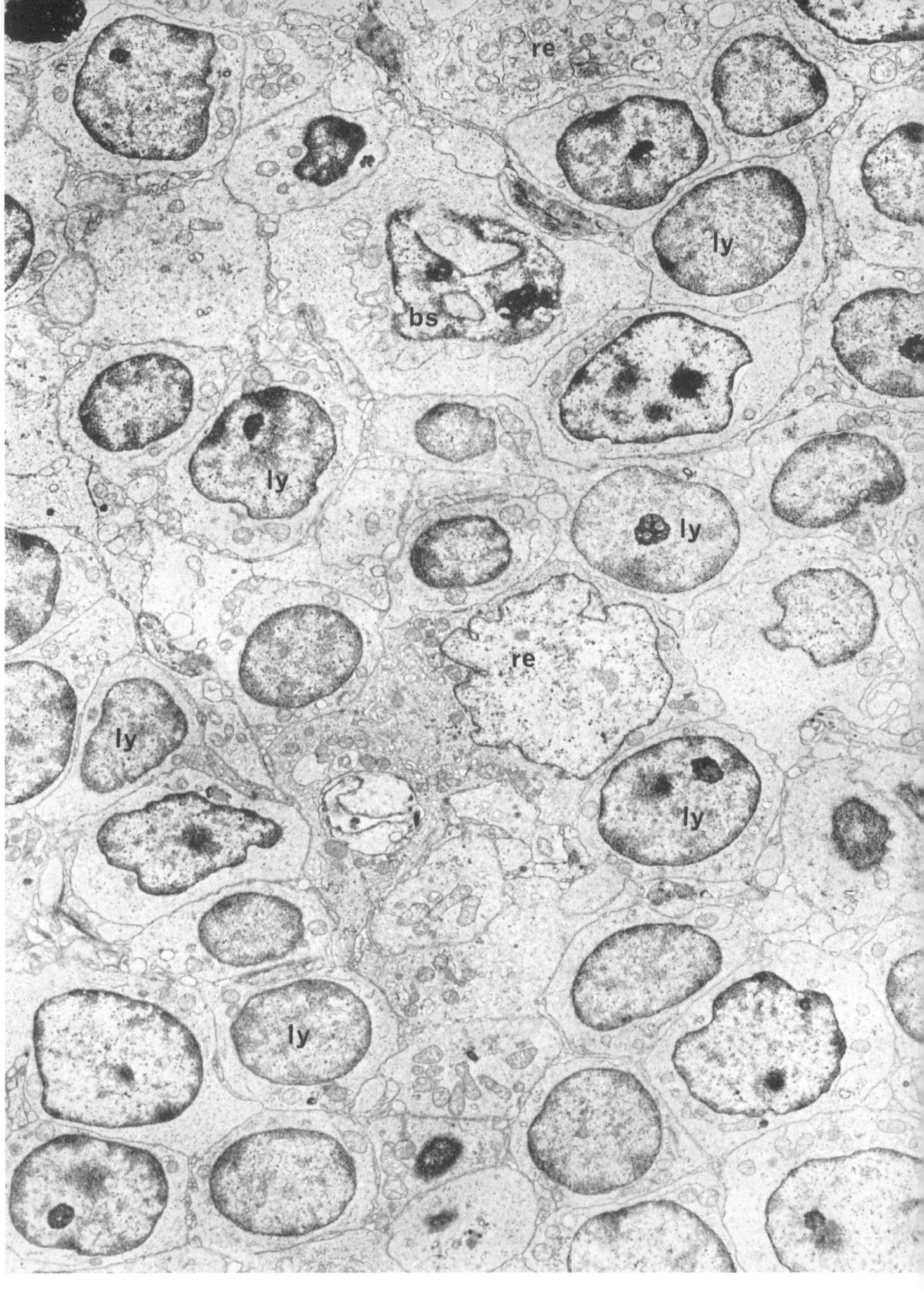

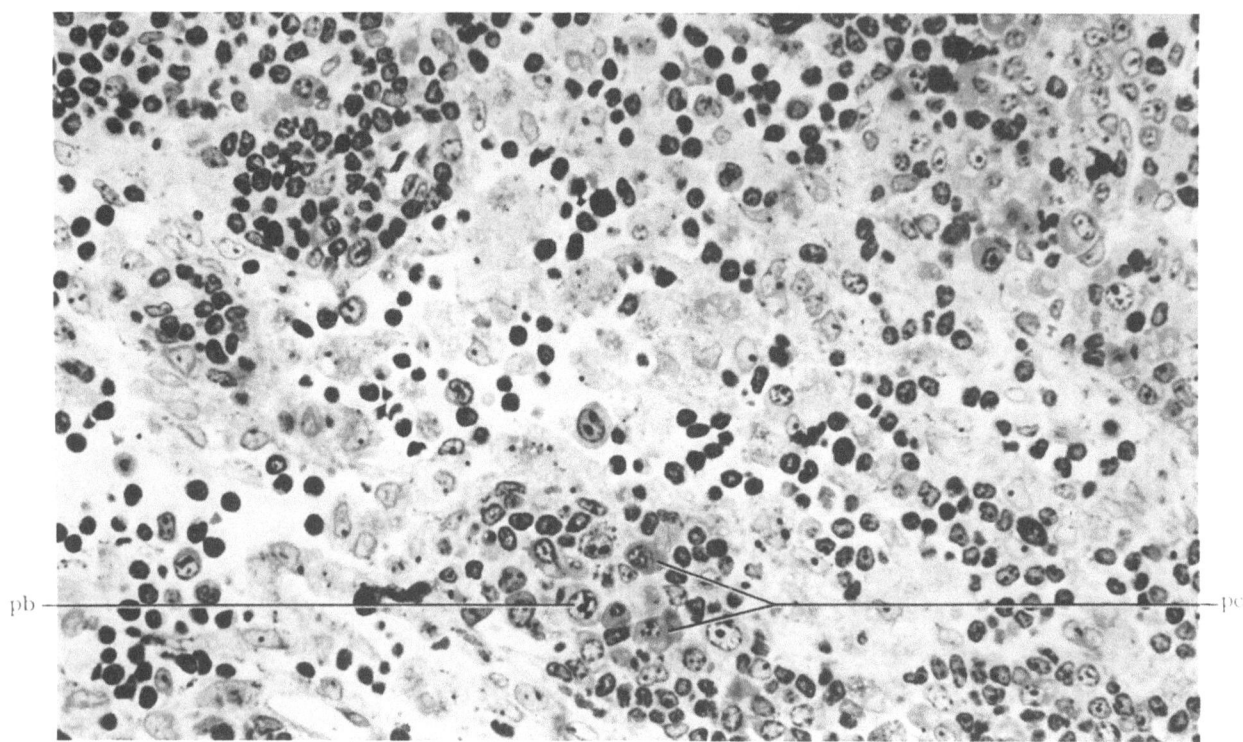

Fig. 35a   Sinus catarrh, medulla of lymph node

The sinus contains many macrophages and lymphocytes and some basophilic cells ("lymphatische Reizformen"). Some plasma cells (*pc*), plasmablasts (*pb*) and a few mast cells are seen in the medulla.

Case 47. Osmium tetroxide fixation. Semithin section. Methylene blue-Azure II.
×400

Fig. 36   Sinus catarrh

Macrophage and retothelial cells with enlarged cytoplasm containing lysosomes. Compare with Fig. 41.

Case 45. ×8,750

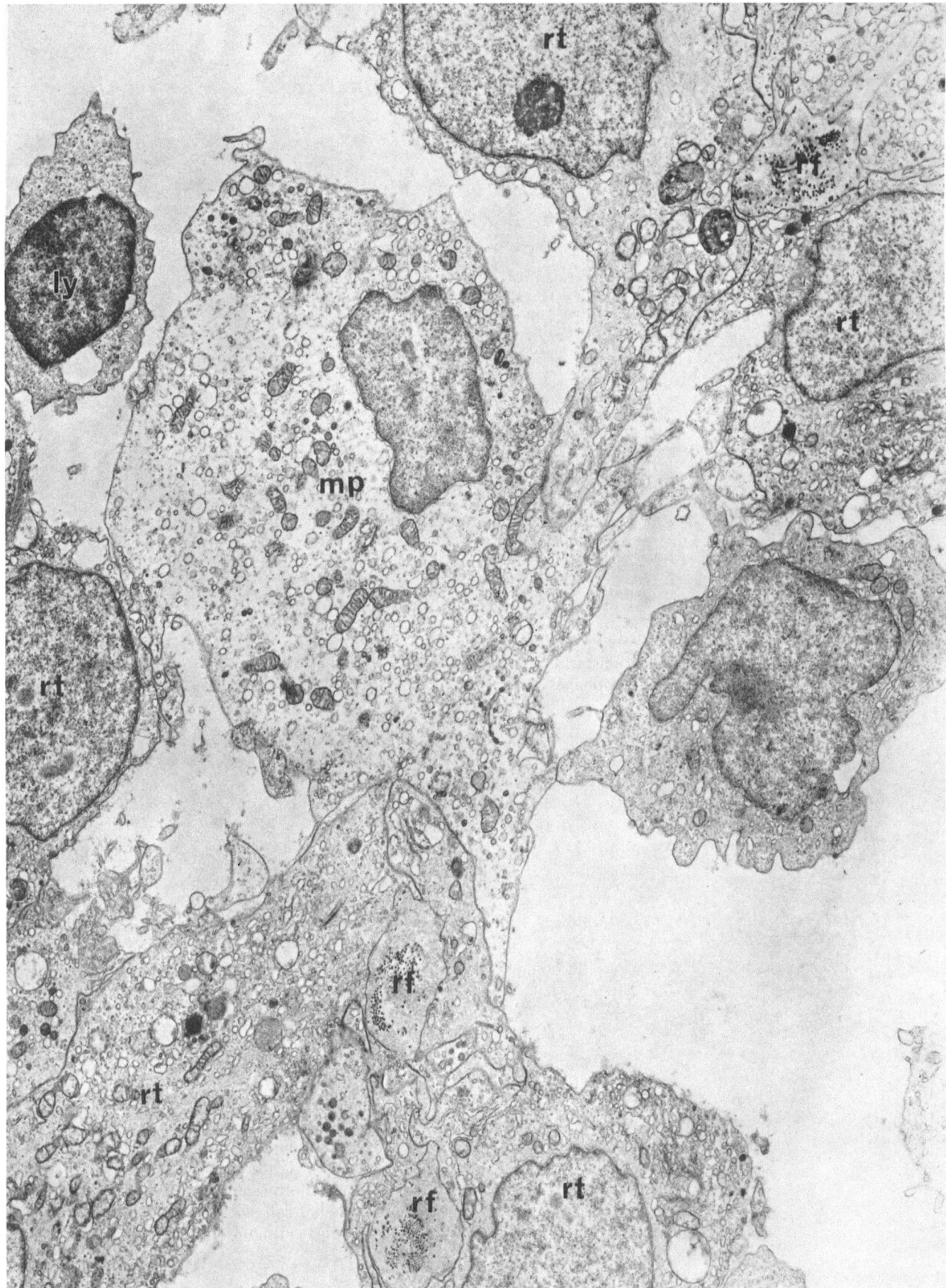

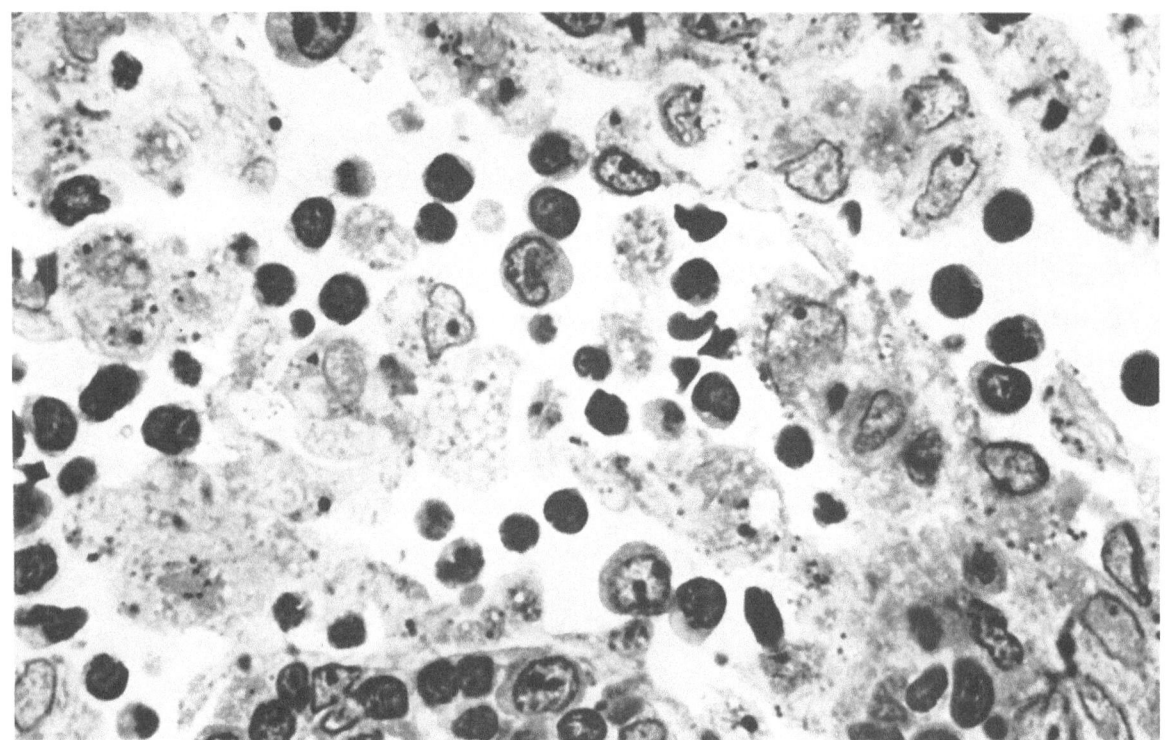

Fig. 35 b   Sinus catarrh, medulla of lymph node.
Higher magnification of field shown in Fig. 35 a. Numerous lysosomes (dark granules) are present in the macrophages.
×1,230

Fig. 37   Macrophage in sinus catarrh
The cytoplasm shows numerous fine processes, and a large amount of finely vesicular and vacuolar endoplasmic reticulum is obvious. In the upper right and lower left corner of the picture retothelial cells of the sinus are seen. Contrast the reticulum fiber in the lower left which is surrounded by the cytoplasm of the retothelial cell with the macrophage that does not envelop such a fiber. Picture was taken from medullary sinus.
Case 45. ×8,750

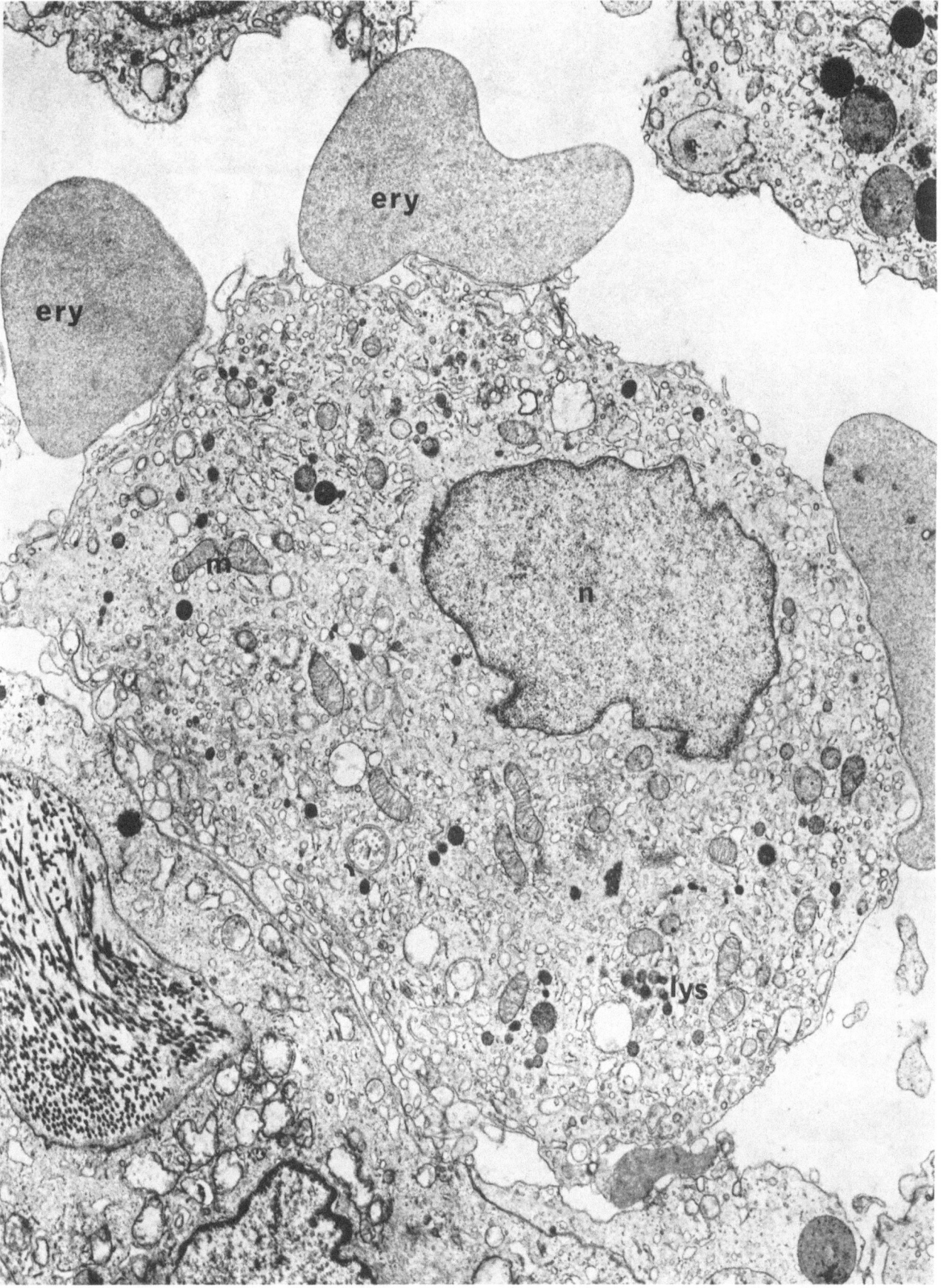

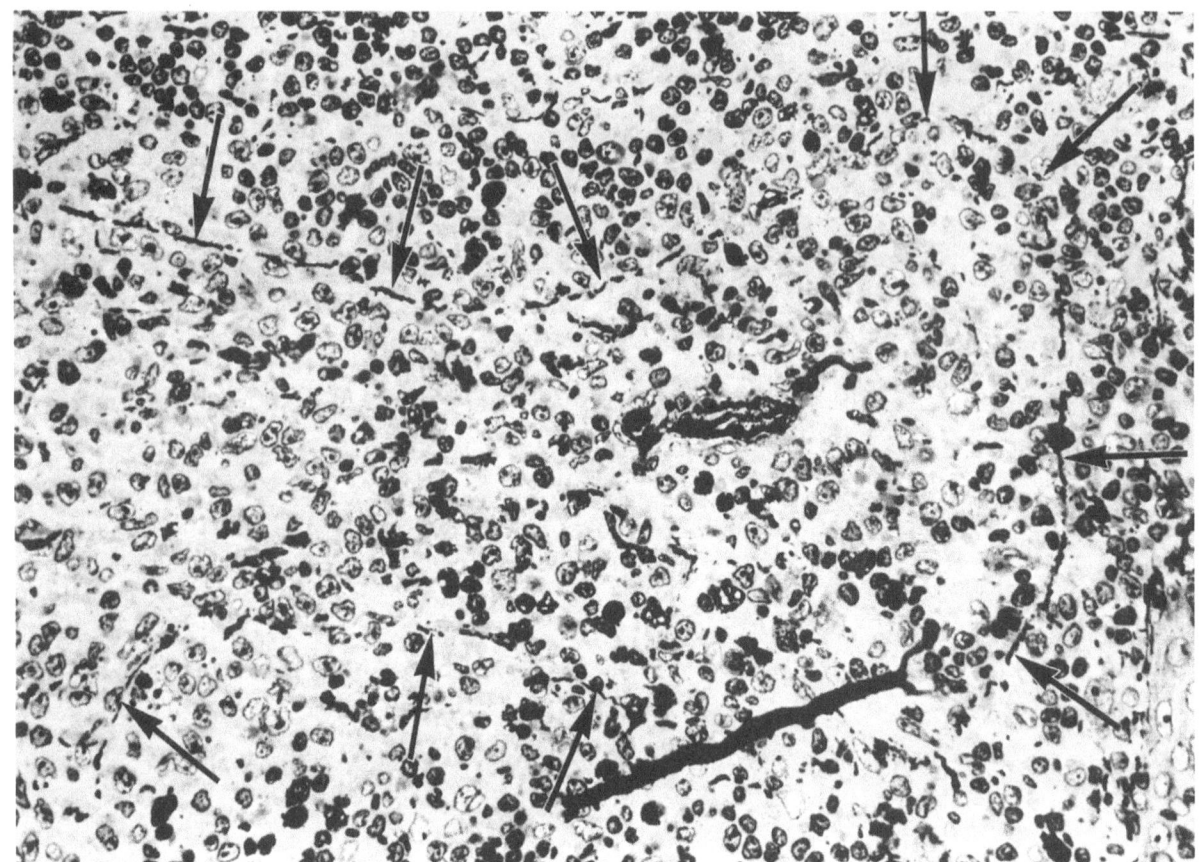

Fig. 38

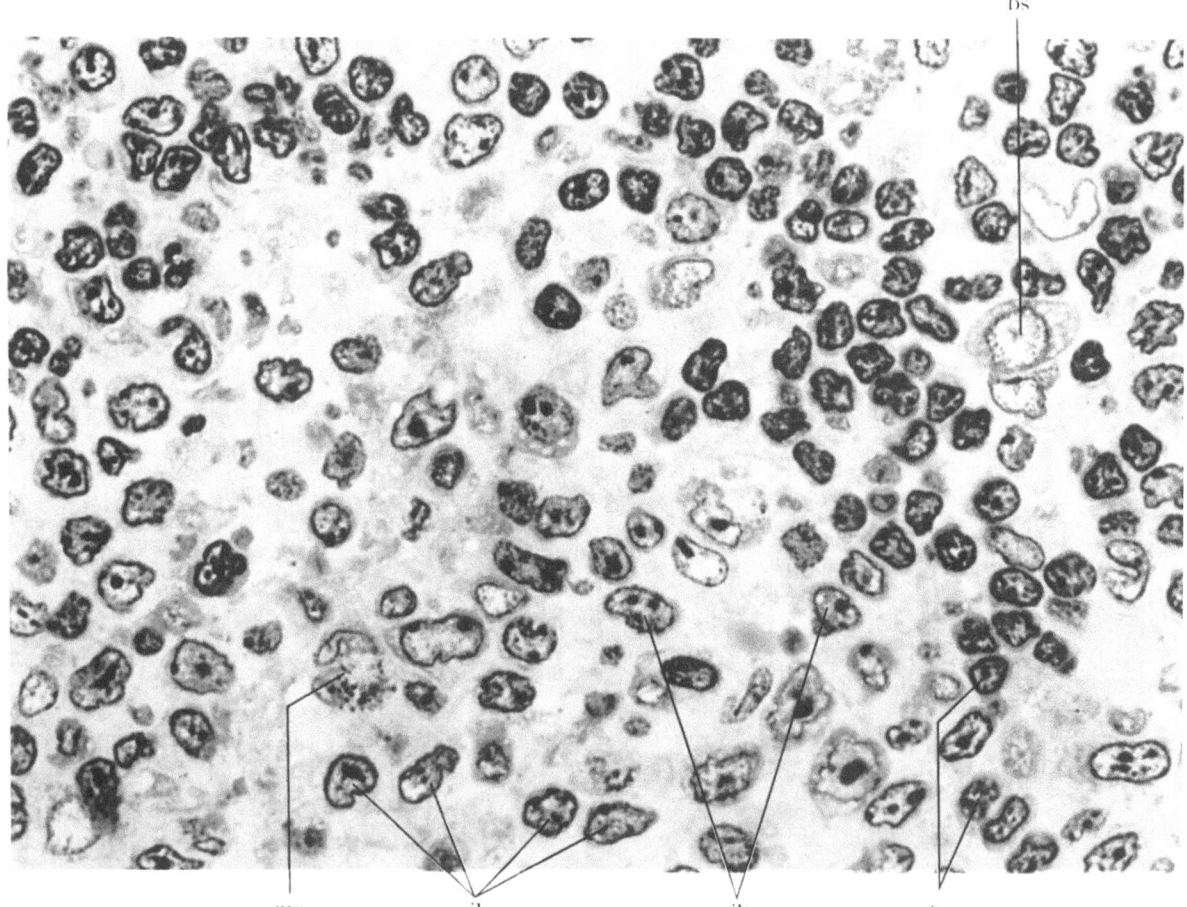

Fig. 39

Fig. 38   Immature sinus histiocytosis in PIRINGER's lymphadenitis (mounted Fig. 40 shows details of same microscopic field depicted in this photomicrograph)
Note reticulum fibers of sinus wall (arrows).
Case 66. Osmium tetroxide fixation. Semithin section. Silver impregnation for reticulin, method of MOVAT. ×400

Fig. 39   Detail from Fig. 38 after application of methylene blue-Azure II stain
Immature sinus histiocytosis (center left), surrounded by lymphocytes (ly). Besides numerous immature histiocytes (ih) a few macrophages (ma) are present. A basophilic stem cell (bs) is seen in the pulp but its large nucleolus is not visible in this plane of section.
Case 66. Osmium tetroxide fixation. Semithin section. Methylene blue-Azure II. ×1,230

Fig. 40   Immature sinus histiocytosis (mounted figure, see next two pages)
The lower third of the picture demonstrates the pulp with the sinus lumen above. The latter is lined (double arrows) by reticulum fibers with collagen bundles (co). The lumen is crowded with immature histiocytes with a slight admixture of lymphocytes and phagocytizing retothelial cells (pre). An angulated lymphocyte is passing through the sinus wall (arrows).
Photomicrograph provided by S. BLÜMCKE and H. R. NIEDORF
Case 66. ×4,500

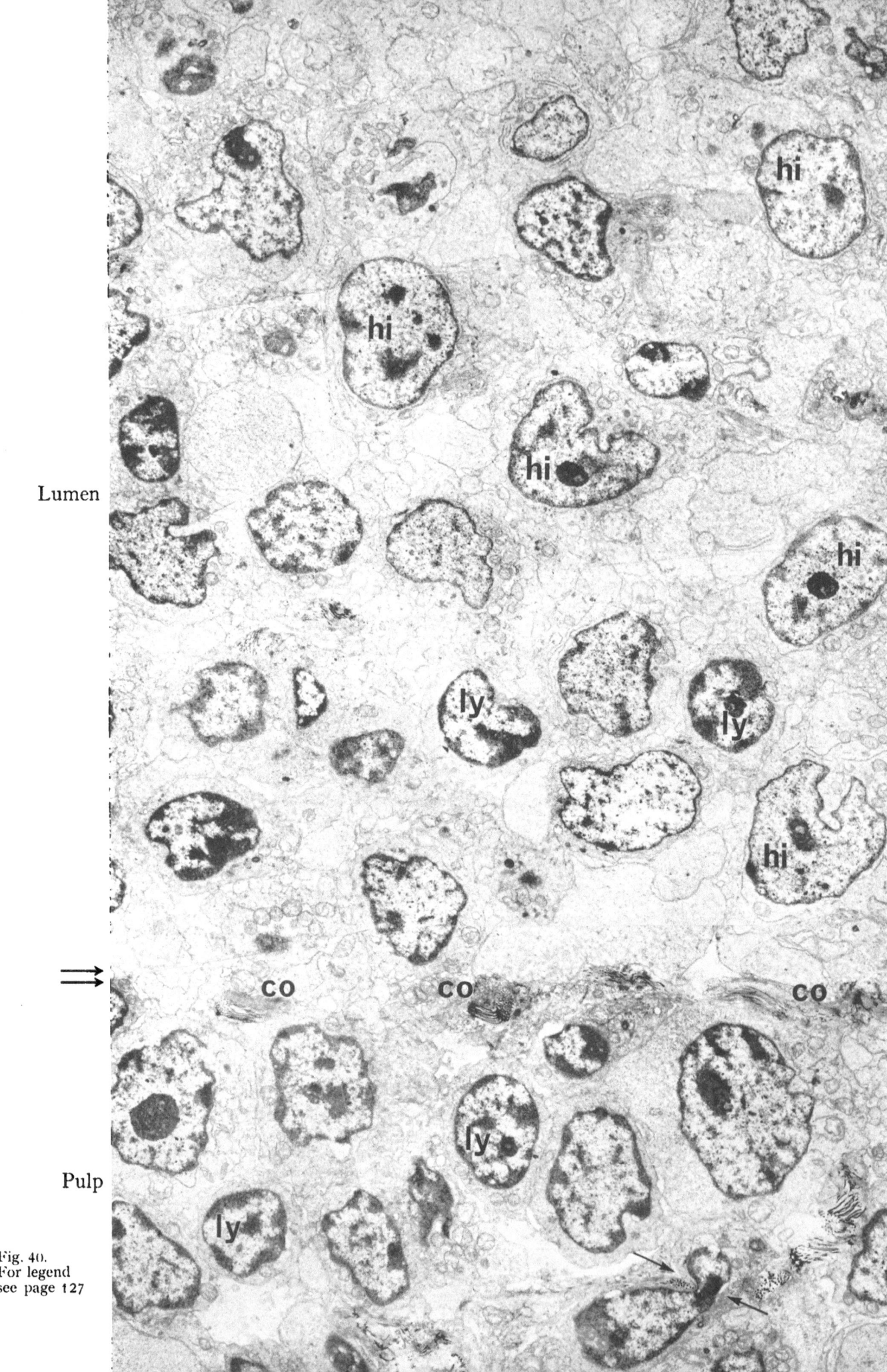

Lumen

Pulp

Fig. 40.
For legend
see page 127

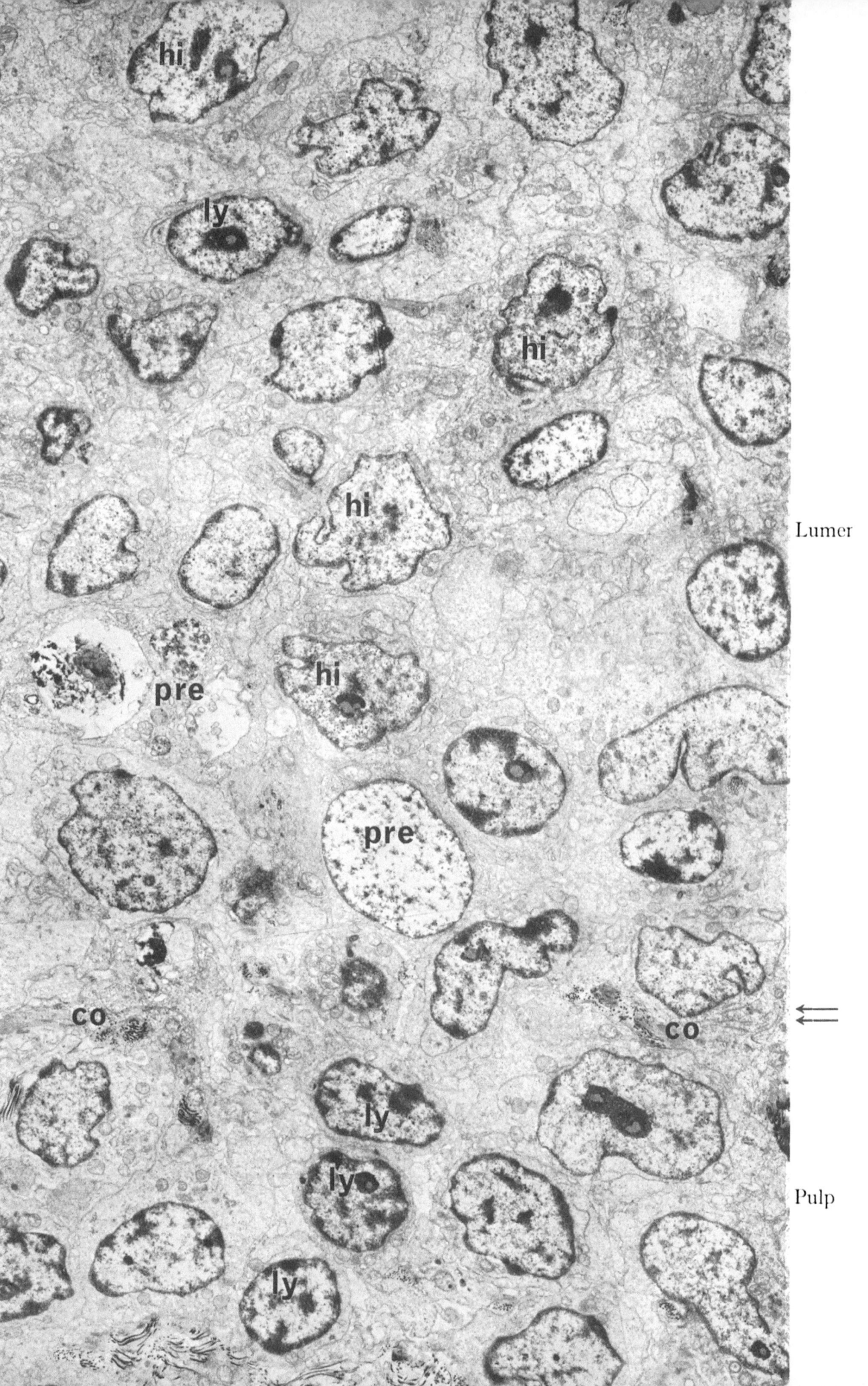

Lumer

Pulp

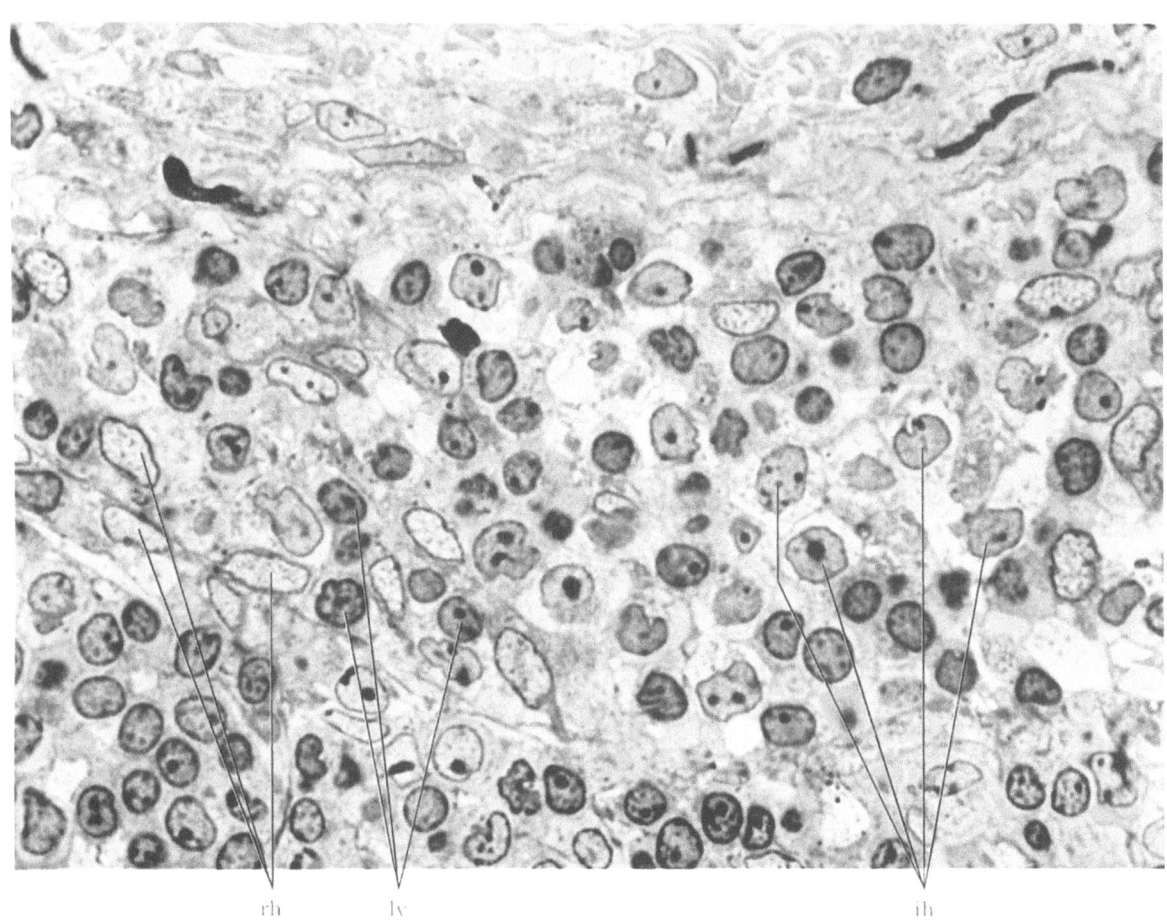

rh   ly   ih

Fig. 41   Immature sinus histiocytosis in PIRINGER's lymphadenitis
Marginal sinus. The capsule of the lymph node is seen in the upper part of the picture.
Note the difference in the nuclear texture of immature histiocytes (*ih*), lymphocytes (*ly*)
and retothelial cells (rh). A portion of the adjacent pulp is present at the bottom.
Case 15. Osmium tetroxide fixation. Semithin section. Methylene blue-Azure II. ×1,230

Fig. 42   Immature sinus histiocytosis
Intraluminal immature histiocytes with pleomorphic nuclei and variable numbers of small
dense bodies (granules).
Case 15. ×8,750

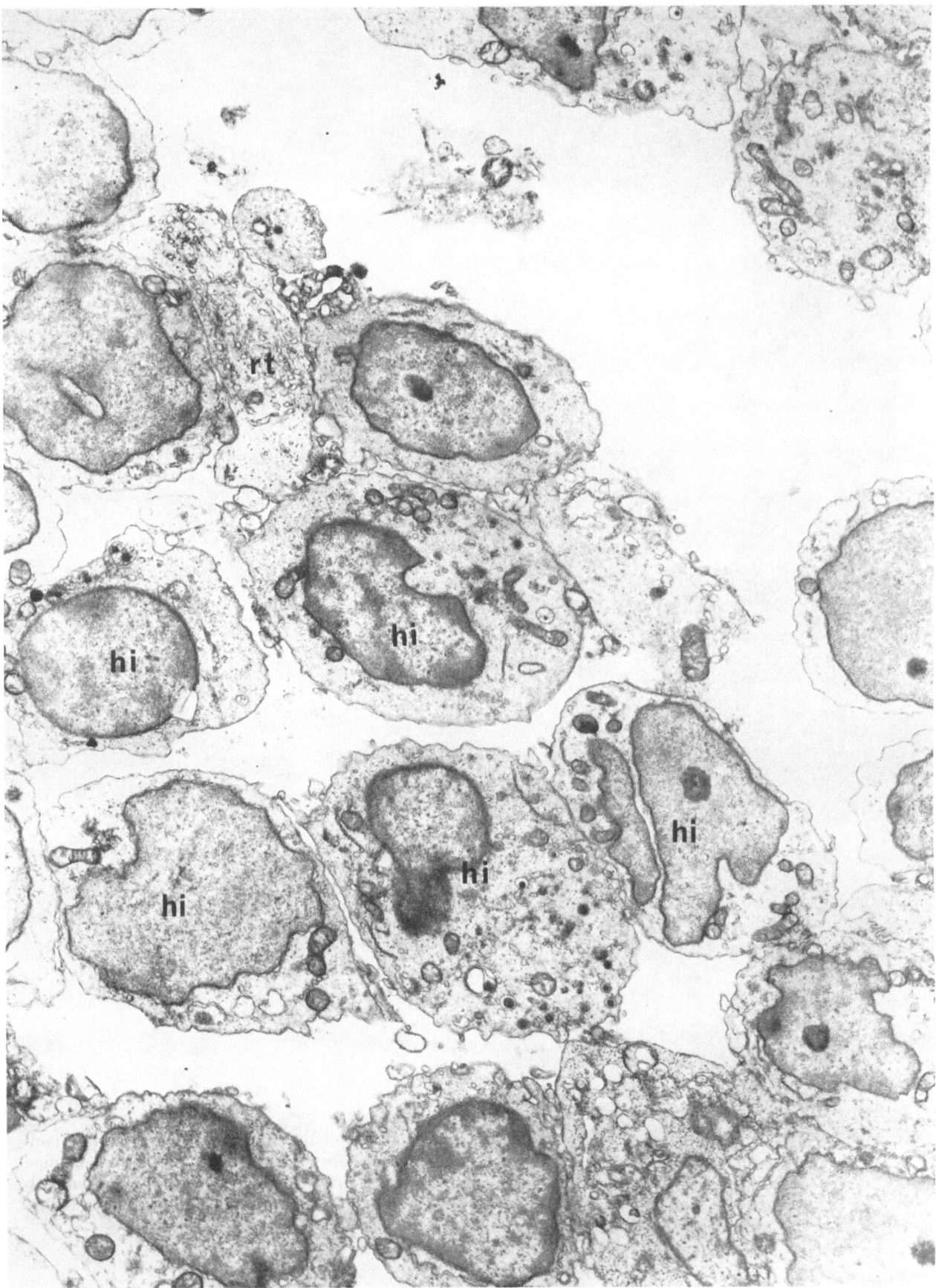

Fig. 43   Immature histiocytes
          Note lobulated or indented nuclei. The cytoplasm of the upper right cell exhibits many
          small dense bodies (granules) resembling "dense core vesicles" (similar to catecholamine
          vesicles). All cells depicted in this micrograph are provided with finely vesicular smooth
          endoplasmic reticulum, very few channels of rough endoplasmic reticulum, abundant ribo-
          somes and occasional polysomes.
          Case 15. ×17,500

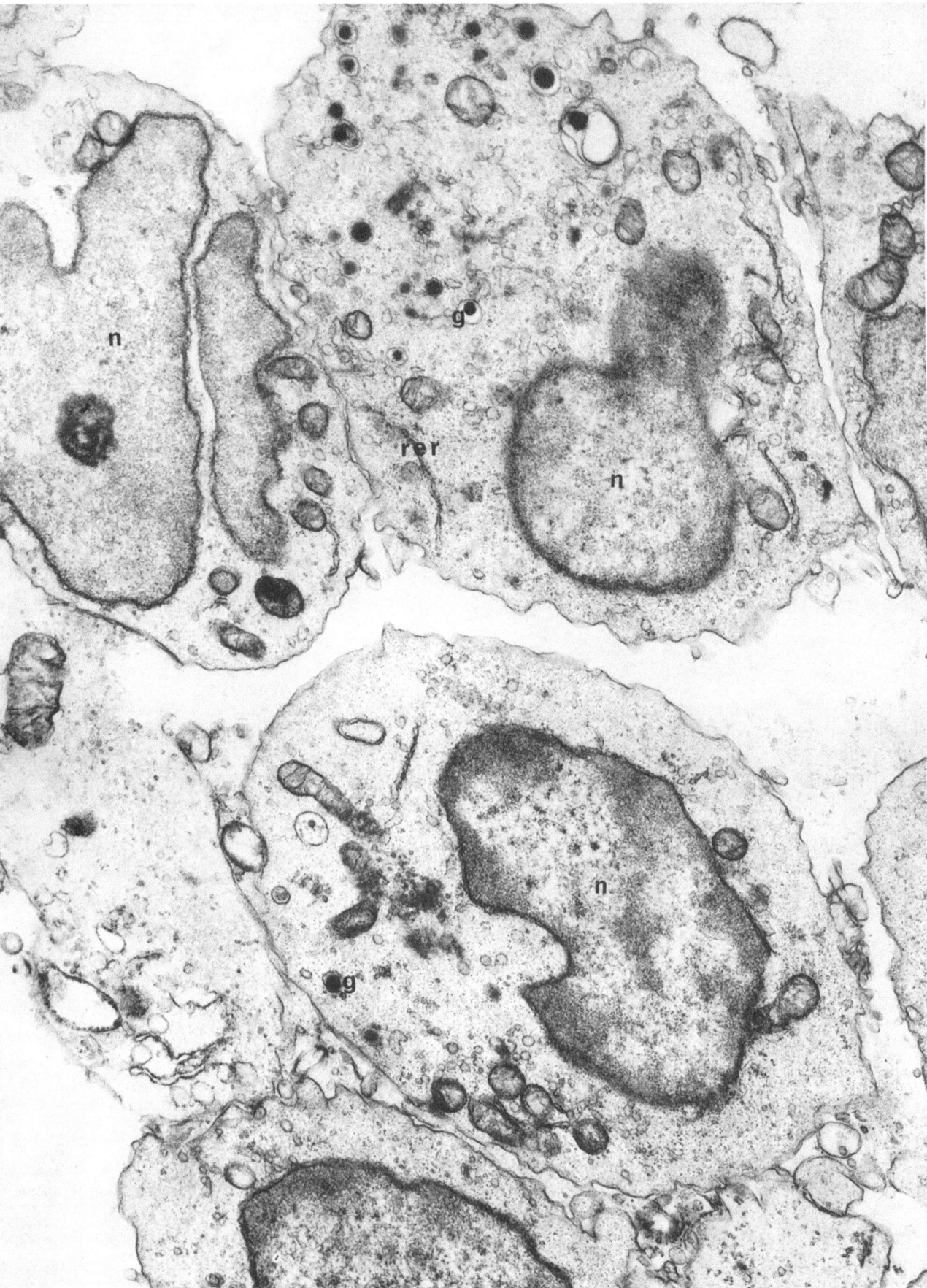

Fig. 44   Immature sinus histiocytosis, case different from that depicted in Fig. 43
Intraluminal histiocytes have more cytoplasm with more vesicular structures than lympho-
cytes. Some neutrophilic granulocytes are also evident.
Case 35. ×8,750

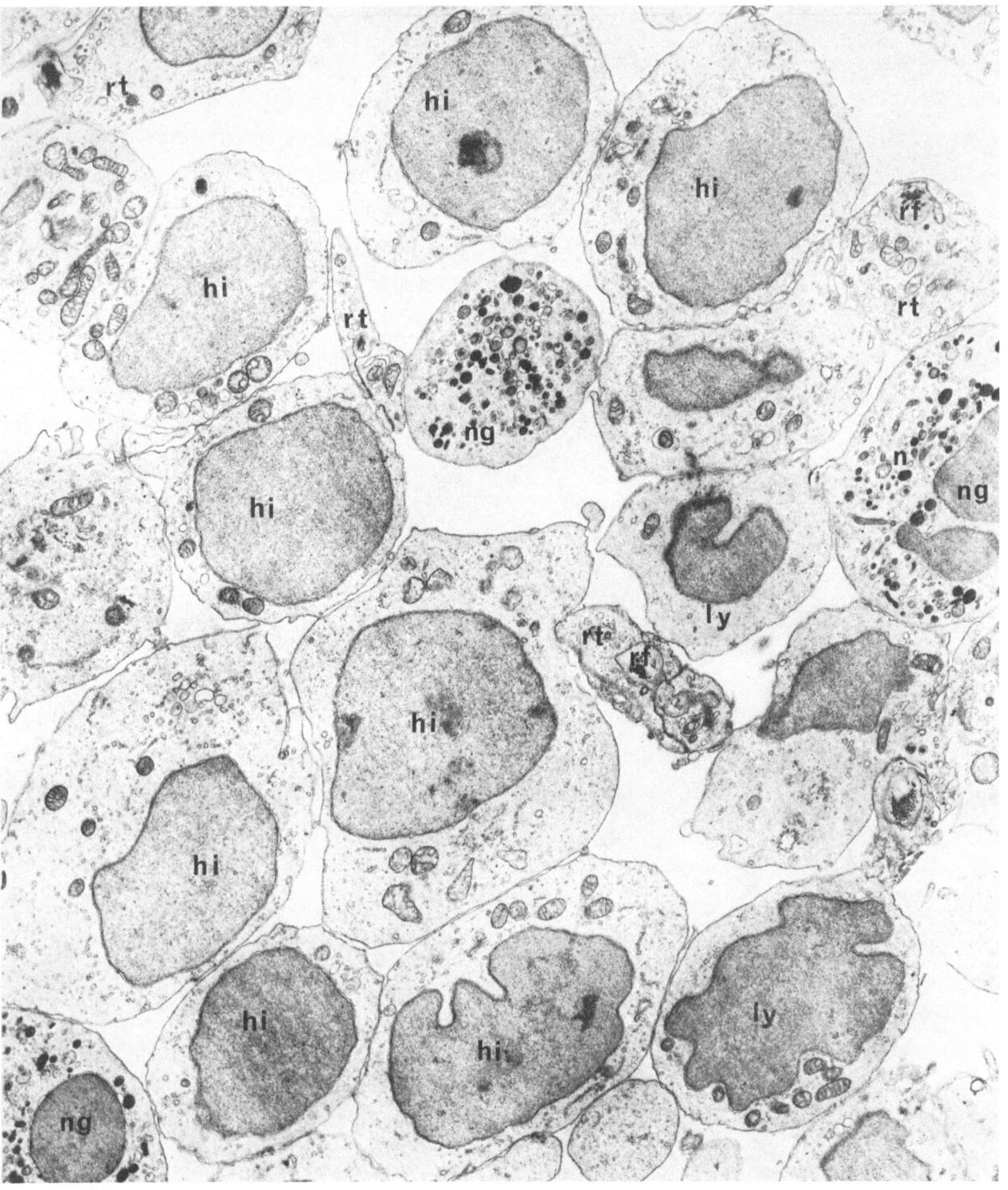

Fig. 45   Immature histiocytes
Immature histiocytes occupy most of the sinus lumen seen in the left. Note long rod-shaped
mitochondria and paucity of rough endoplasmic reticulum. Dense vesicles are absent
(compare with Fig. 43).
Case 35. ×17,500

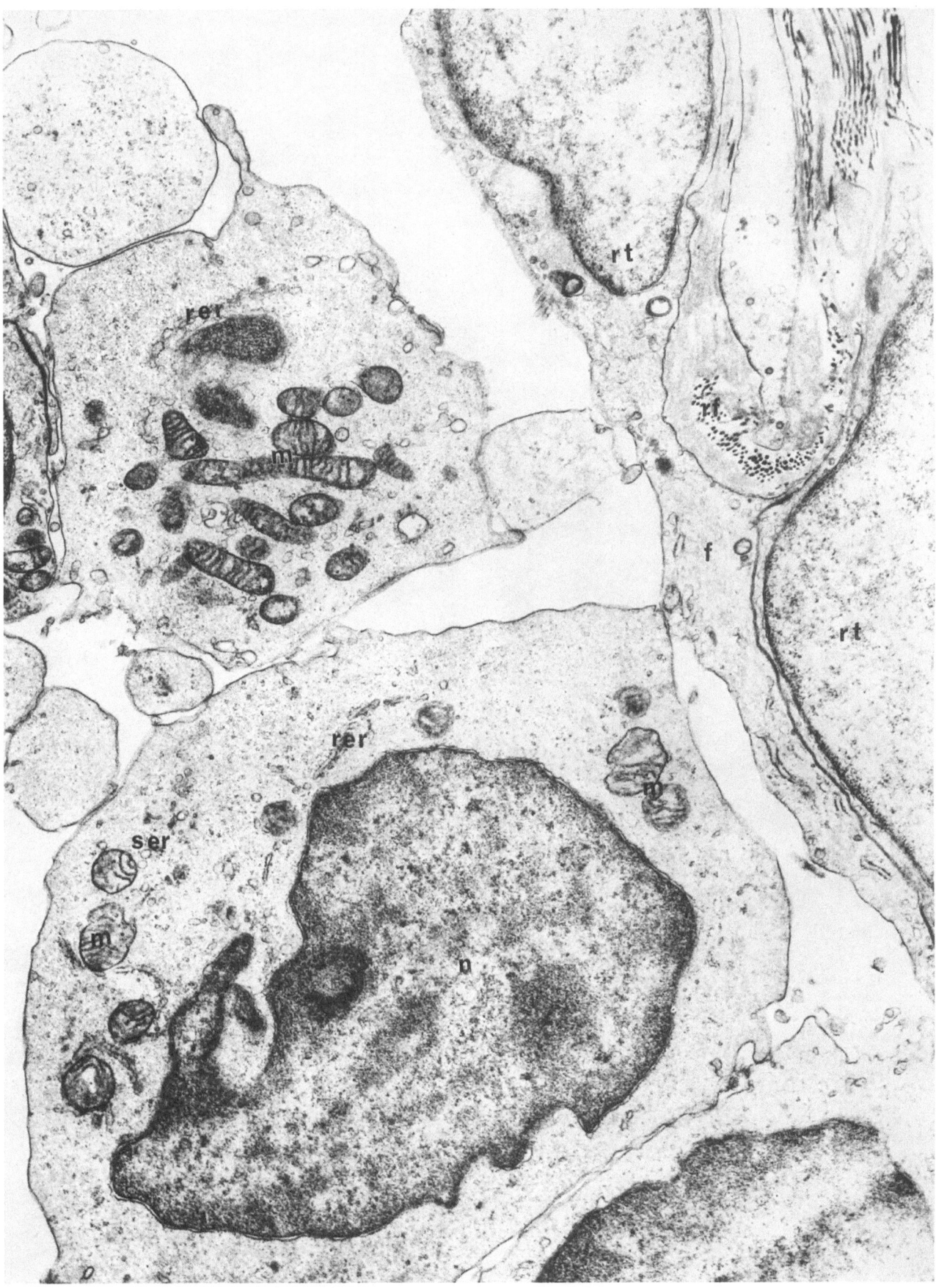

Tuberculosis
Sarcoidosis
Toxoplasmosis (Piringer's Lymphadenitis)

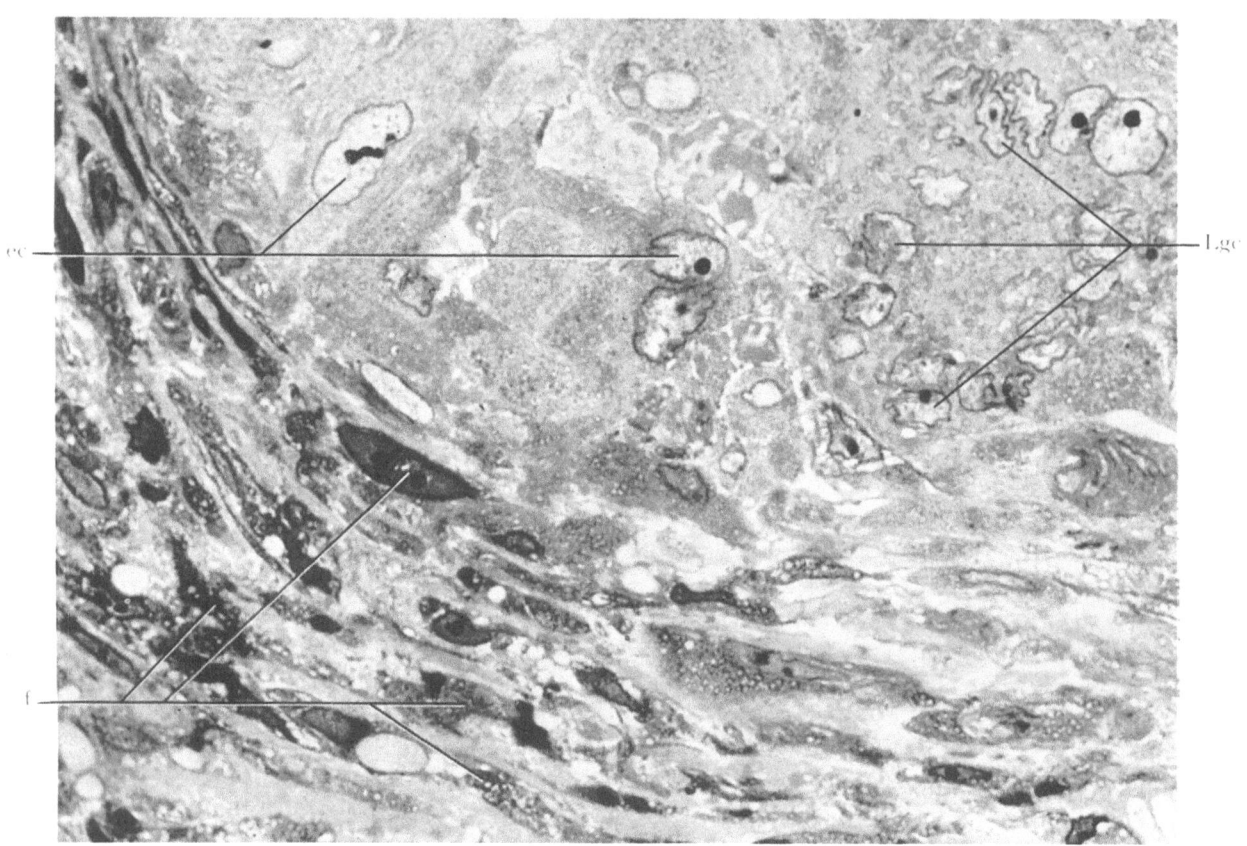

Fig. 46   Tuberculosis with epithelioid cell pattern

Tubercle with a Langhans' type giant cell (*Lgc*) and some epithelioid cells (*ec*). The tubercle is surrounded by numerous fibroblasts (*f*) with basophilic vacuolated cytoplasm (dilated cisternae of endoplasmic reticulum). Note the slender nuclei of the fibroblasts and the succulent appearing nuclei of the epithelioid cells.

Case 52. Osmium tetroxide fixation. Semithin section. Methylene blue-Azure II. ×1,230

Fig. 47   Epithelioid cell in tuberculosis ("preepithelioid" cell)

The cell appears somewhat "immature" and conforms to what YAMORI has called "pre-epithelioid" cell. Four small Golgi fields are visualized. The cytoplasm is filled with numerous vesicles and vacuoles with varied contents and many small mitochondria. Typical interdigitations of the cellular surface are lacking.

Case 52. ×8,750

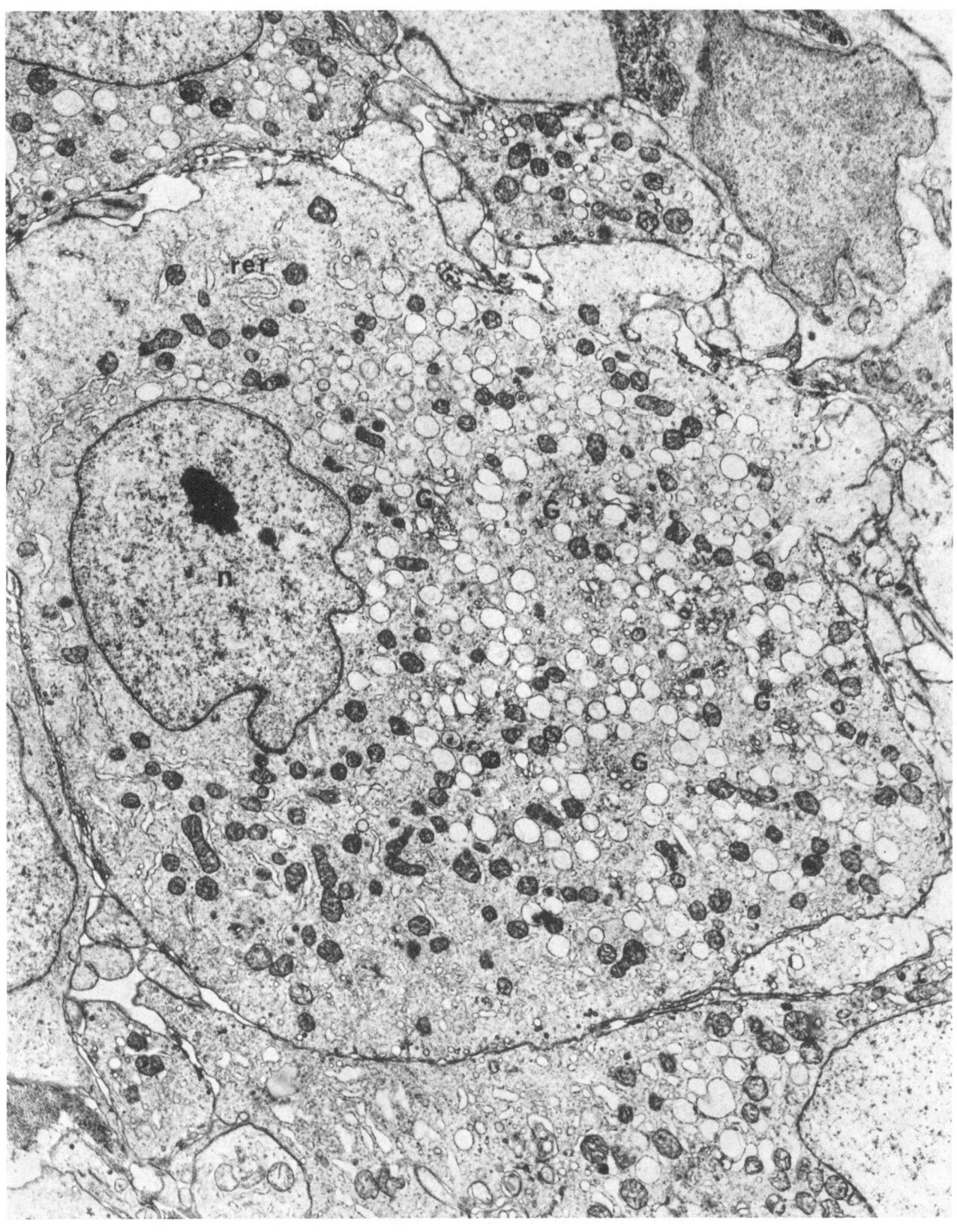

Fig. 48   Epithelioid cell in tuberculosis, higher magnification
The central portion of the cell has many vesicles and vacuoles enclosing substances of varied electron density. Numerous ribosomes and a number of polysomes are seen.
Case 52. ×35,000

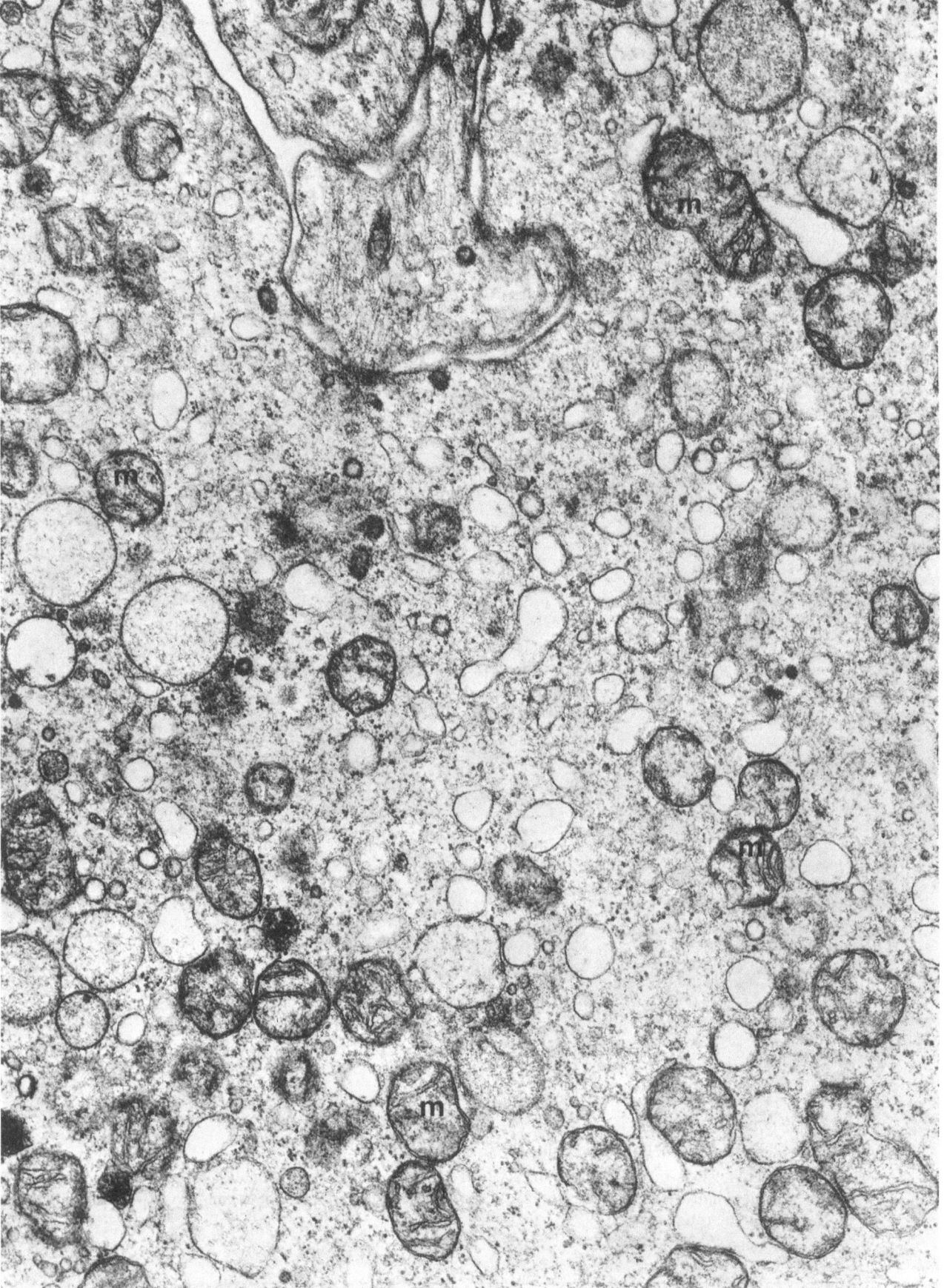

Fig. 49   Epithelioid cell in tuberculosis ("mature" type)
This cell is representative of YAMORI'S "typical" epithelioid cell. Note the characteristic interdigitations of the cytoplasmic processes. The endoplasmic reticulum is predominantly of the rough variety.
Case 52. ×35,000.

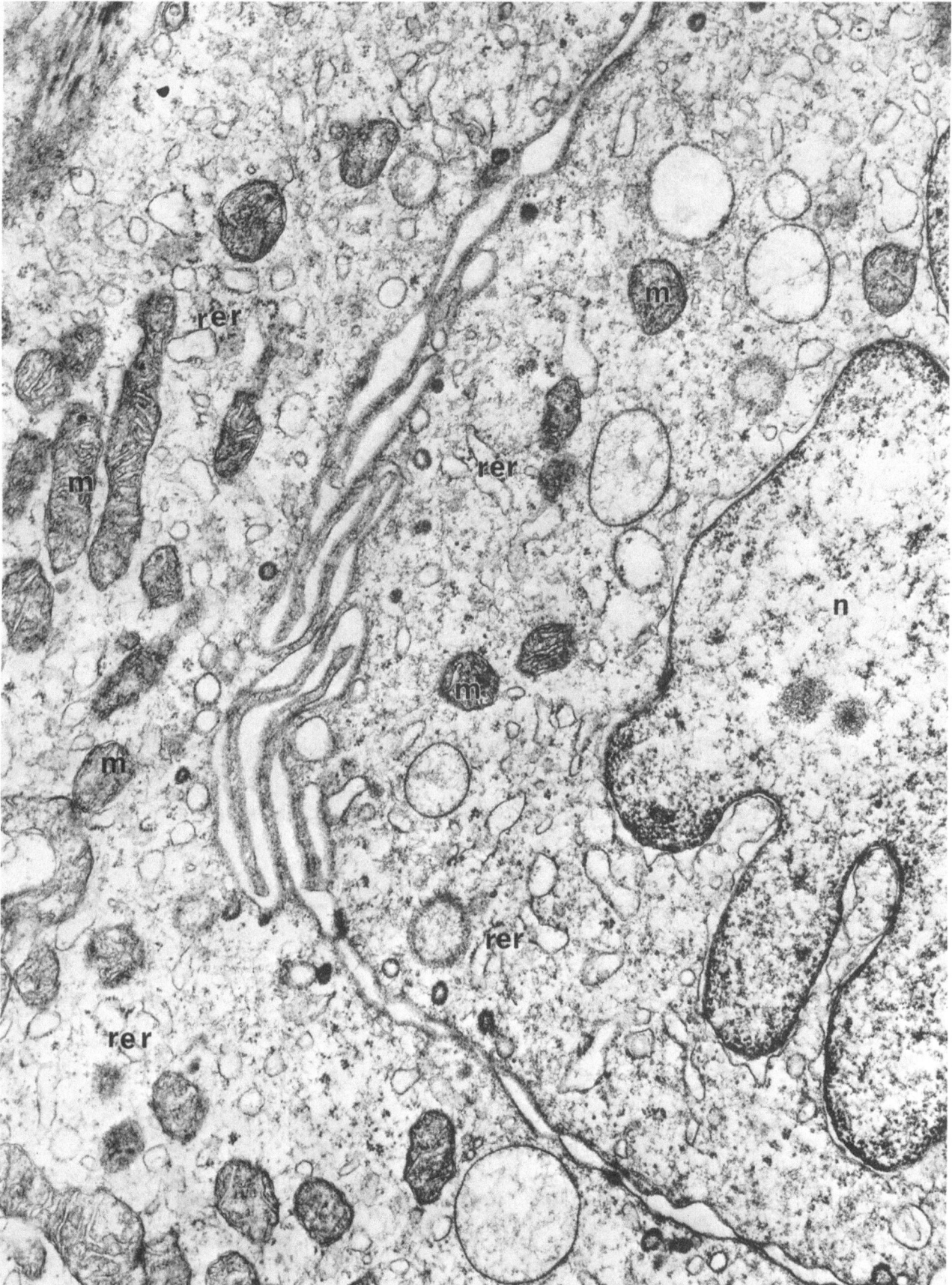

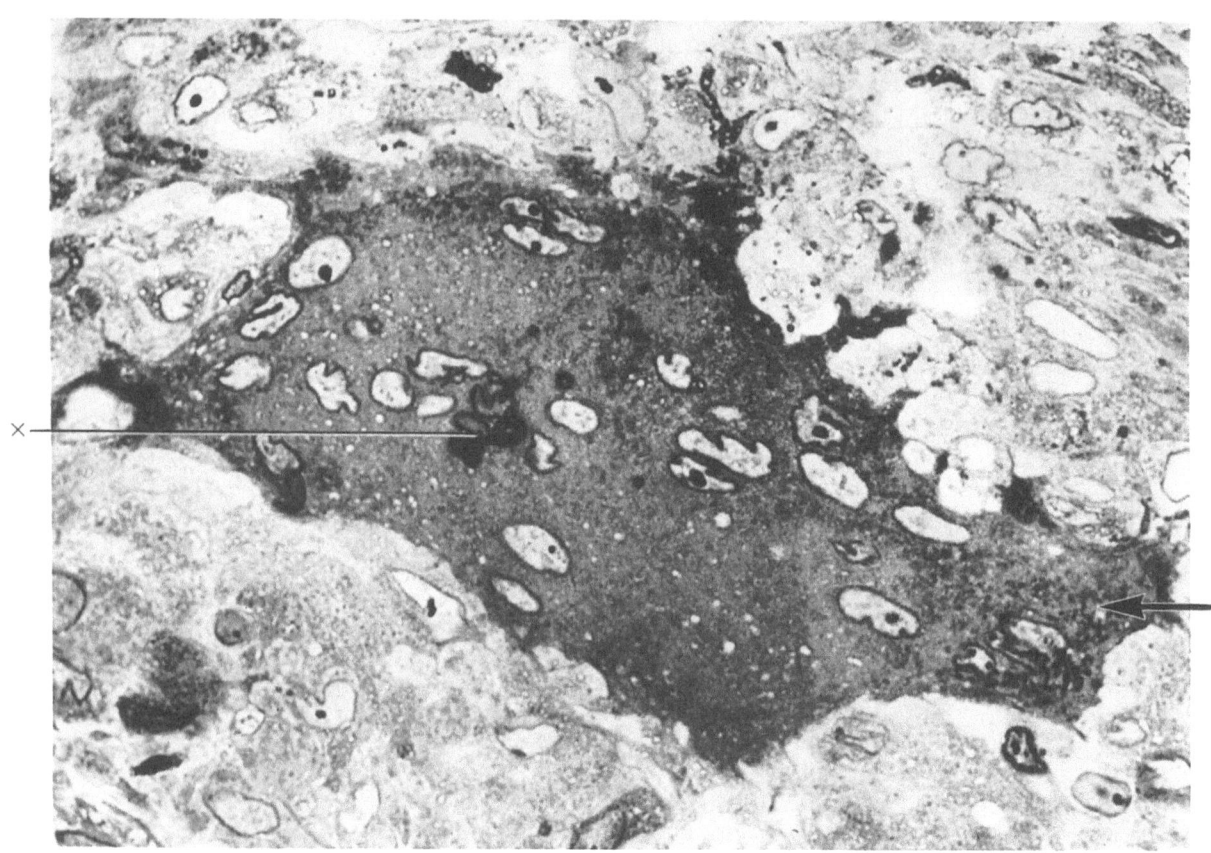

Fig. 50   Degenerating Langhans' giant cell

Note pleomorphic nuclei, numerous cytoplasmic vacuoles, a few lysosomes (arrow) and central "inclusions" (×) (? myelin figures, ? small Schaumann bodies). Many epithelioid cells are present in the vicinity.

Case 52. Osmium tetroxide fixation. Semithin section. Methylene blue-Azure II. ×1,230

Fig. 51   Langhans' giant cell in tuberculosis

The fine structure closely approximates that of the "mature" epithelioid cell. In addition to the interdigitations with adjacent cells (single arrows) note the villous folds (double arrow) imparting a characteristic corrugated appearance on the cellular limit. (For details see Fig. 52).

Case 52. ×8,750

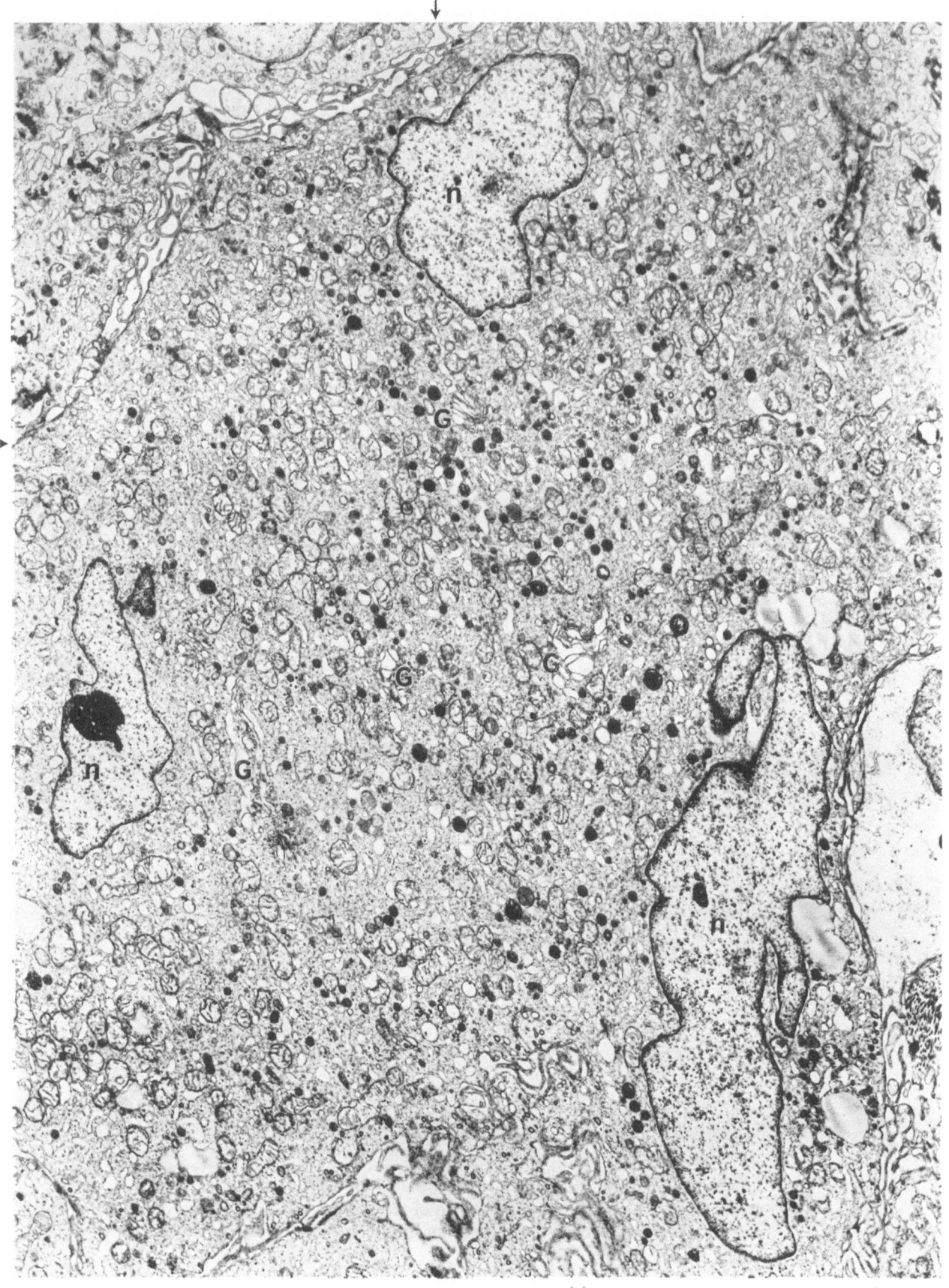

Fig. 52   Langhans' giant cell in tuberculosis, higher magnification
Again, note the corrugated appearance of the cellular limit (double arrow). A number of lysosome-like dense bodies and electron-lucent vacuoles are scattered in the cytoplasm. Case 52. ×17,500

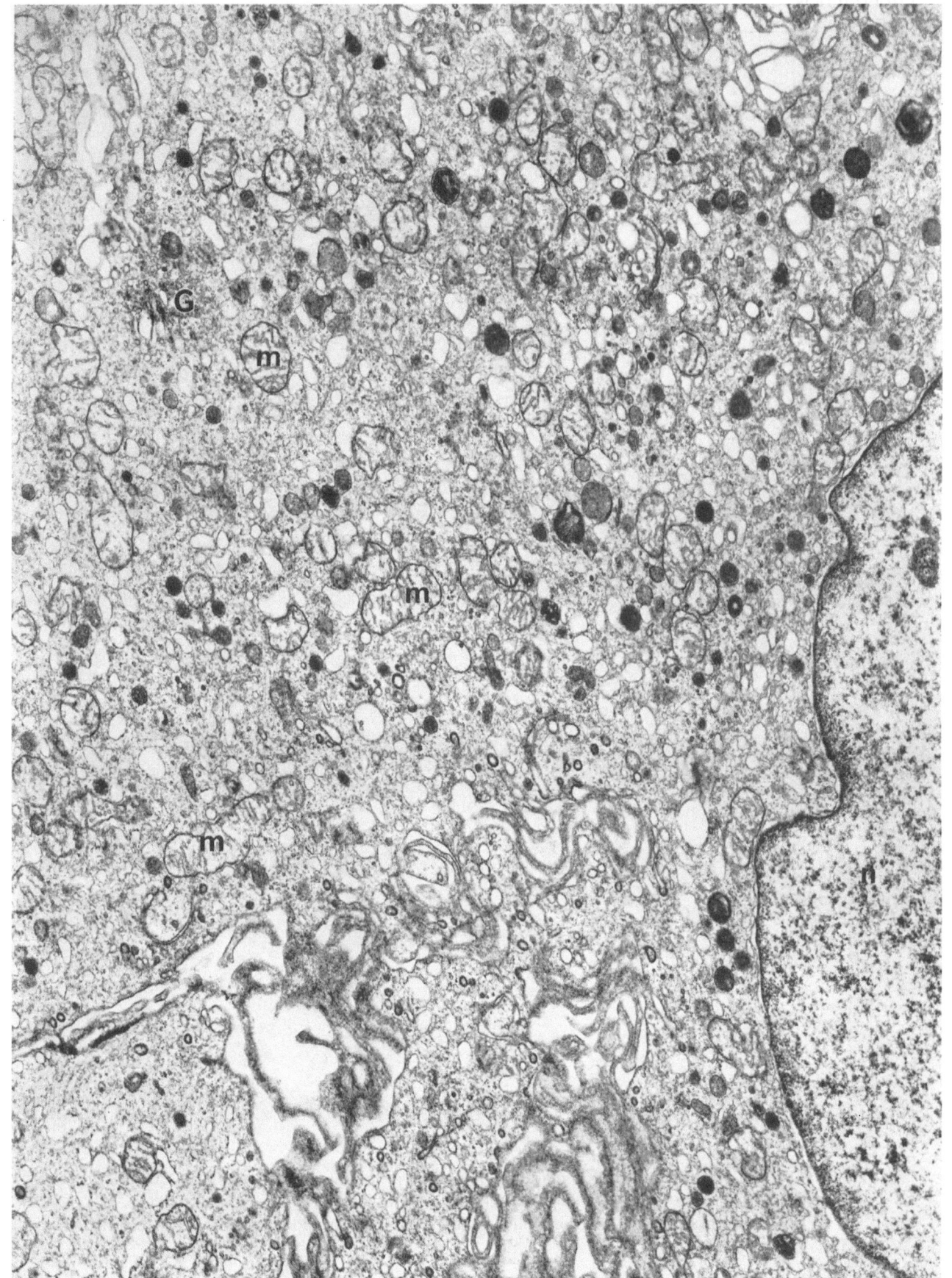

Fig. 53   Schaumann body in tuberculosis
The Schaumann body is made up of a series of laminated electron-dense calcium deposits
(arrows). Note myelin figure in upper center.
Case 18. Formalin fixation followed by osmium tetroxide fixation. Lead citrate stain.
×19,000

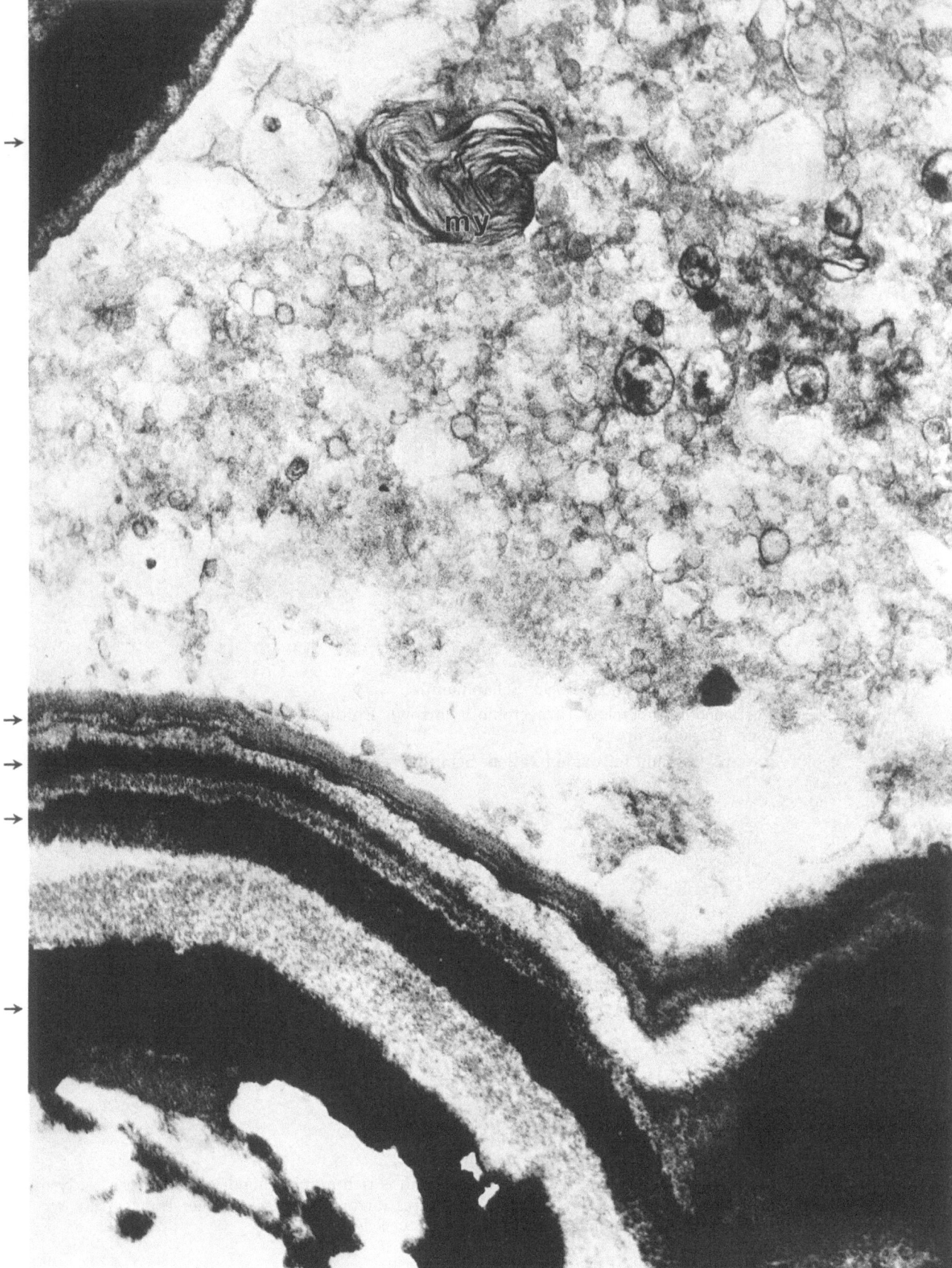

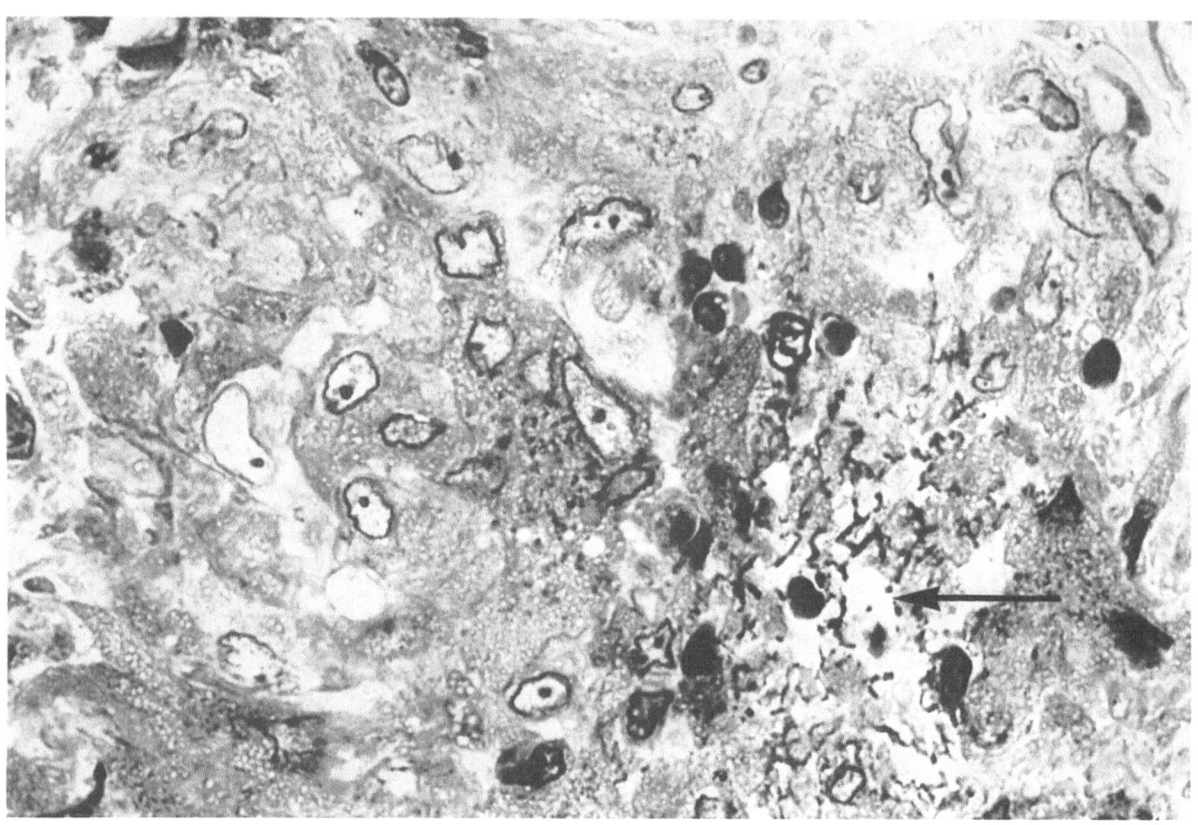

Fig. 54   Sarcoidosis (M. Besnier-Boeck-Schaumann)
                  Portion of a tubercle with necrobiosis (arrow). Epithelioid cells with many vacuoles and
                  some lysosomes are seen.
                  Case 12. Osmium tetroxide fixation. Semithin section. Methylene blue-Azure II. ×1,230

Fig. 55   Epithelioid cell in sarcoidosis ("mature" type)
                  The cytoplasm has long slender processes and is rich in mitochondria, small vesicles, lyso-
                  somes and vacuoles. The vesicular structures (arrows) represent either transversely sec-
                  tioned tubular invaginations or pinocytotic vesicles.
                  Case 12. ×17,500

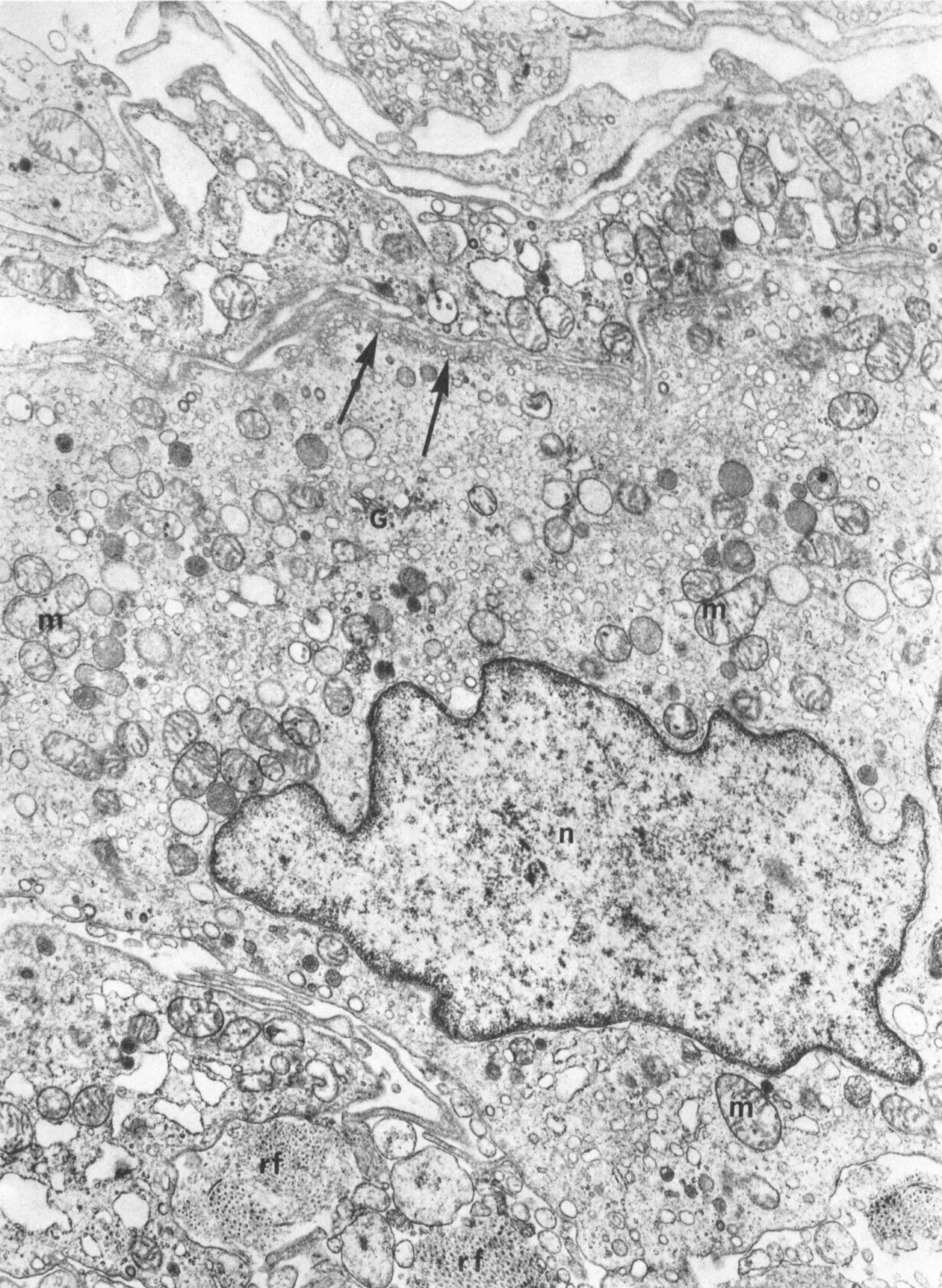

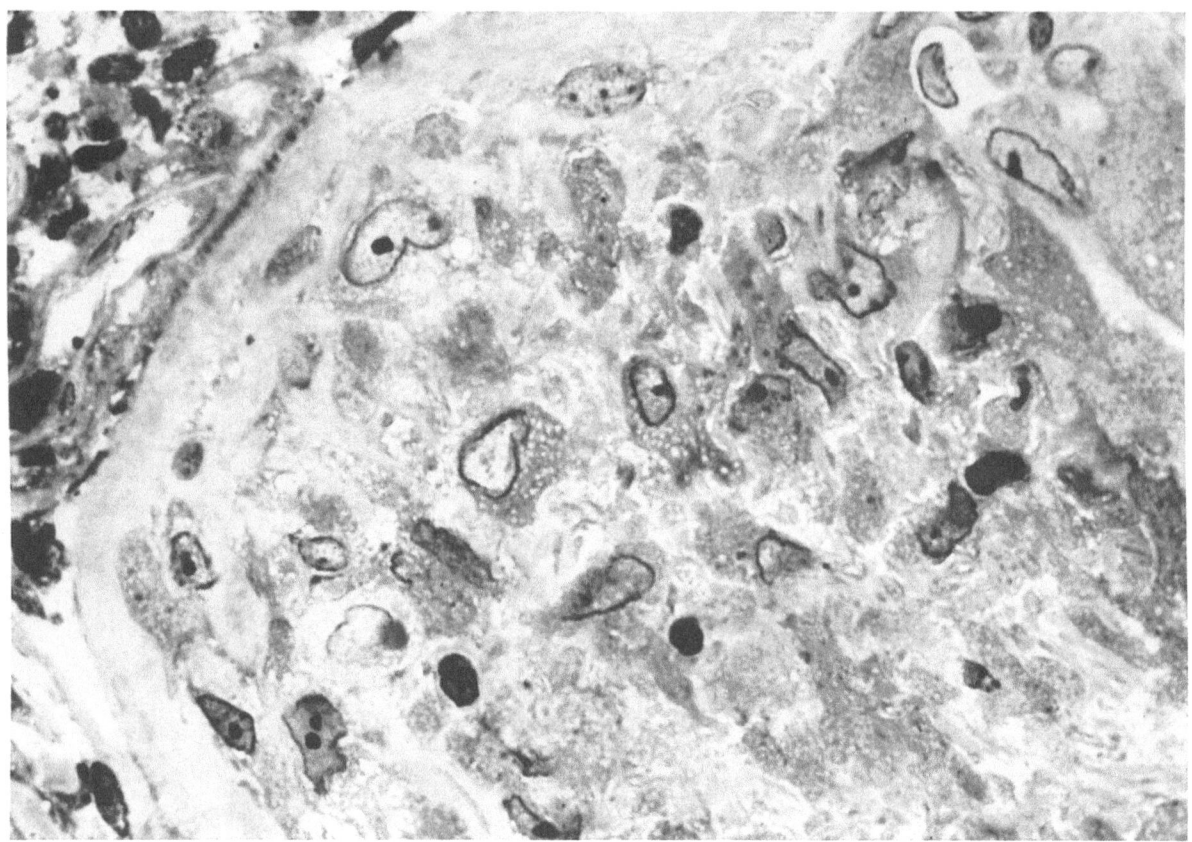

Fig. 56   Portion of a tubercle in sarcoidosis
               Vacuoles and lysosomes are prominently displayed in the epithelioid cells. Hyaline sub-
               stances are conspicuous at the border (left) of the tubercle.
               Case 12. Osmium tetroxide fixation. Semithin section. Methylene blue-Azure II. ×1,230

Fig. 57   Epithelioid cell in sarcoidosis
               Many dense bodies (? lysosomes) are interspersed with numerous mitochondria that vary
               in size and shape. The dense bodies show gradual transitions to Golgi vesicles and larger
               bodies of lesser electron density (upper right). The latter may correspond to the PAS-
               positive structures as seen with the light microscope.
               Case 12. ×45,000

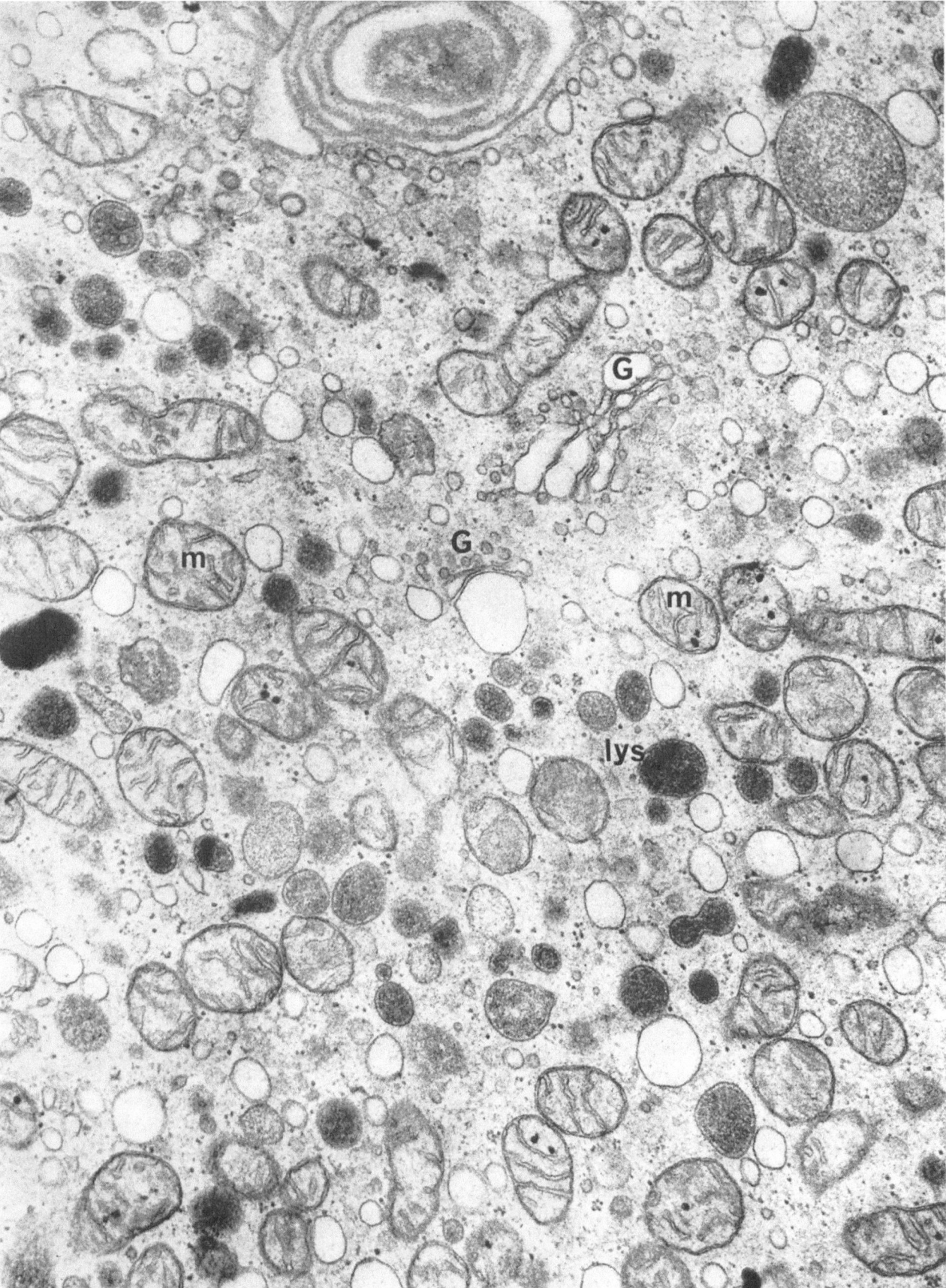

Fig. 58   Epithelioid cell in sarcoidosis ("mature" type)
Note long slender interdigitating cytoplasmic processes. The cytoplasmic fine structure closely approximates that depicted in the previous picture, but channels of rough endoplasmic reticulum are somewhat more distended.
Case 12. ×35,000

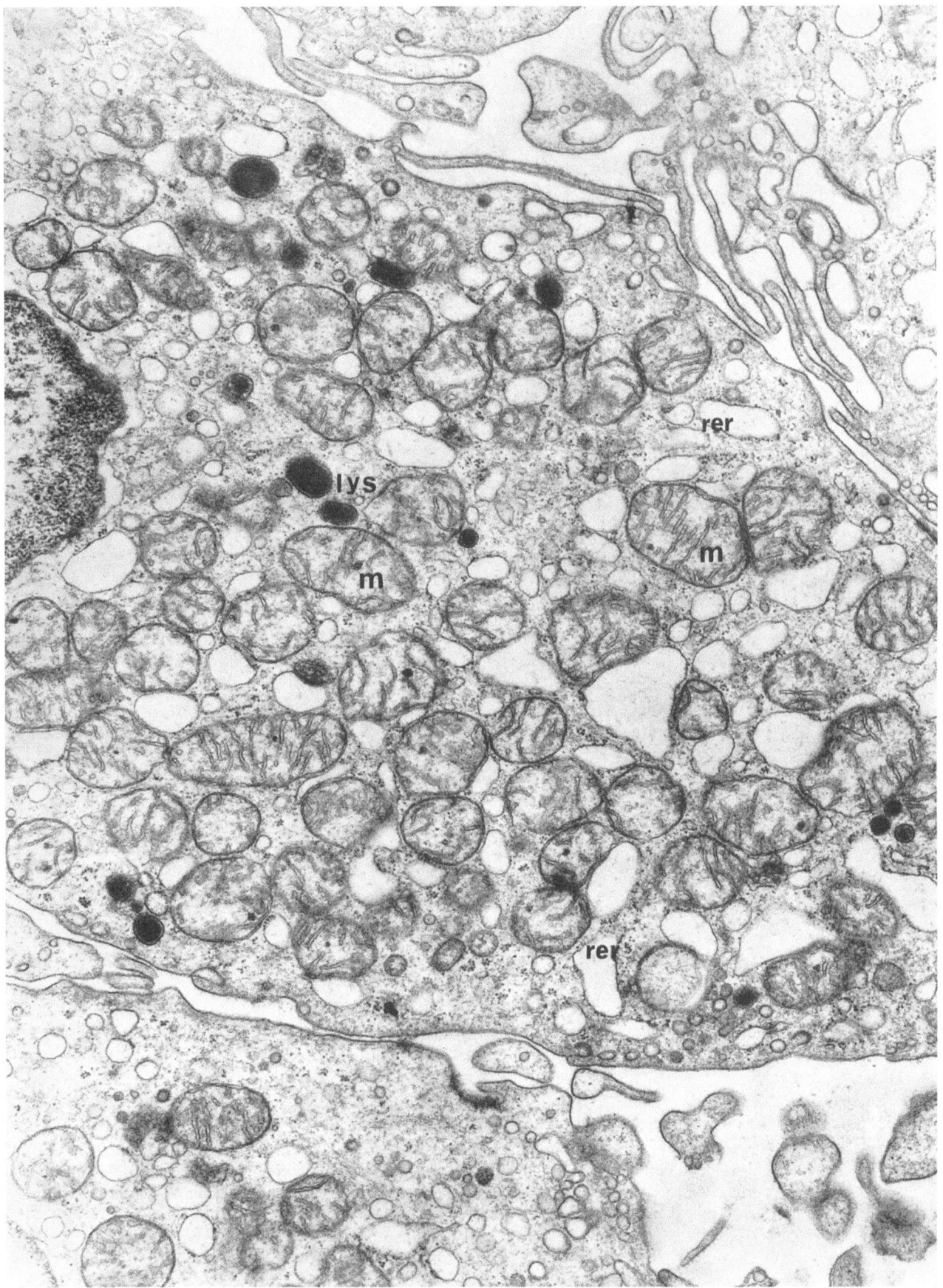

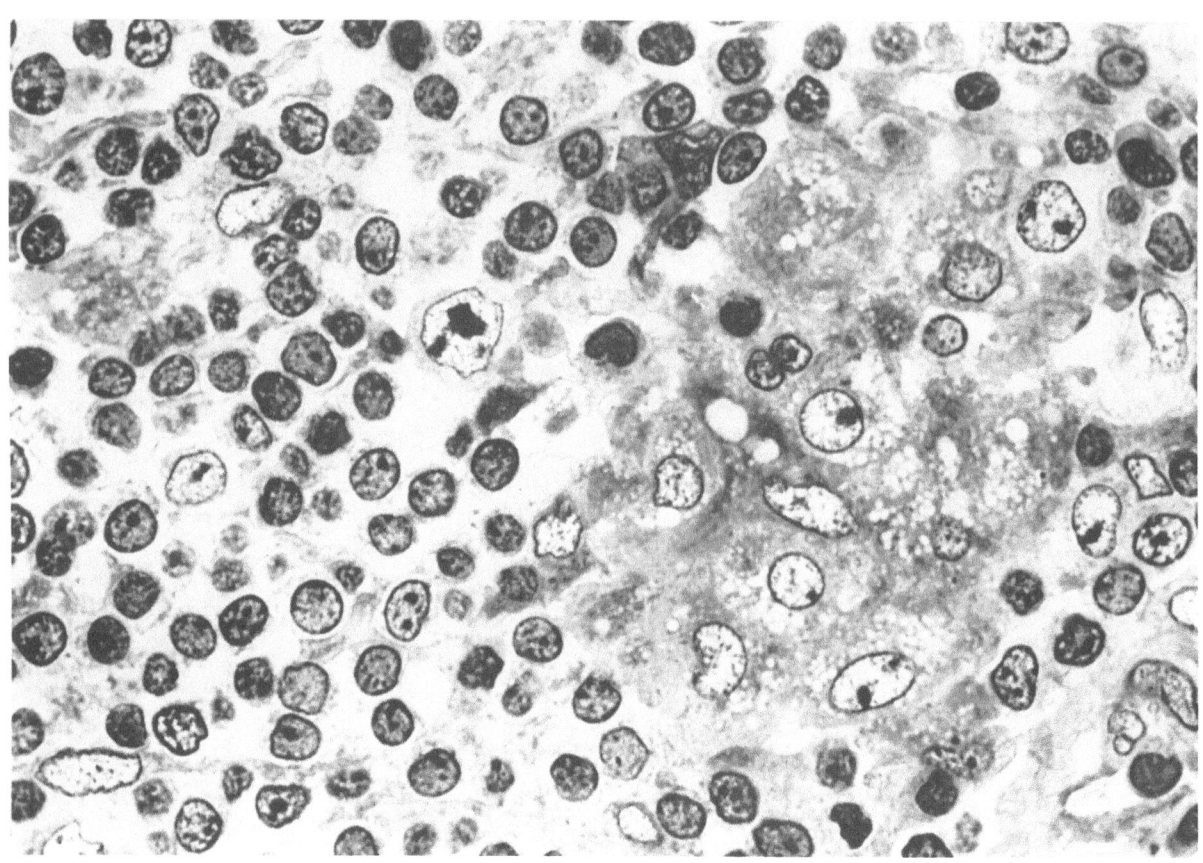

Fig. 59   Toxoplasmosis
               Cluster of epithelioid cells (right) with numerous lysosomes and vacuoles. Basophilic stem
               cell (center left).
               Case 15. Osmium tetroxide fixation. Semithin section. Methylene blue-Azure II. ×1,230

Fig. 60   Epithelioid cell in toxoplasmosis ("preepithelioid cell")
               Numerous electron-lucent small vesicles are scattered throughout the cytoplasm. Well-
               developed lamellar ergastoplasm is accumulated in the lower portion of the cytoplasm. A few
               phagosomes are also present; one of them probably represents toxoplasma gondii (arrow).
               No typical interdigitations of the cellular surface are seen.
               Case 15. ×8,750

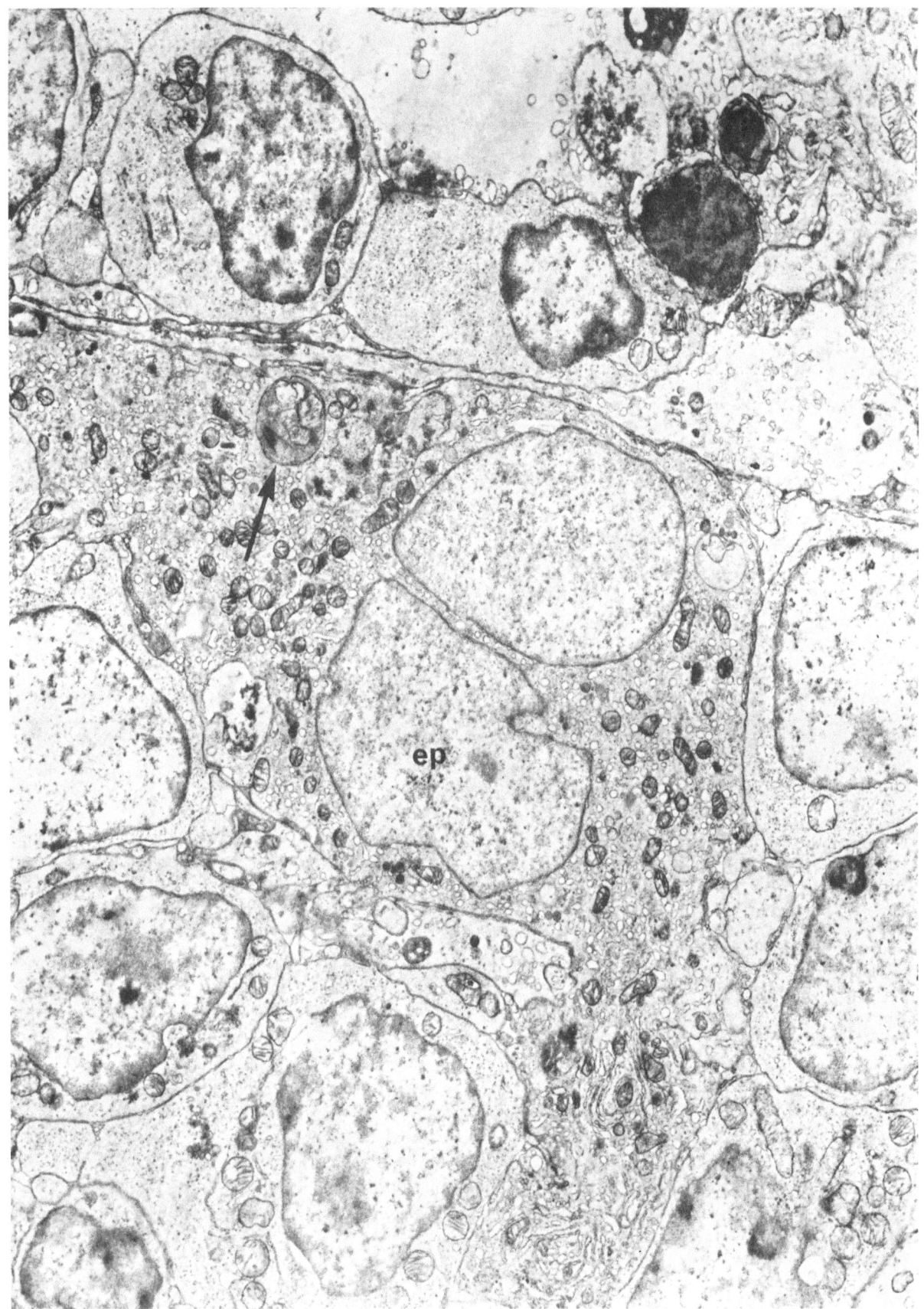

Fig. 61   Central area of epithelioid cell in toxoplasmosis
                 Note distinct accumulation of somewhat distended tubules of rough endoplasmic reticulum.
                 Case 15. ×35,000

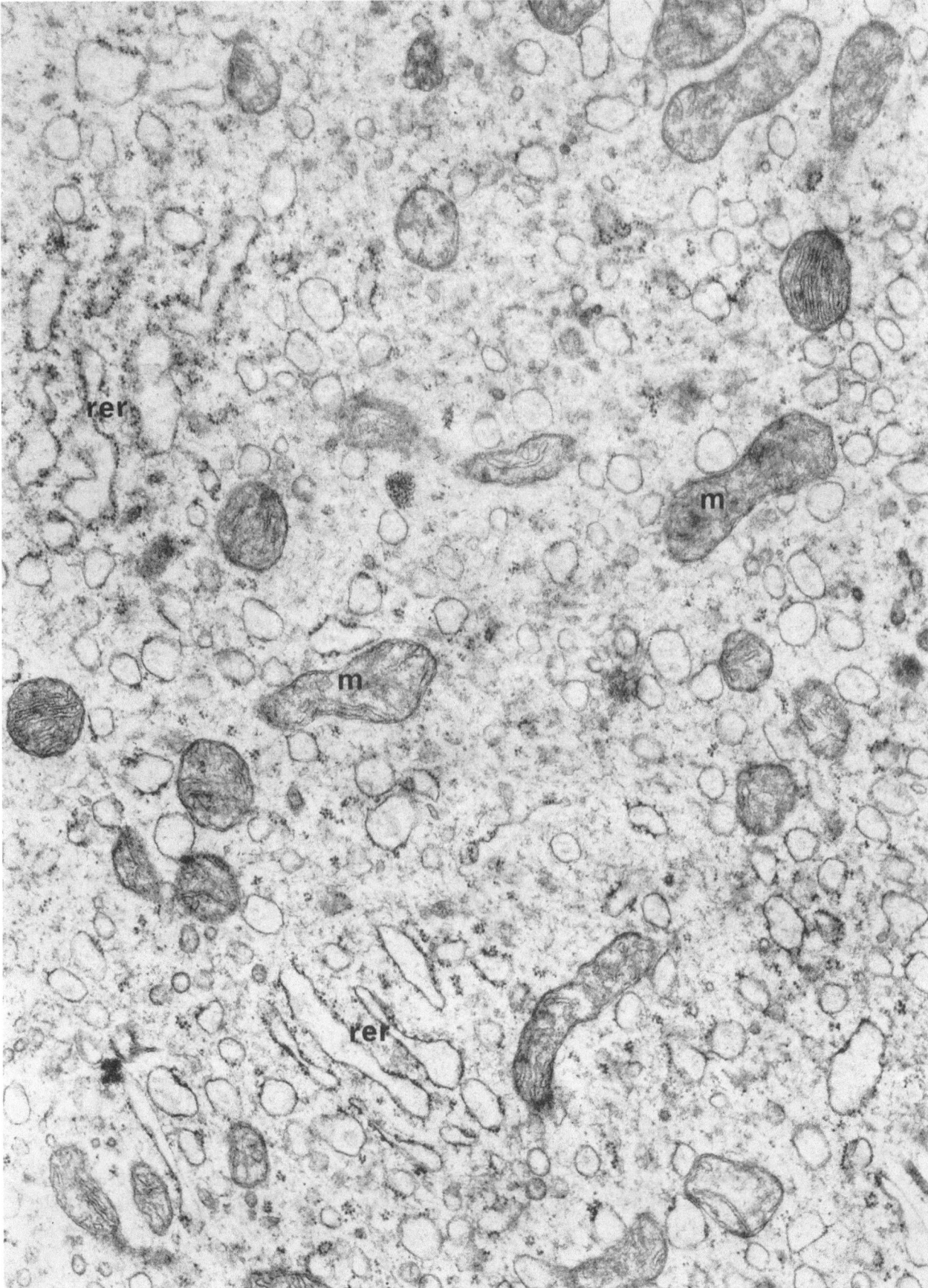

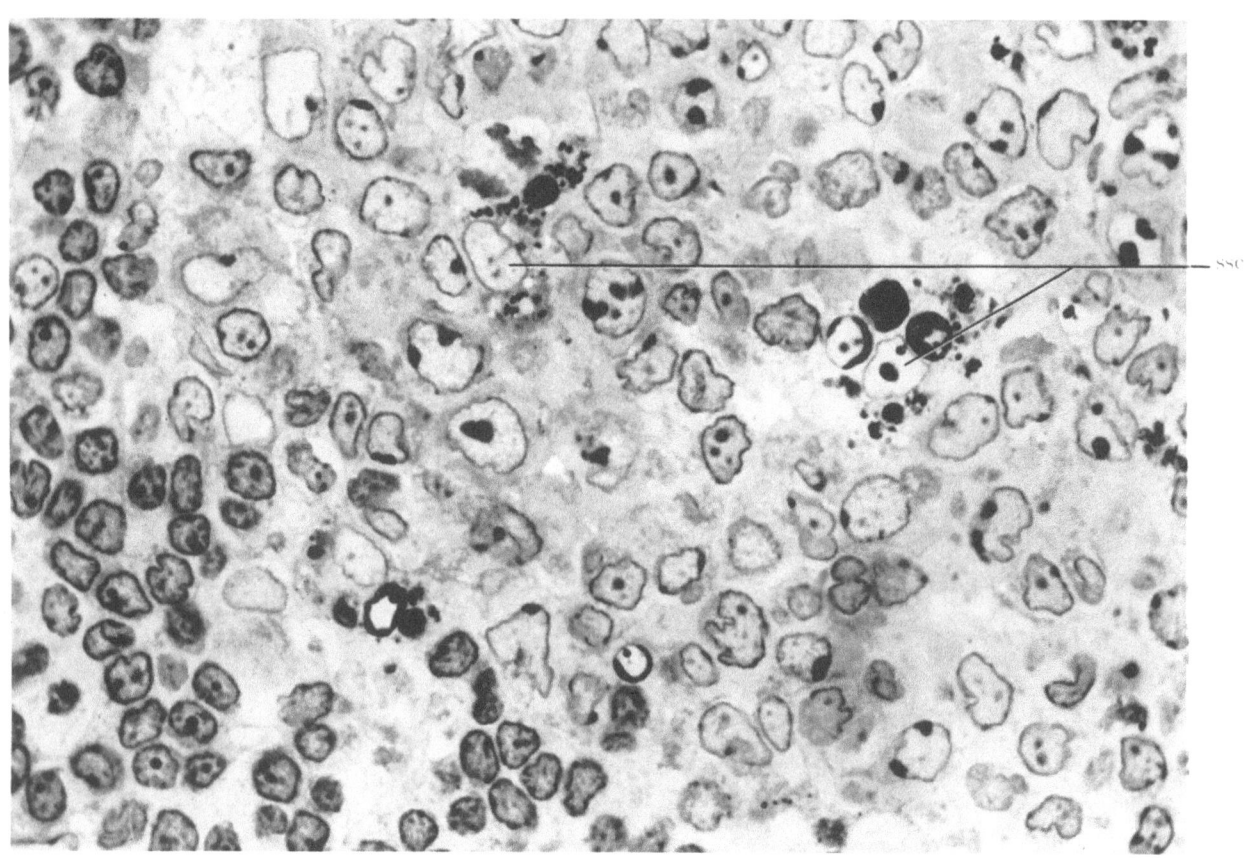

ssc

Fig. 62   Germinal center with a few "starry sky cells" (*ssc*)
Degenerating macrophages are filled with cellular debris. Germinoblasts and germinocytes
are common. Peripheral lymphocytic mantle is seen at left lower corner.
Case 35. Osmium tetroxide fixation. Semithin section. Methylene blue-Azure II. ×1,230

Fig. 63   Tingible body macrophage
Phagosomes and residual bodies are of varying size and composition. Some of them contain
lamellar structures (arrows), others contain cellular debris or granular material of variable
electron density.
Case 35. ×17,500

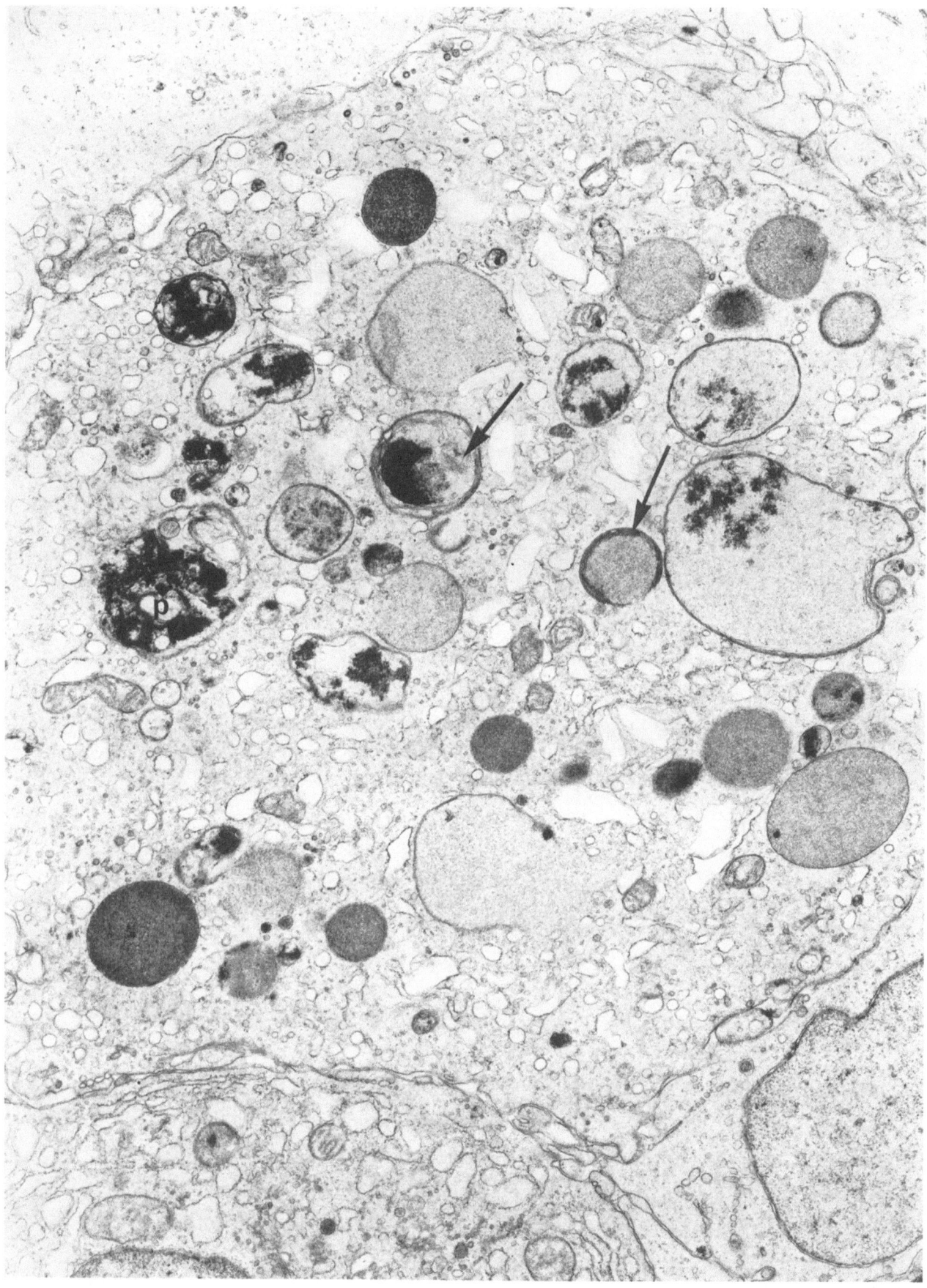

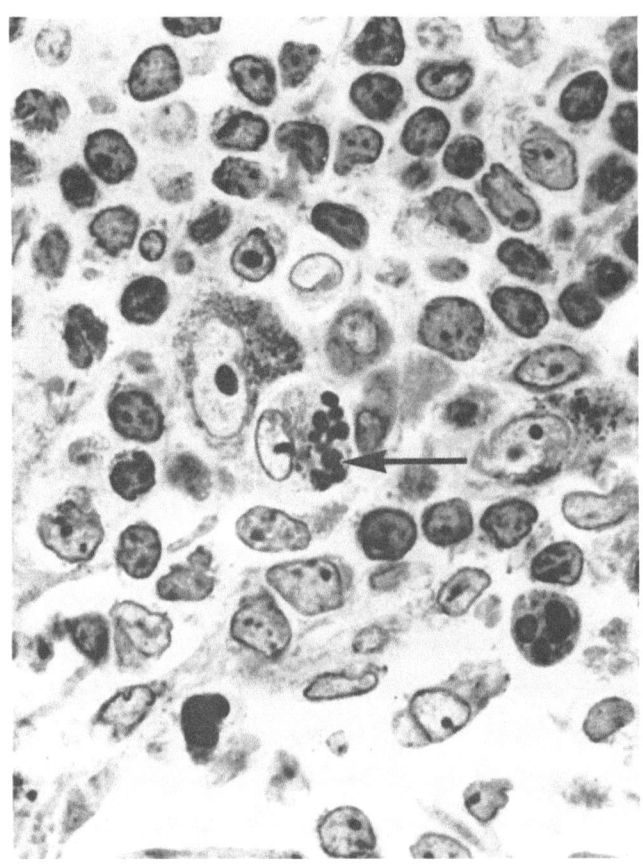

Fig. 64   Toxoplasmosis
                 Macrophages (? in a sinus), one with quite a few round to oval inclusions of equal size
                 (probably toxoplasma organisms, proliferating type, arrow); one with numerous lysosomes.
                 Case 35. Osmium tetroxide fixation. Semithin section. Methylene blue-Azure II. ×1,230

Fig. 65 a—c   Toxoplasma gondii engulfed by cytoplasm of macrophage
                 a) and c) proliferating type, well preserved
                 b)   proliferating type (? disintegrating)
                 Case 35. a) ×17,500   b) ×35,000   c) ×35,000

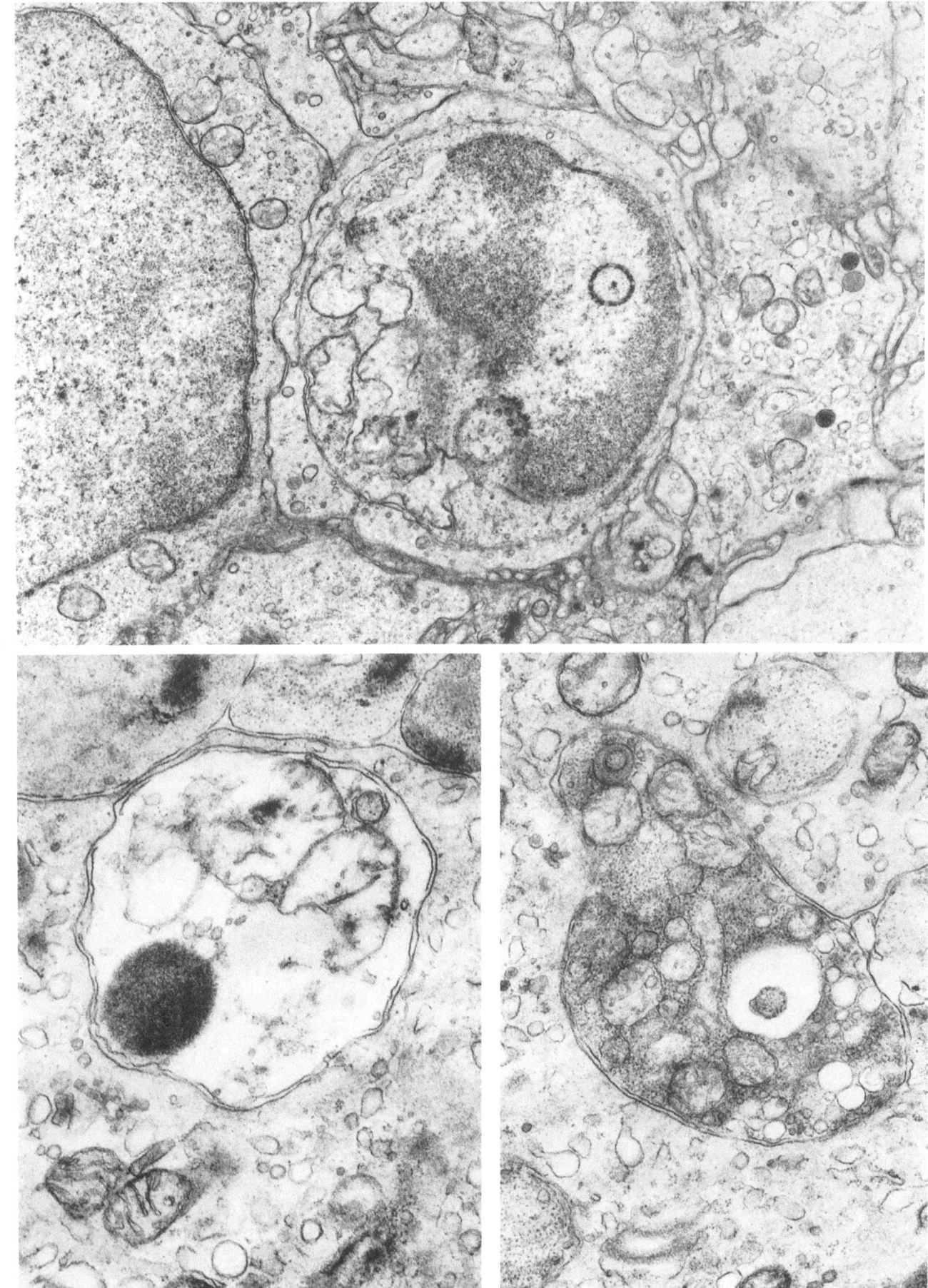

Mixed Type
Nodular Sclerosing Type (Lukes)
Epithelioid Cellular
Lymphogranulomatosis

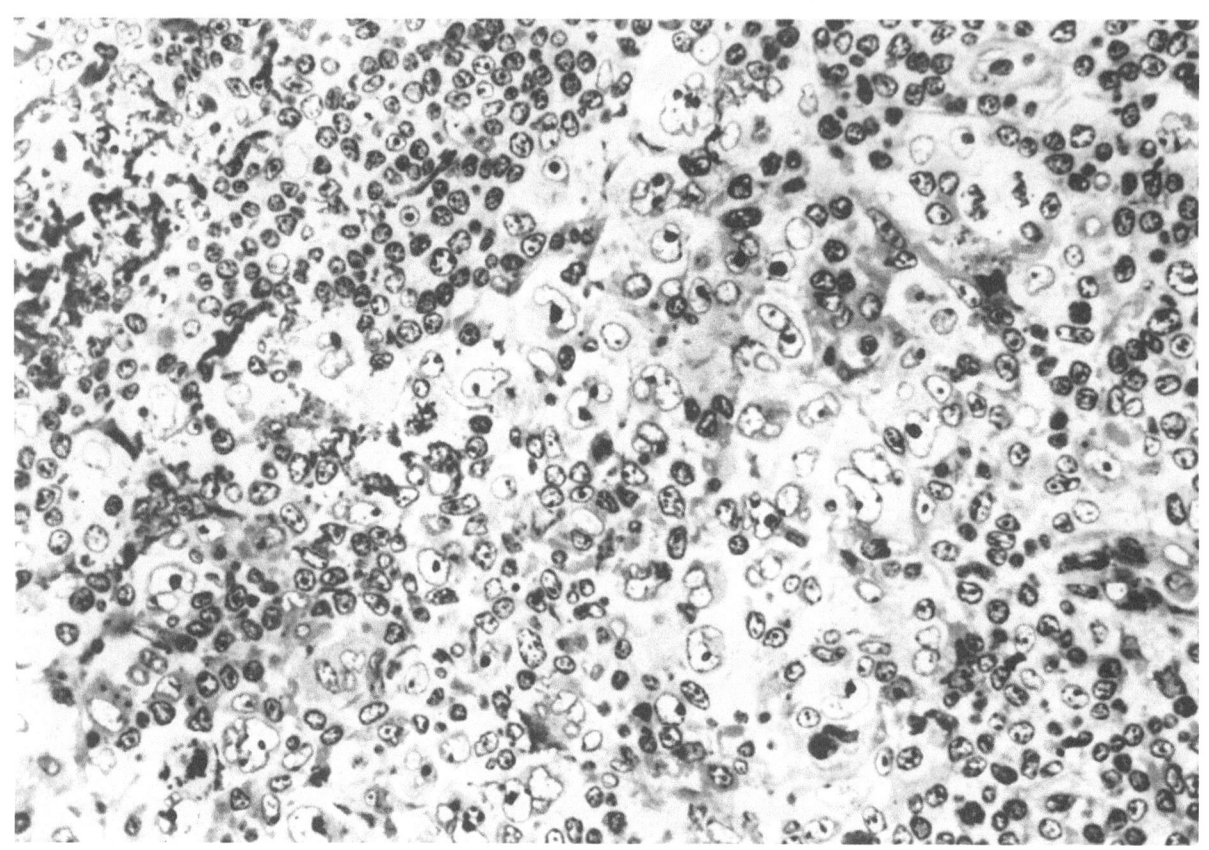

Fig. 66   HODGKIN'S disease, mixed type

Area with many Hodgkin and Sternberg-Reed cells. Note lack of fibrosis. Numerous lymphocytes and reticulum cells and some plasma cells are seen.

Case 36. Osmium tetroxide fixation. Semithin section. Methylene blue-Azure II. ×400

Fig. 67   Sternberg-Reed giant cell

Typical lobulated nuclei with prominent nucleoli are illustrated. The endoplasmic reticulum is conspicuous and appears vesicular or to consist of narrow channels.

Case 36. ×8,750

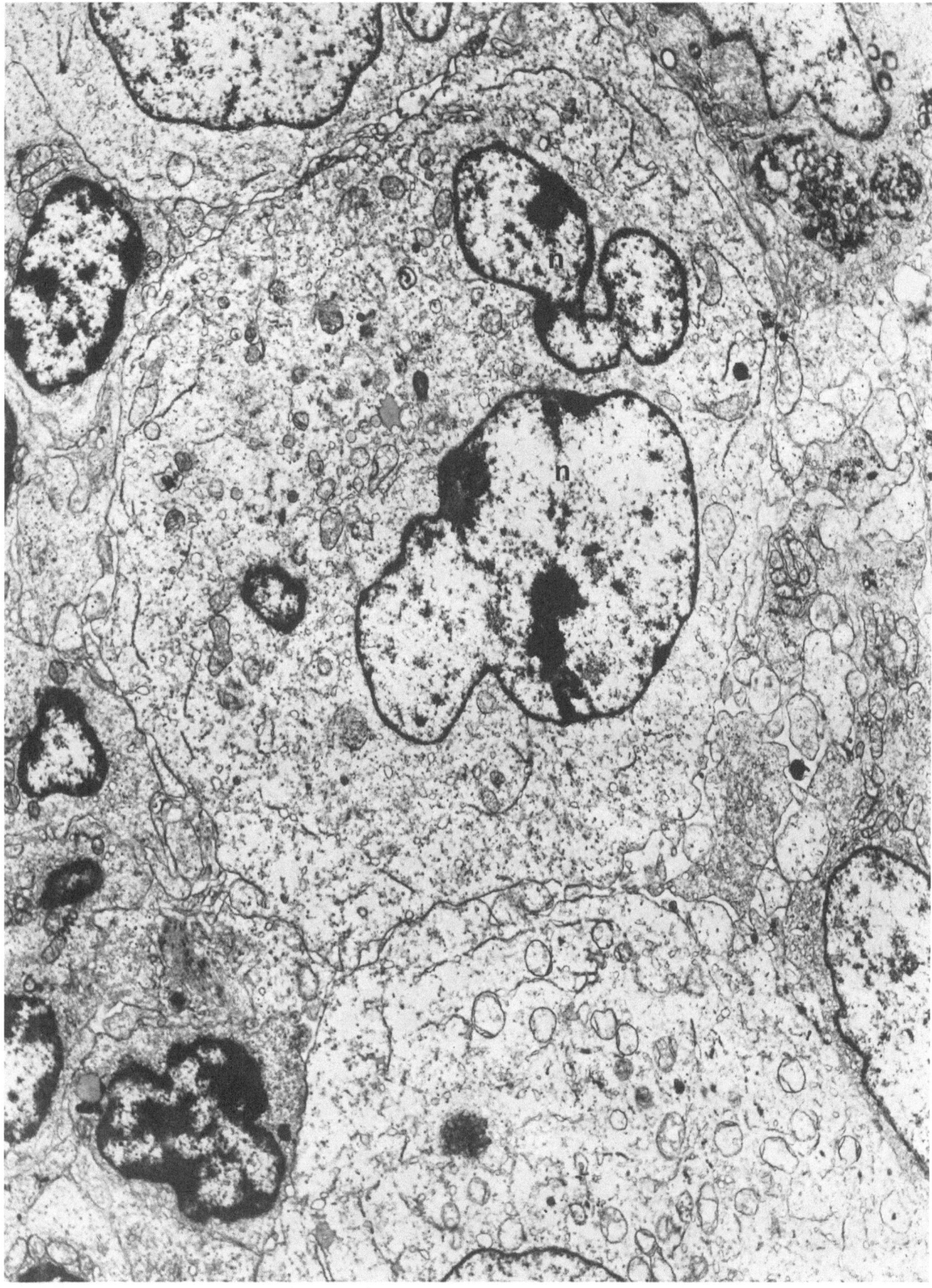

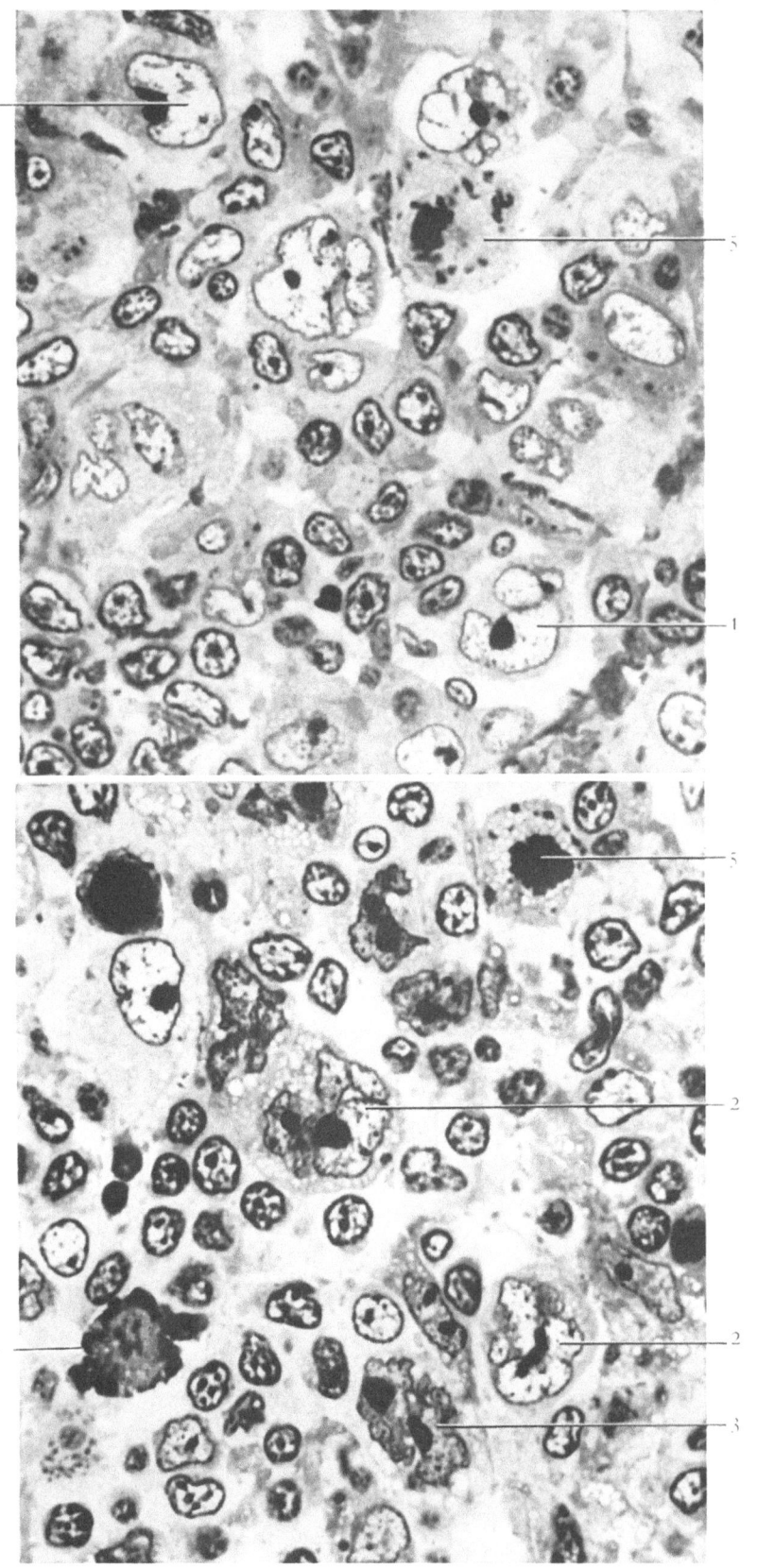

Fig. 68
HODGKIN's disease,
mixed type.
Sternberg-Reed giant
cells in all stages of
development:
1 well preserved
2 markedly vacuolated
3 slightly degenerated
  (pyknotic)
4 completely pyknotic
5 atypical mitoses

Case 36. Osmium
tetroxide fixation.
Semithin section.
Methylene blue-Azure II.
×1,230

Fig. 69
Hodgkin cell

Numerous free ribosomes
and bundles of filaments
(arrows). Small rod-
shaped mitochondria and
small amount of vesicular
structures. Giant nucle-
olus.
Mixed type of HODGKIN's
disease.

Case 10. ×17,500

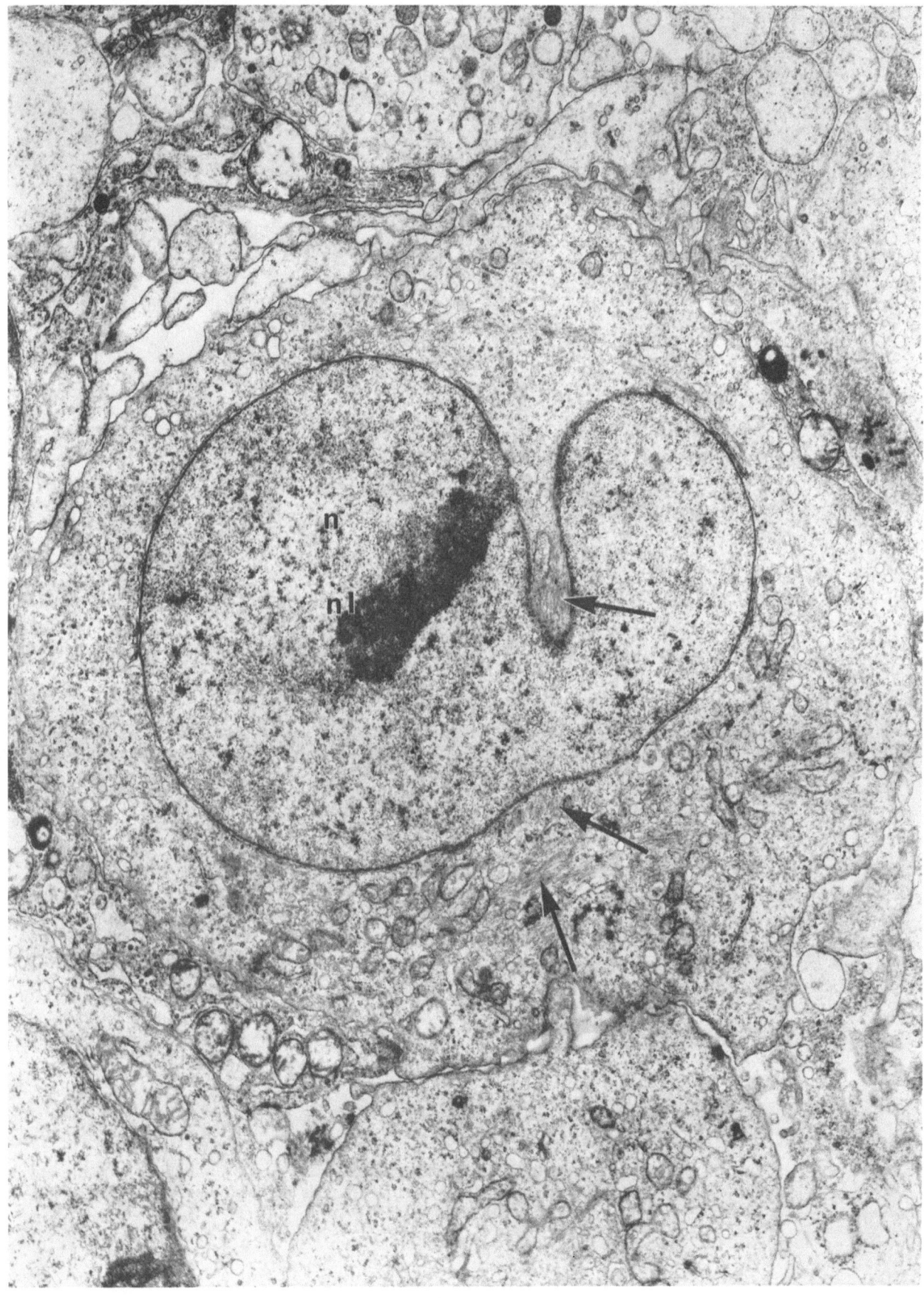

Fig. 69

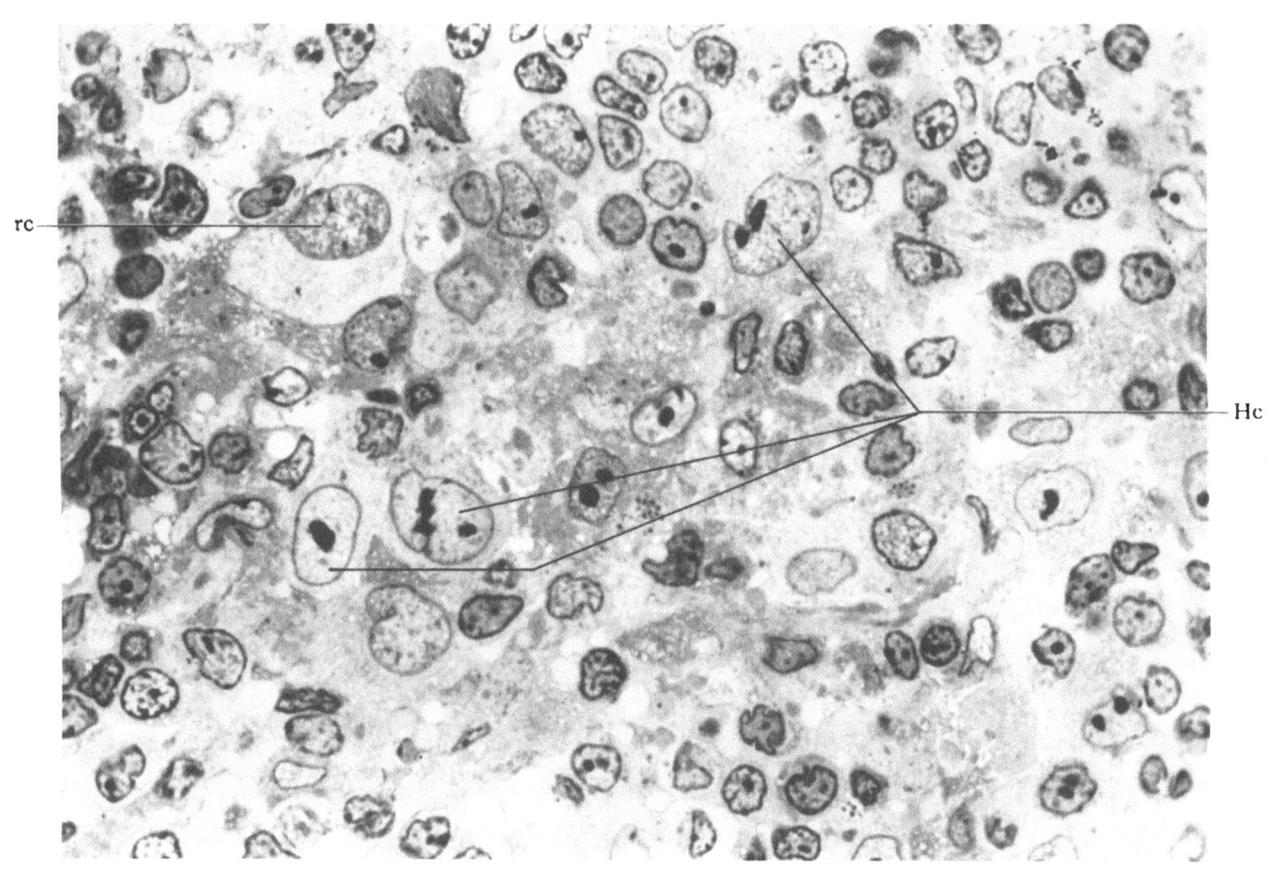

Fig. 70   Hodgkin's disease, nodular sclerosing type (Lukes)

Three so-called Hodgkin cells (*Hc*) differ slightly from typical Hodgkin cells as seen in the mixed type. Reticulum cell with abundant cytoplasm (*rc*) corresponds to reticulum cells with a clear halo as they appear in paraffin sections.

Case 10. Osmium tetroxide fixation. Semithin section. Methylene blue-Azure II. ×1,230

Fig. 71   Hodgkin cell

Swollen mitochondria and vacuoles occur in the cytoplasm. Three large nucleoli are present in the nucleus. The morphologic features of Hodgkin cells are variable even in a single case. Nodular sclerosing type of Hodgkin's disease.

Case 10. ×17,500

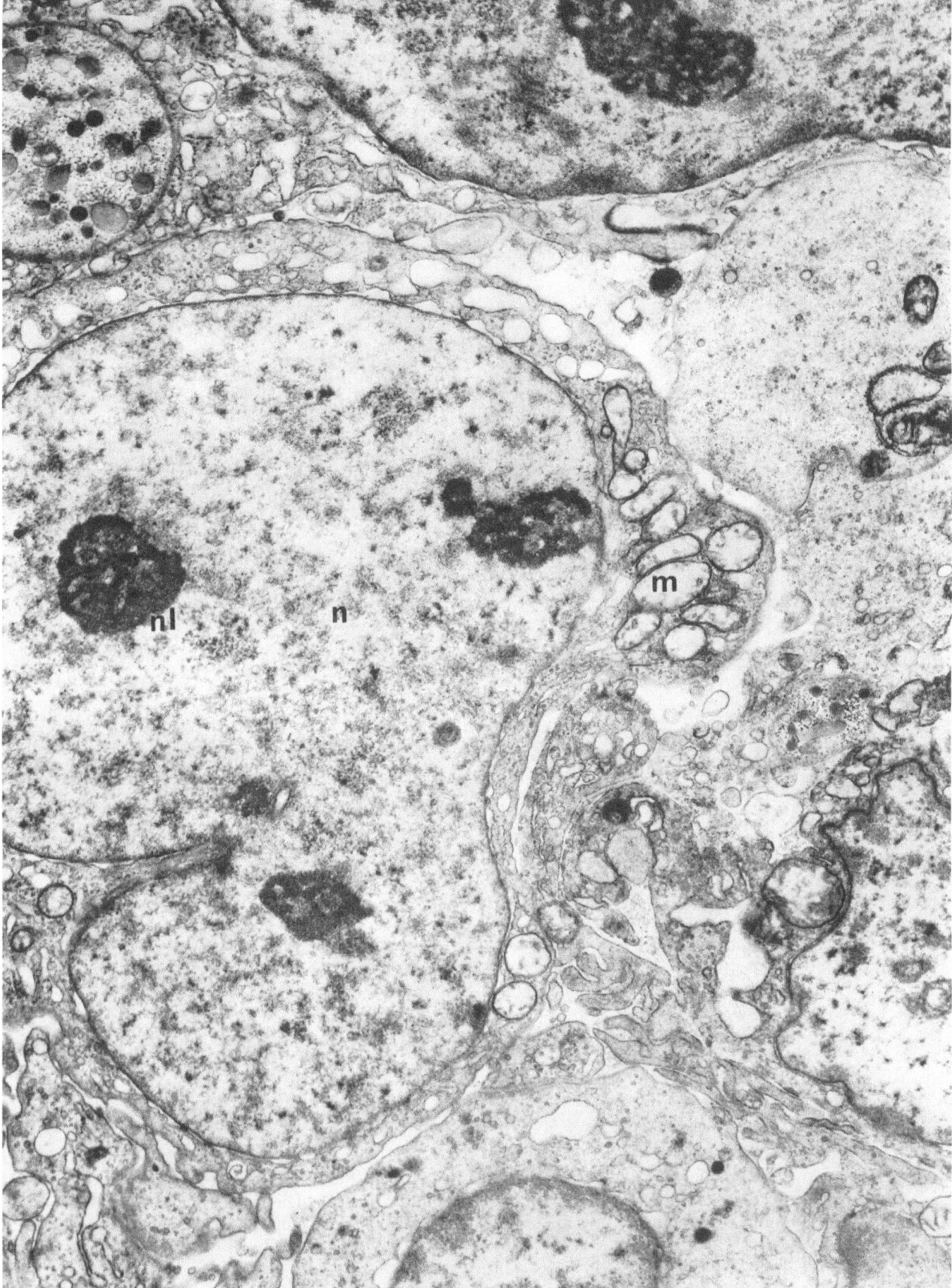

Fig. 72   Sternberg-Reed giant cell
               Typical binucleated form with large reticulated nucleoli. The Golgi apparatus is prominent
               and vesicular structures are seen. Nodular sclerosing type of HODGKIN's disease.

               Case 10. ×17,500

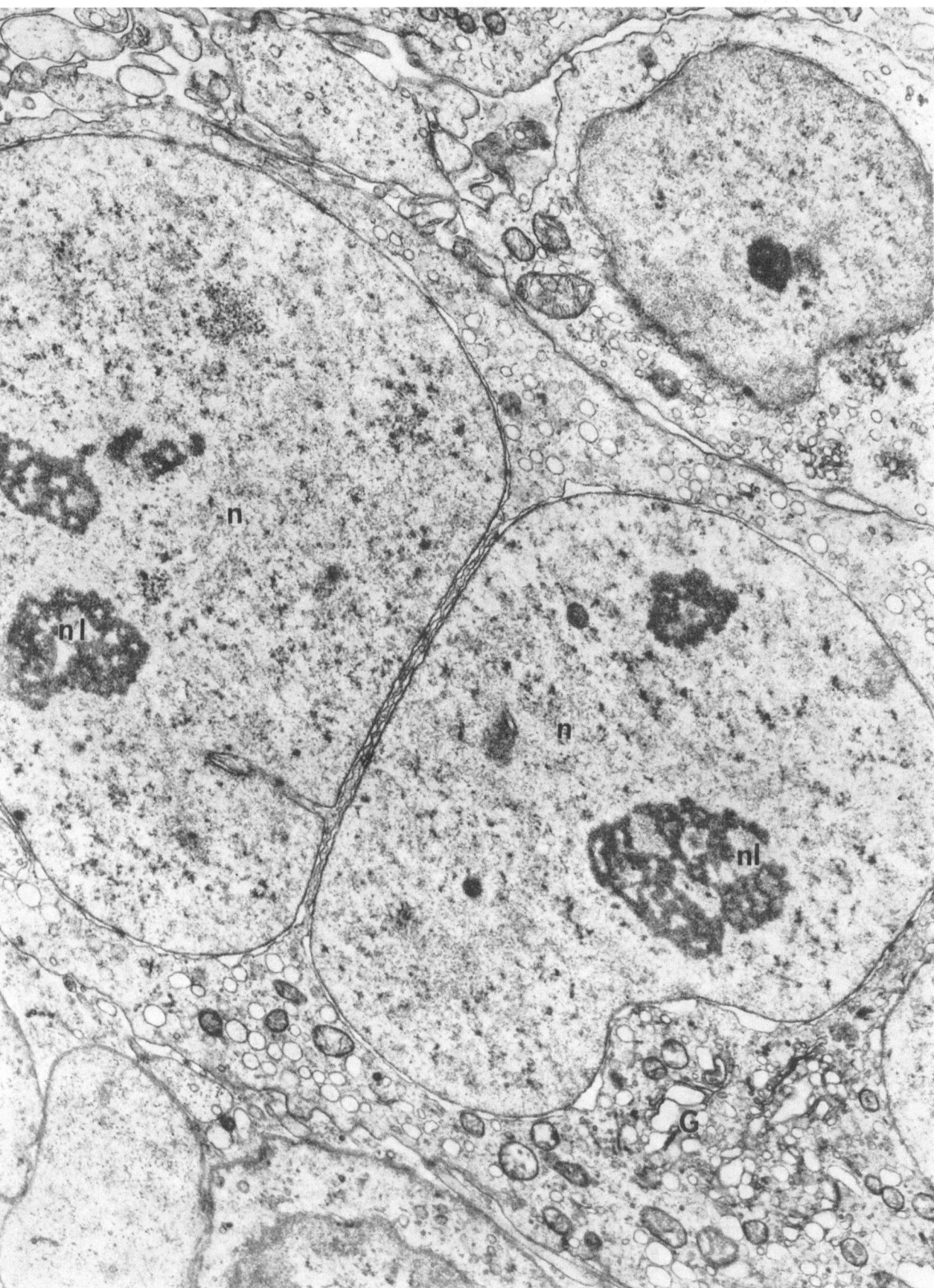

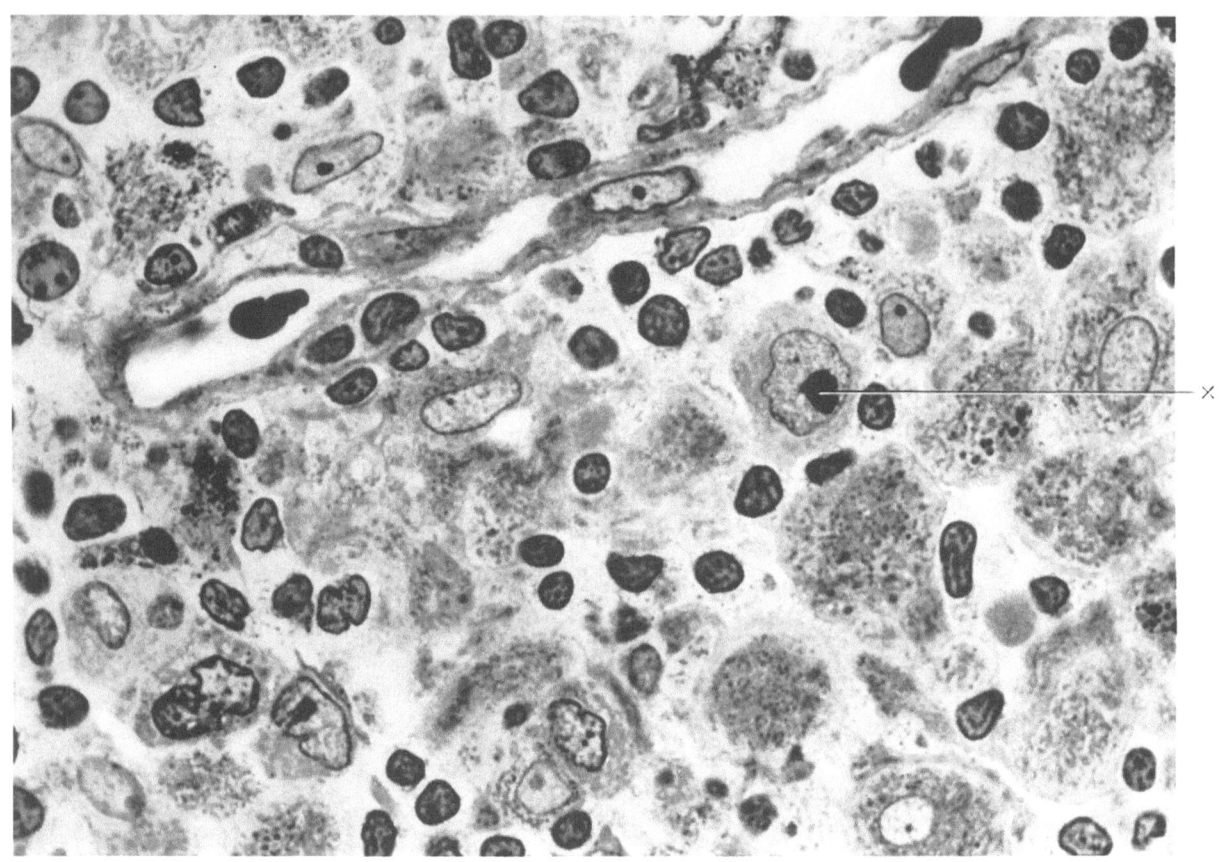

Fig. 73   Epithelioid cellular lymphogranulomatosis
                 Epithelioid cells with many lysosomes and one atypical cell with some morphologic charac-
                 teristics of a Hodgkin cell (×) are seen.

                 Case 26. Osmium tetroxide fixation. Semithin section. Methylene blue-Azure II. ×1,230

Fig. 74   Epithelioid cell in HODGKIN's disease (epithelioid cellular lymphogranulomatosis)
                 Note prominent Golgi apparatuses, unevenly scattered smooth and rough endoplasmic
                 reticulum and many cytoplasmic processes. Many lysosome-like dense bodies of variable
                 electron-opacity are discernible between the numerous mitochondria.

                 Case 26. ×17,500

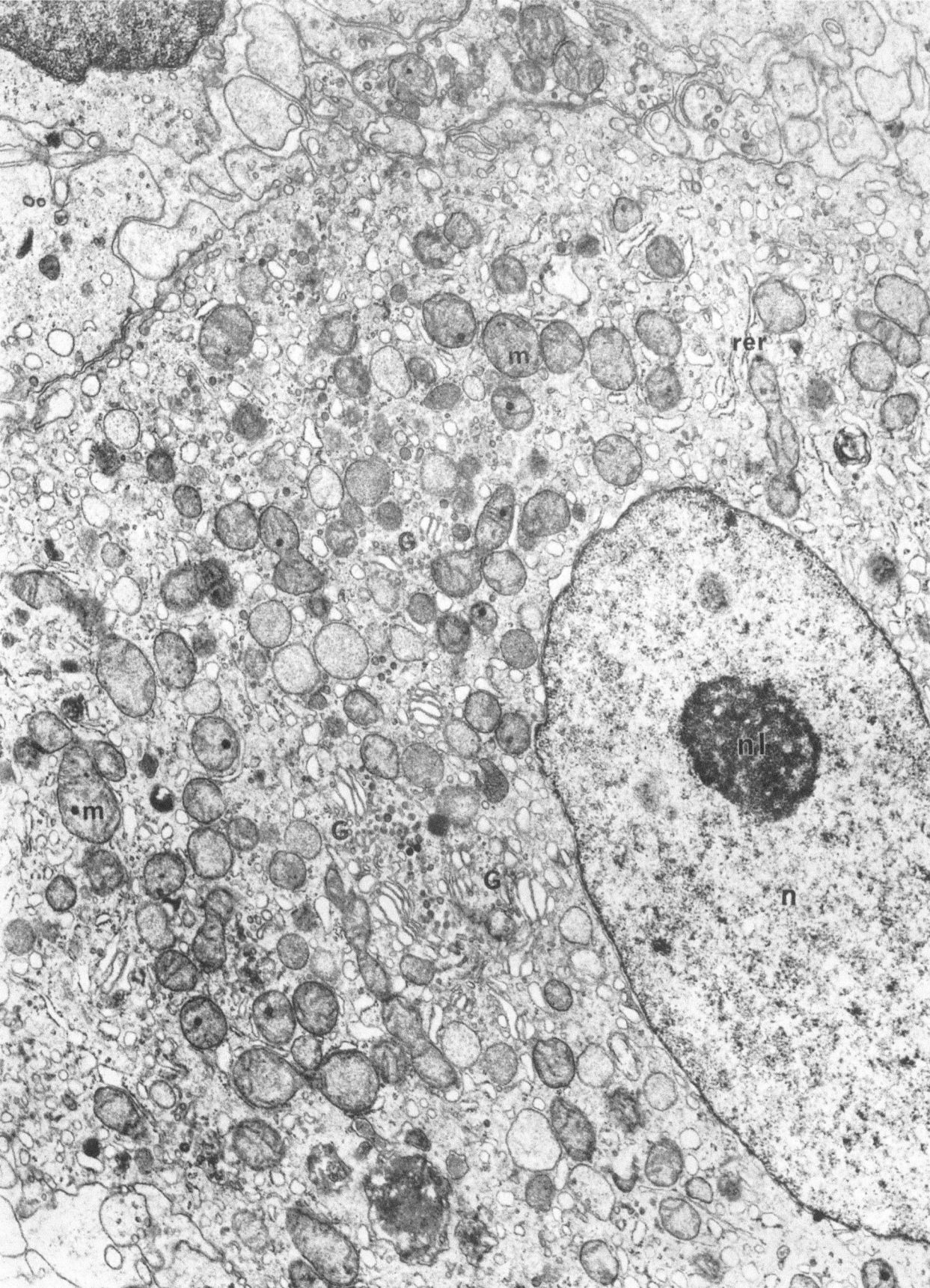

Fig. 75   Perinuclear area of epithelioid cell in epithelioid cellular lymphogranulomatosis
Conspicuous numbers of lysosome-like bodies of variable size and density are interspersed
between the numerous mitochondria. The smaller bodies show transitional forms to dense
Golgi vesicles. A questionable nuclear body is contained in the part of the nucleus seen at
the lower left.

Case 26. ×35,000

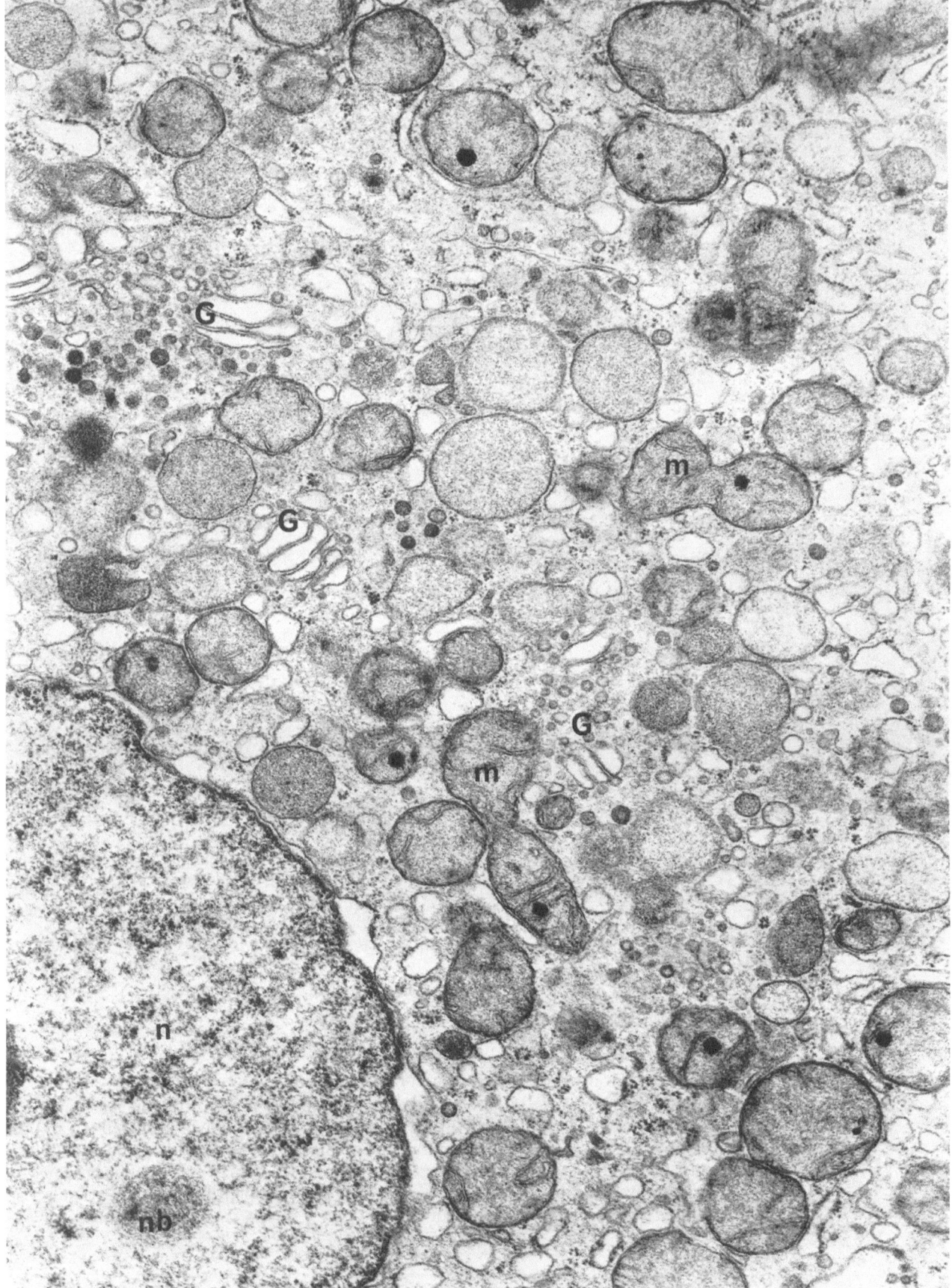

Fig. 76   Portion of cytoplasm of epithelioid cell in epithelioid cellular lymphogranulomatosis

This photomicrograph depicts the perinuclear area and a small part of the nucleus in the right lower corner. Numerous lysosome-like bodies of variable size and electron density and large mitochondria are distributed throughout the cytoplasm. Well-developed rough endoplasmic reticulum is evident (upper center and upper right). A large rather homogeneous appearing structure is visible (center right). It contains evenly distributed finely granular material and is invested in a distinct membrane.

Case 26. ×35,000

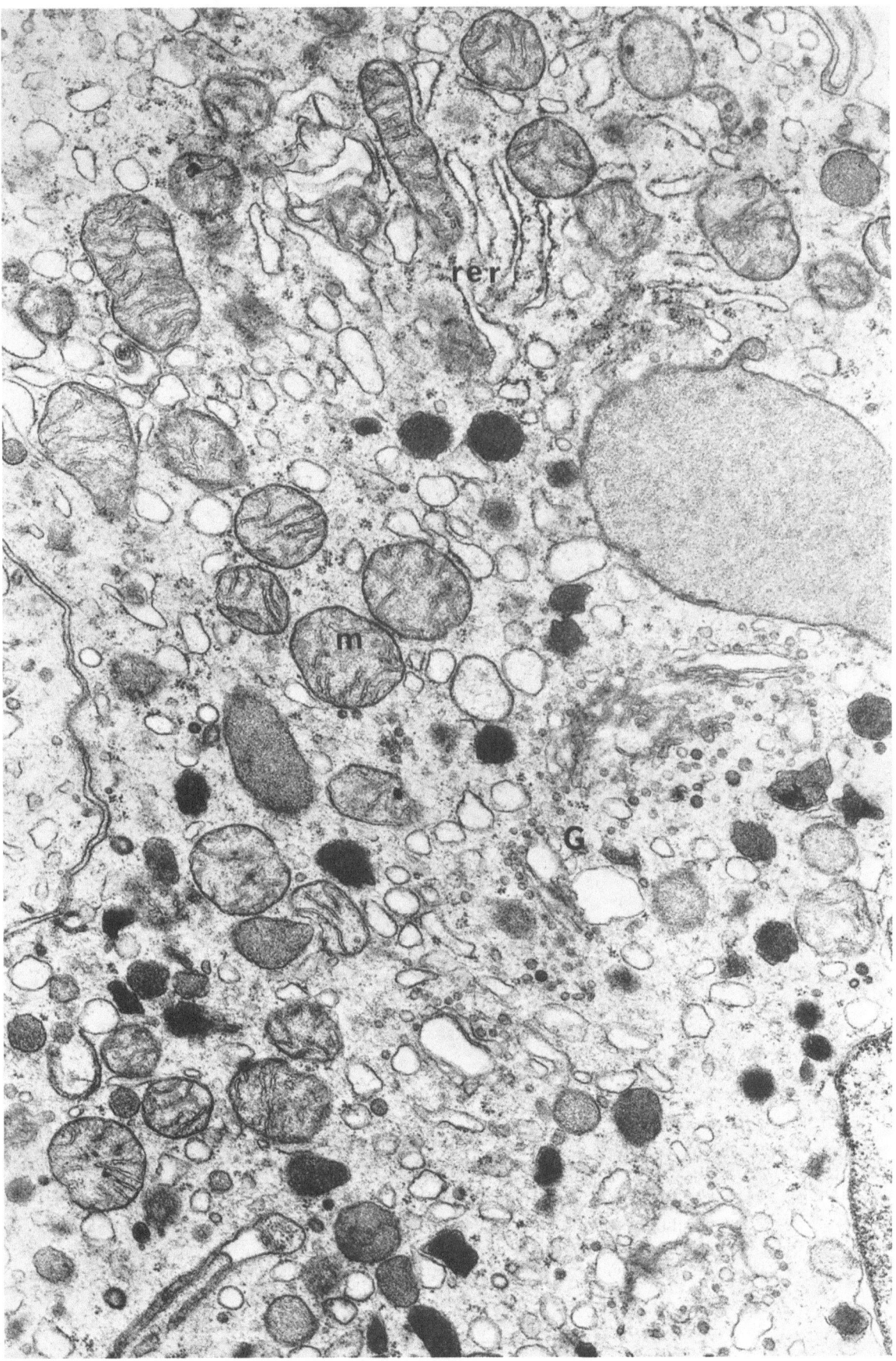

Fig. 77a and b   Two epithelioid cells of variegated appearance in epithelioid cellular lymphogranulomatosis

a) A number of lysosomes or phagolysosomes of high electron density gather in the vicinity of the nucleus.

b) Note prominence of ergastoplasm arranged in parallel fashion.

Case 26.   a) ×20,500   b) ×21,500

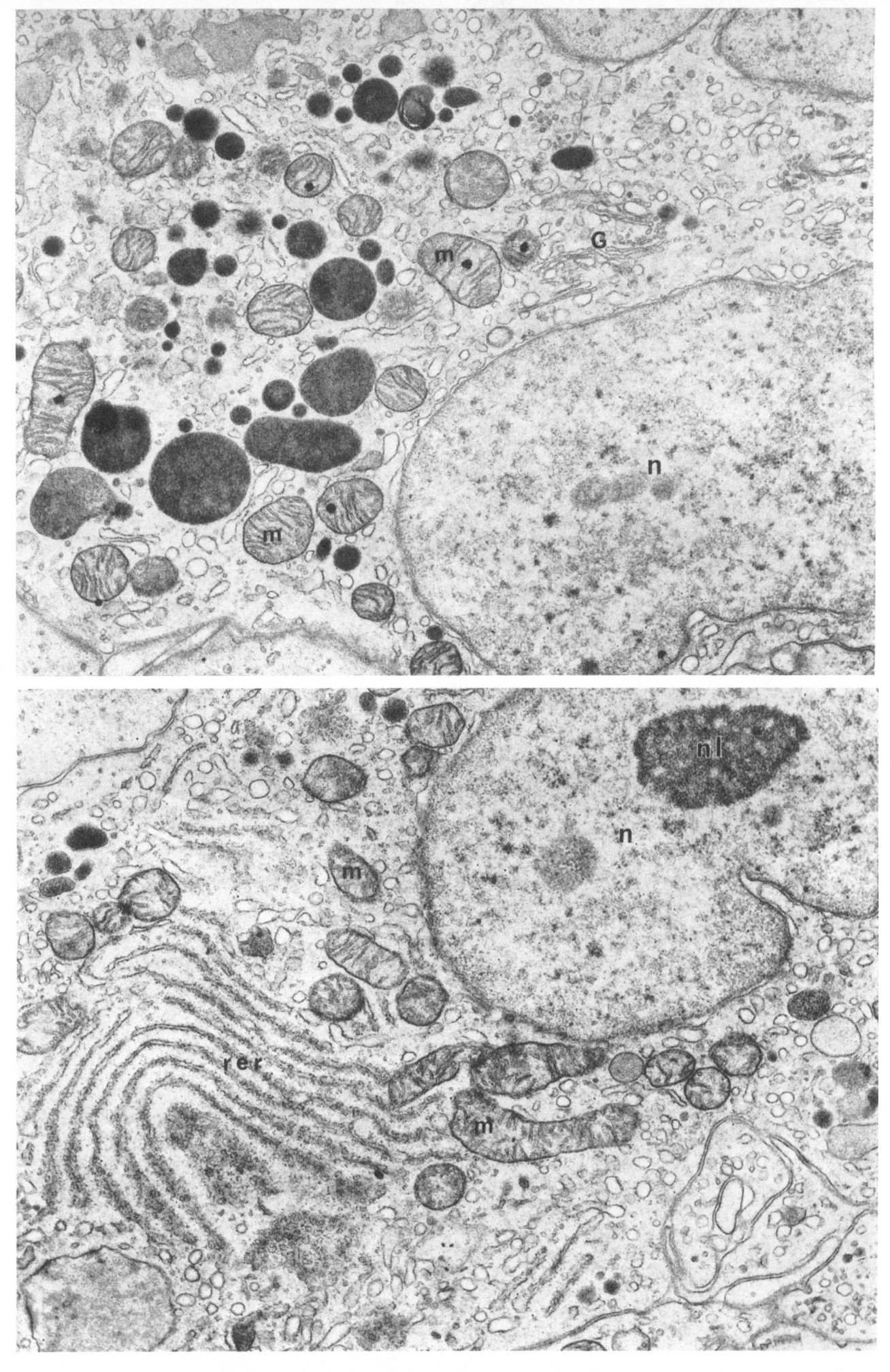

a

b

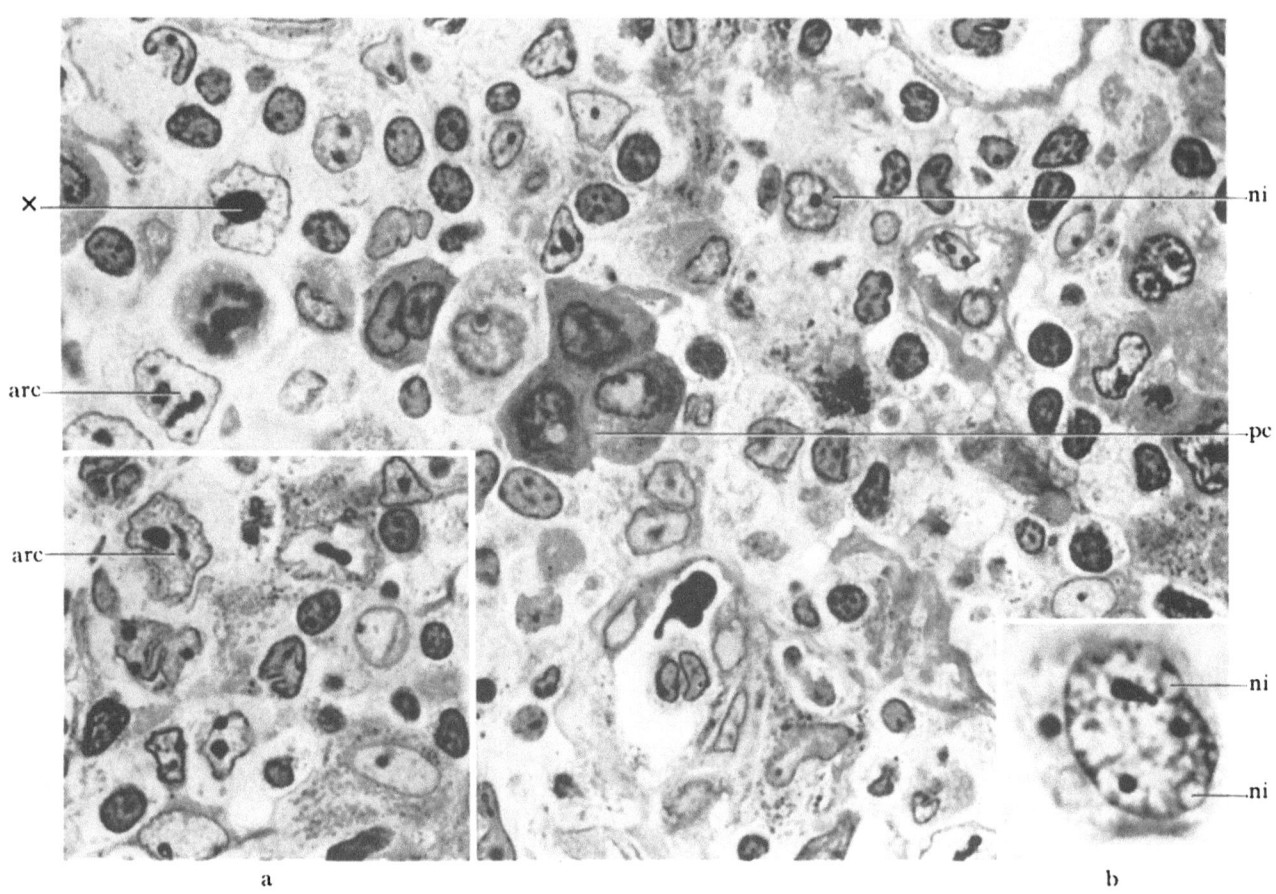

a   b

Fig. 78   Epithelioid cellular lymphogranulomatosis
                   Plasma cells (*pc*), atypical reticulum cells (*arc*) and one cell with features similar to Hodgkin
                   cell (*x*) are visualized. Note nuclear inclusions in atypical cells (*ni*) and in plasma cells (*ni*).

Inset a   Few atypical cells and one epithelioid cell
Inset b   Nuclear inclusions in a large basophilic cell

                   Case 26. Osmium tetroxide fixation. Semithin section. Methylene blue-Azure II. ×1,230

Fig. 79a—c   Nuclear pockets and intranuclear protein inclusion in plasma cells (epithelioid cellular
                   lymphogranulomatosis)
                   a)   One protein inclusion and one nuclear pocket containing cytoplasmic components
                   b) and c) Nuclear pockets filled with many vesicular and vacuolar structures and delimited
                             by a four-layer nuclear membrane

                   Case 26.   a) ×20,000   b), c) ×25,000

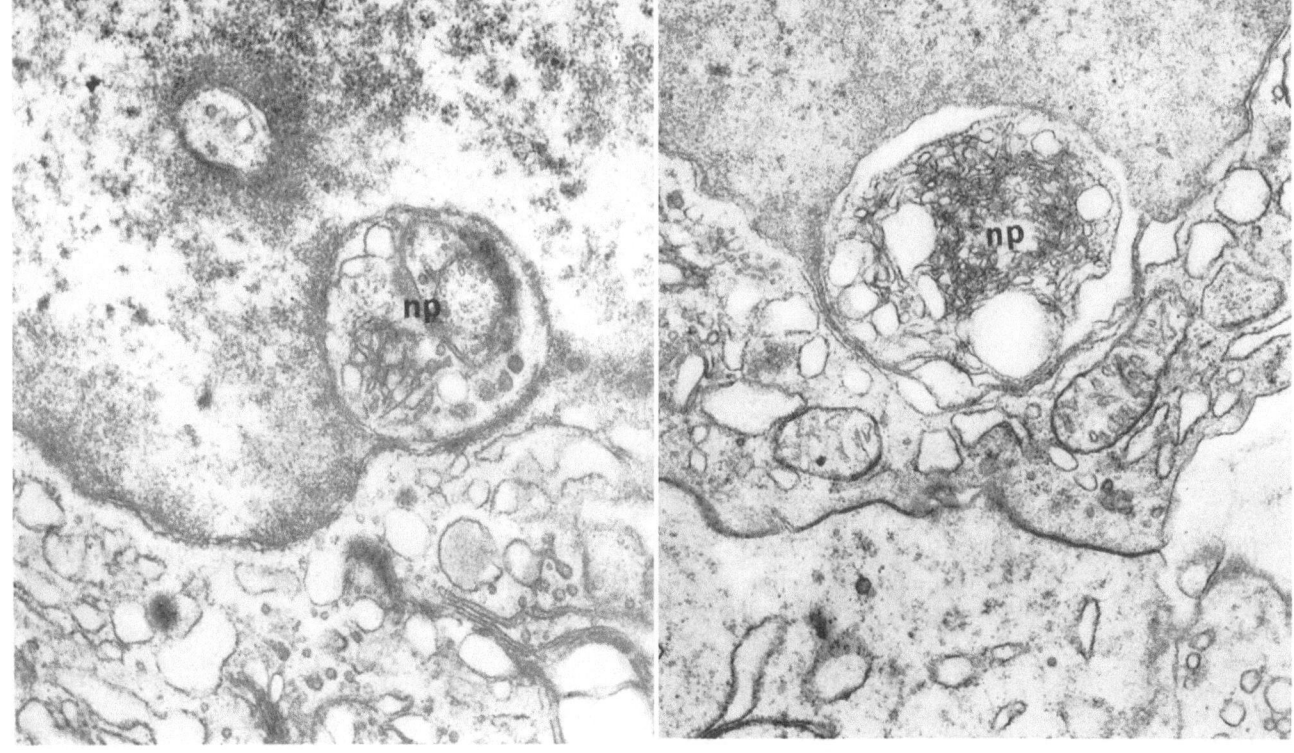

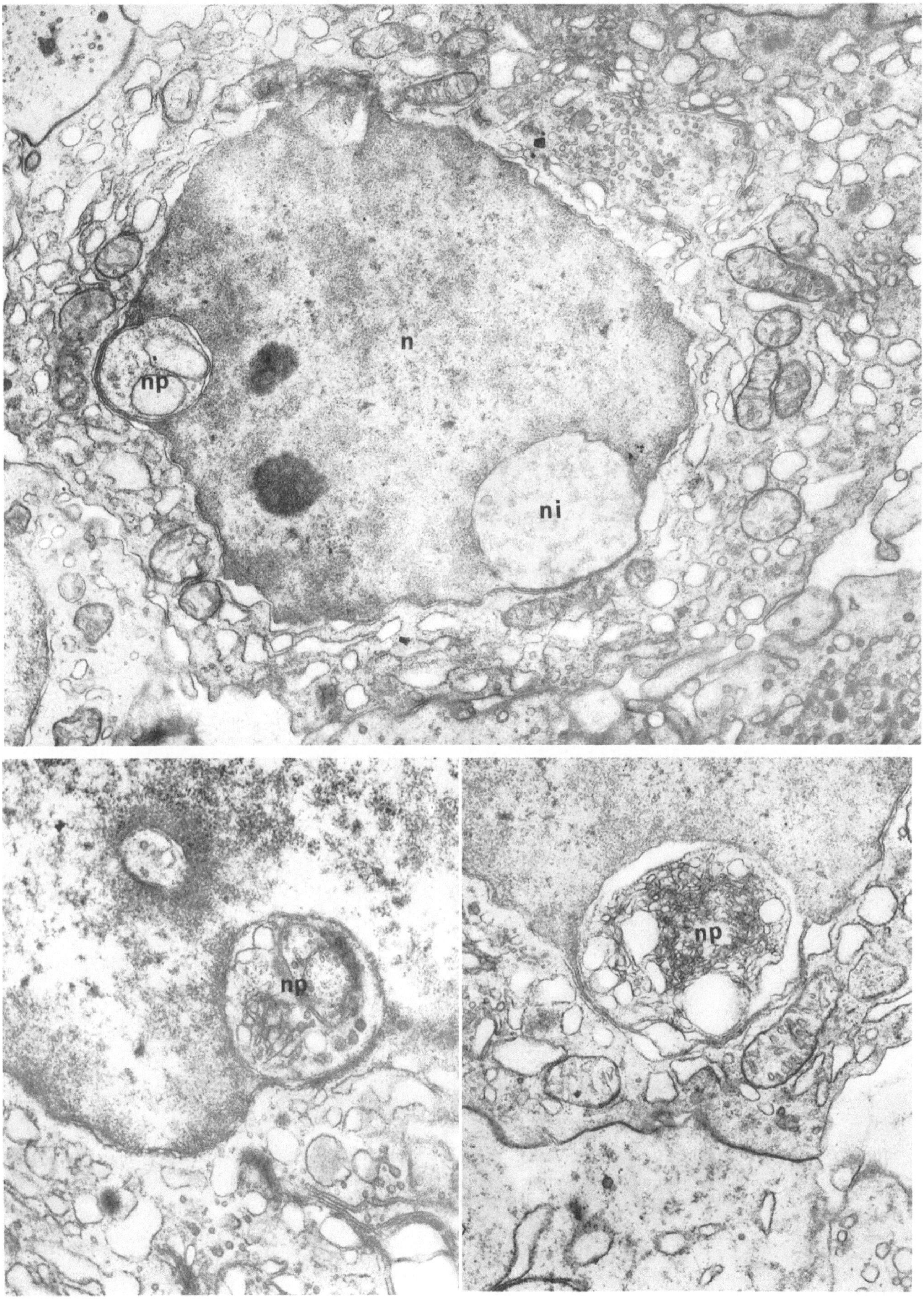

a

b   c

Fig. 80   Intranuclear protein inclusion in plasma cell (epithelioid cellular lymphogranulomatosis)
                 The protein inclusion occupies a large fraction of the nuclear area. Note concentric runs of
                 well-developed rough endoplasmic reticulum in the cytoplasm.
                 Case 26. ×42,000

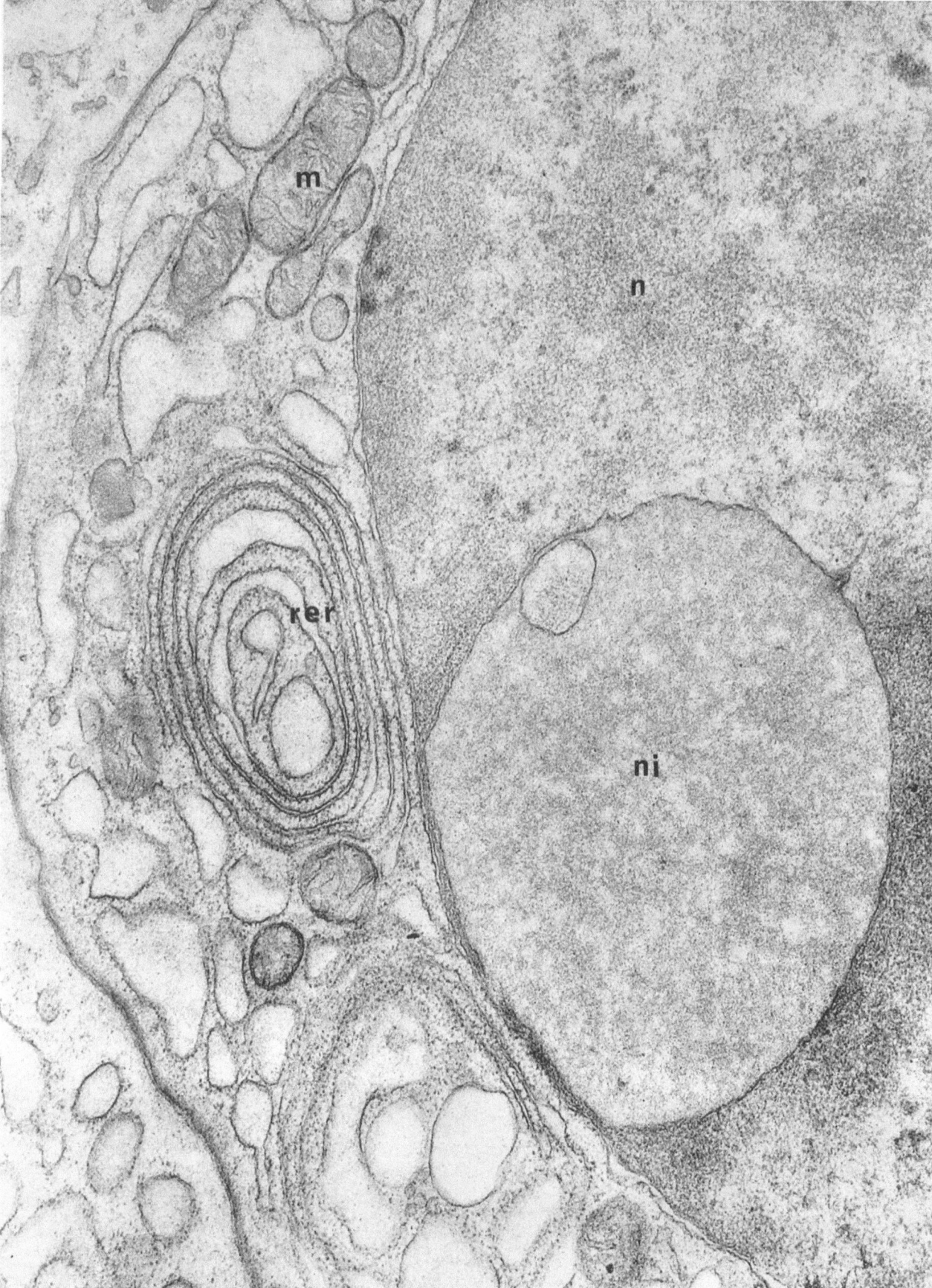

Germinoblastoma (Follicular or Nodular Lymphoma, M. Brill-Symmers)

Lymphosarcoma

Reticulum Cell Sarcoma

Burkitt's Tumor

Macroglobulinemia of Waldenström

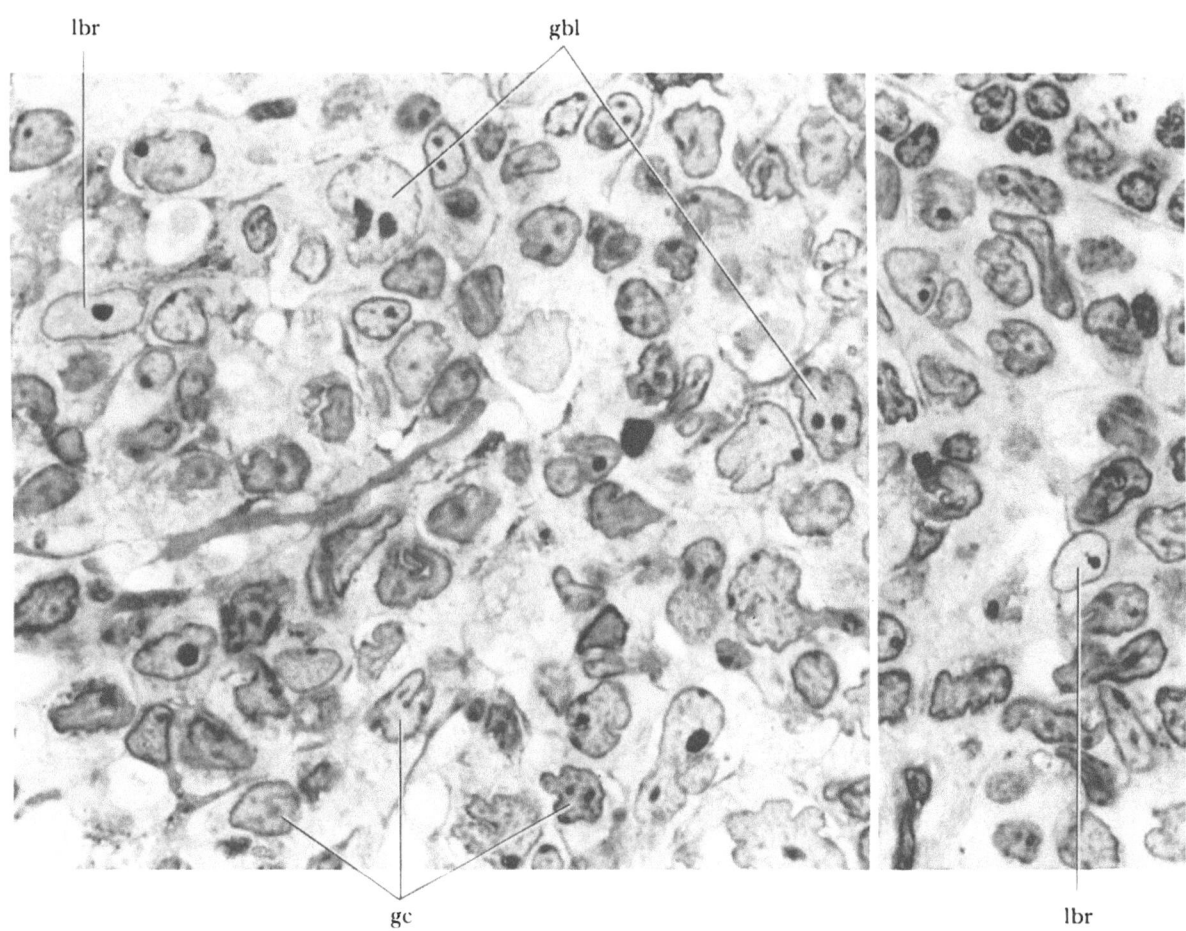

Fig. 81   Follicular lymphoma (germinoblastoma)

Germinoblasts (*gbl*), germinocytes (*gc*) and long-branching reticulum cells (*lbr*) of germinal center are shown.

Case 21. Osmium tetroxide fixation. Semithin section. Methylene blue-Azure II. ×1,230

Fig. 82   Follicular lymphoma (germinoblastoma)

The nucleus in the center exhibits two nuclear pockets (arrows). These cells show ultrastructural features similar to normal germinocytes or germinoblasts. Few tubules of endoplasmic reticulum with only occasional ribosomes attached are seen. Mitochondria are rod-shaped. The Golgi apparatus is poorly developed. Note paucity of ribosomes and polysomes. The cytoplasmic matrix is unusually pale.

Case 21. ×17,500

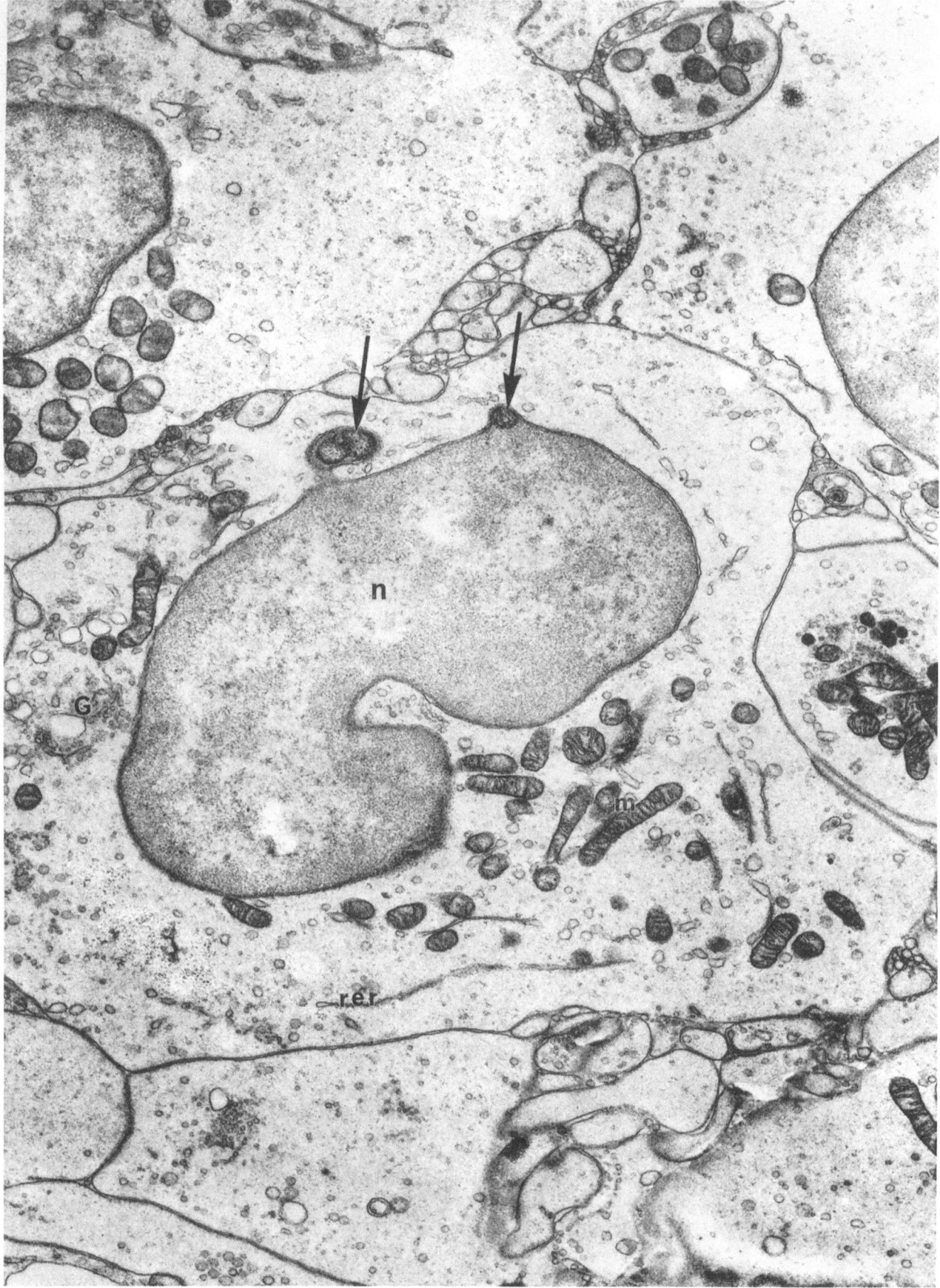

Fig. 83   Follicular lymphoma (germinoblastoma), higher magnification
Note the nuclear pocket in the center containing one mitochondrion, a few vacuoles and ribosomal granules. The cytoplasm contains some weakly electron-dense vesicles, ovoid or elongated mitochondria and scant endoplasmic reticulum associated with only a few ribosomes.

Case 21. ×42,000

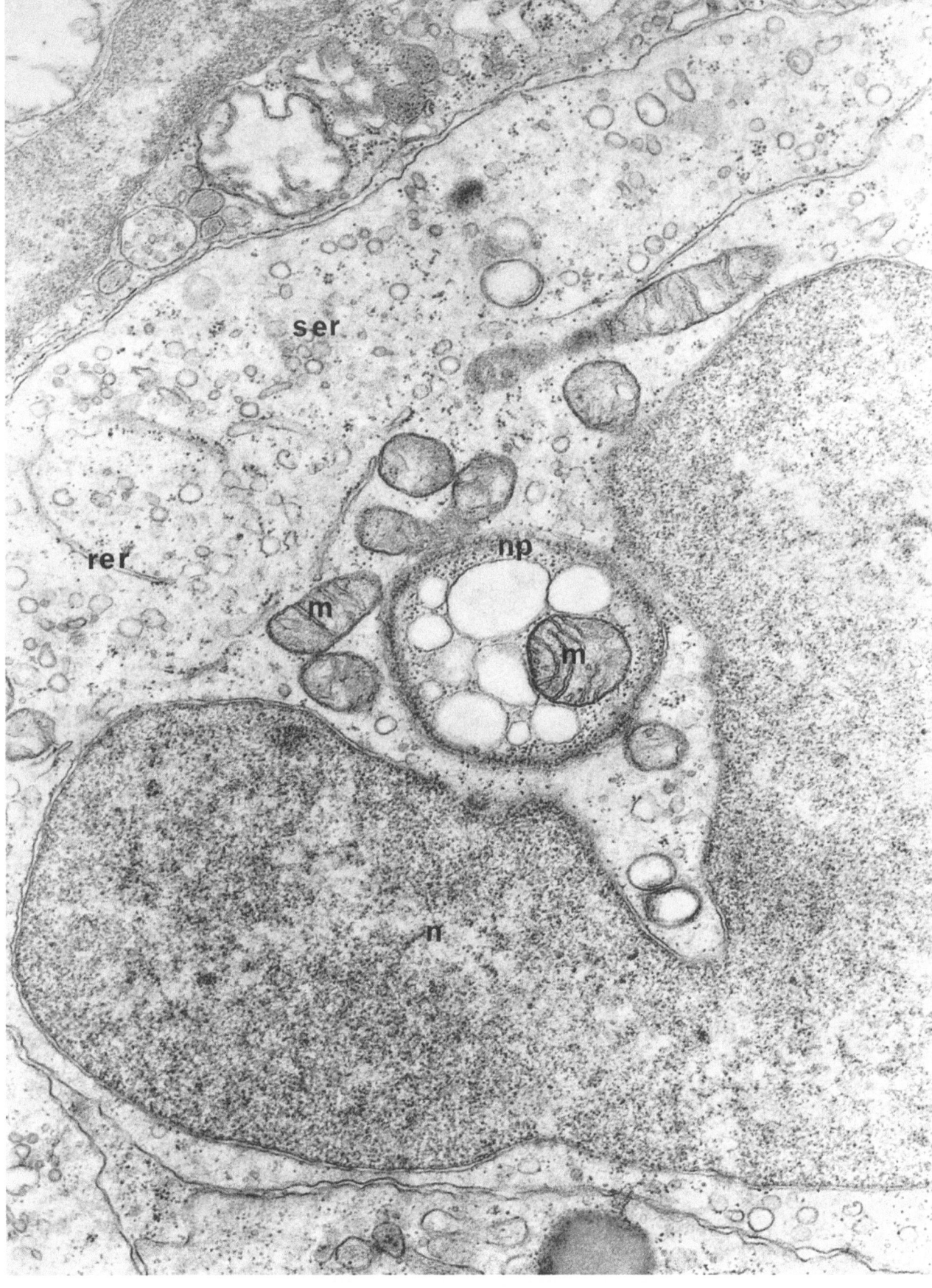

Fig. 84   Nucleus of follicular lymphoma (germinoblastoma) cell
Peculiar nuclear pockets enclosing cytoplasmic components are demonstrated. They are limited by a series of membranes made up of two layers of the cytoplasmic and a reduplicated nuclear membrane. Chromatin material is observed in the space between the two layers of the nuclear membrane.

Case 21. ×35,000

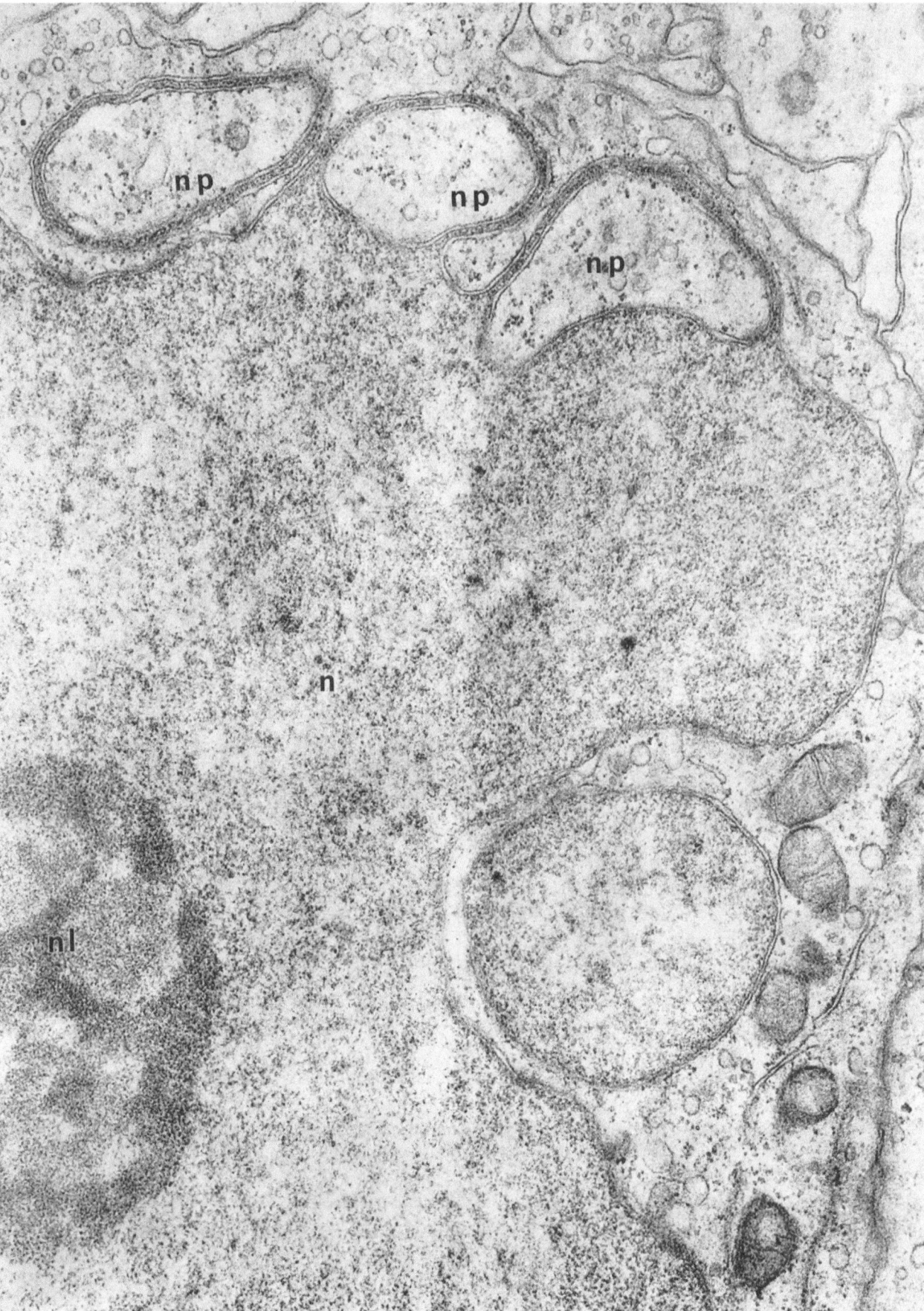

Fig. 85 a—c   Several types of nuclear pockets in follicular lymphoma

a) The nuclear pocket appears to be budding from the nucleus into the cytoplasm. It is delimited by a typical 4-layer structure (compare with Fig. 84). The centrally placed mitochondrion is encircled by a series of cytoplasmic membranes, making an onion peel-like body.

b) Nuclear pocket crowded with cytoplasmic vacuoles

c) Inverted nuclear pocket with cisternae in concentric parallel arrangement

Case 21.   a) ×35,000   b) ×42,000   c) ×28,000

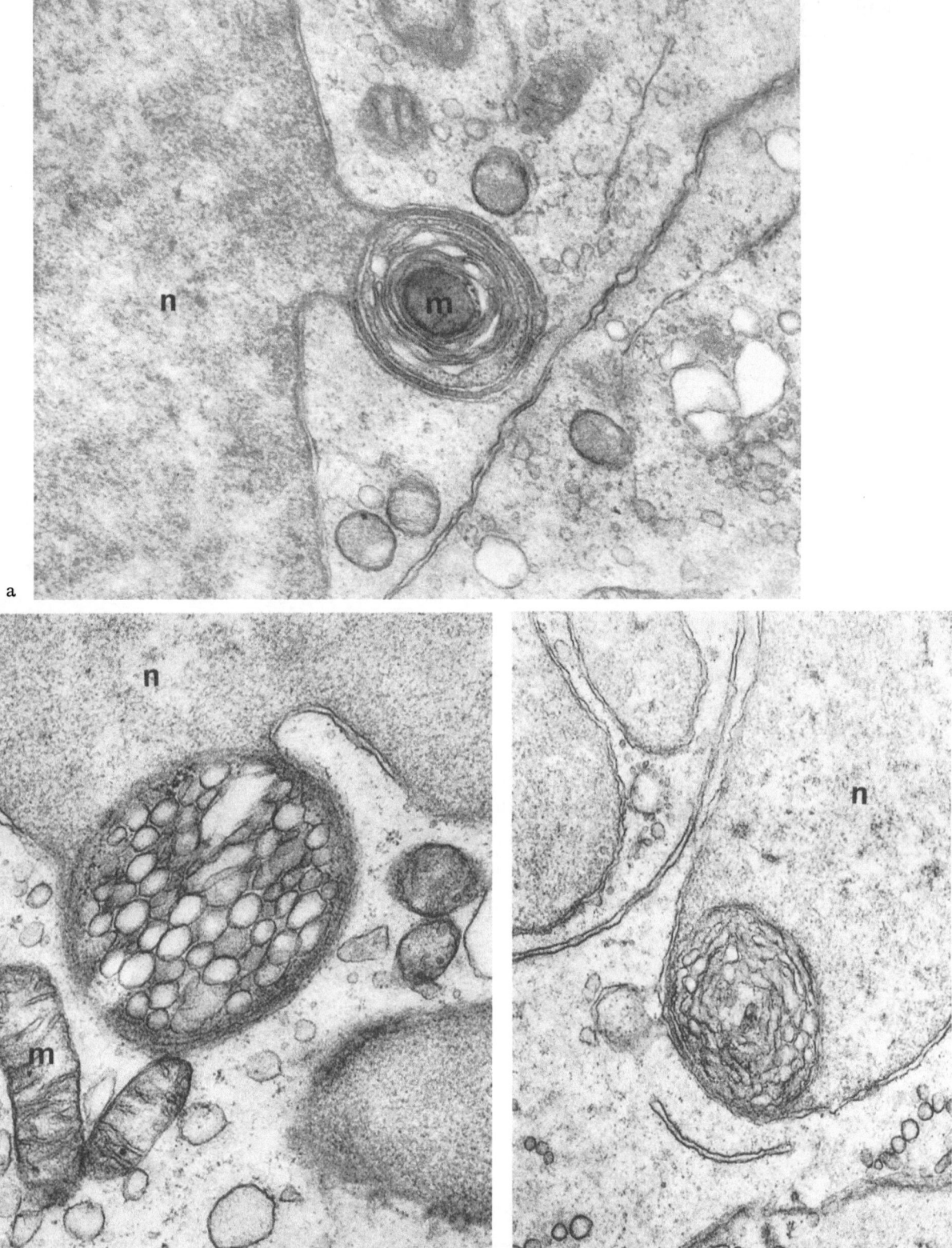

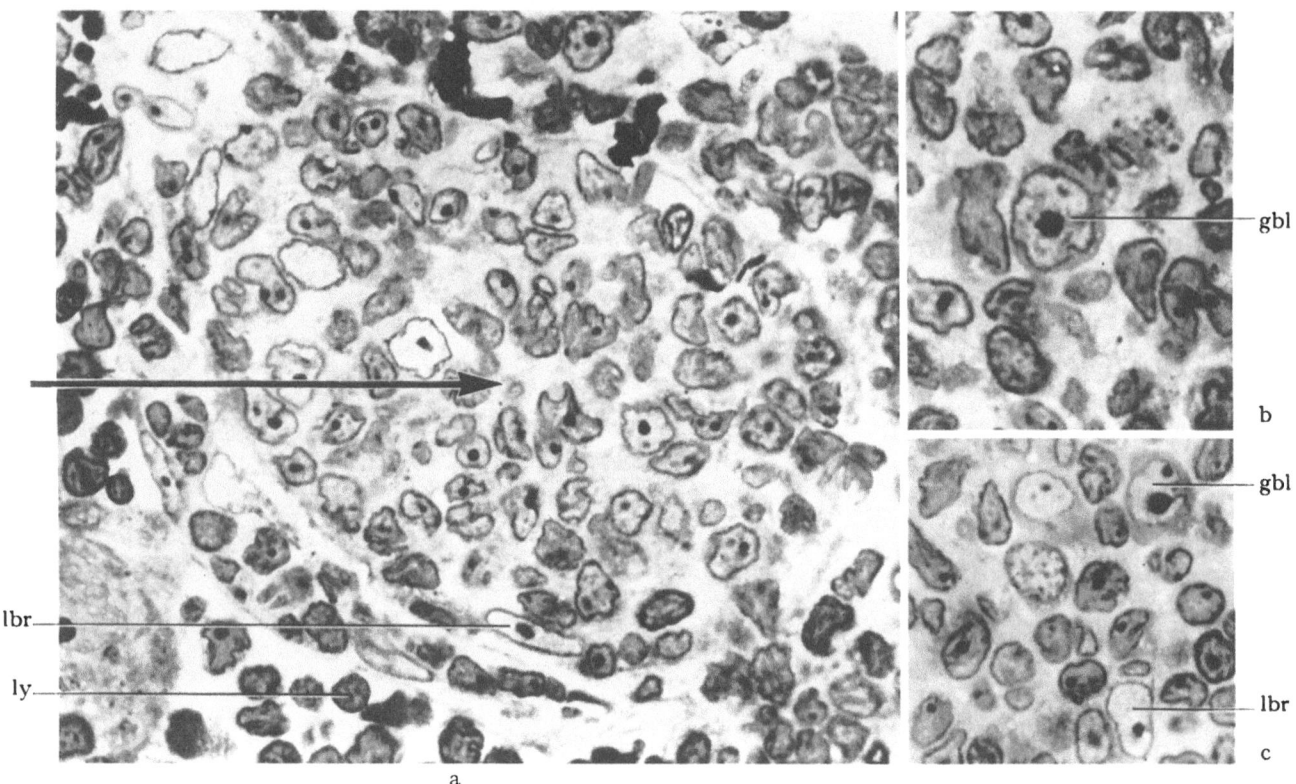

Fig. 86a—c   Follicular lymphoma (germinoblastoma)

a) Germinal center (arrow) with many germinocytes, a few germinoblasts and one typical long-branching reticulum cell (*lbr*)

b) Large germinoblast from another germinal center (*gbl*)

c) Germinoblast (*gbl*) and probably long-branching reticulum cell (*lbr*)

Case 32. Osmium tetroxide fixation. Semithin section. Methylene blue-Azure II. ×1,230

Fig. 87   Follicular lymphoma (germinoblastoma)

Note bizarre nuclear pocket enclosing cytoplasmic materials and prominent irregularly shaped vesicles. A distended Golgi apparatus, a distinct centriole, a multivesicular body and a residual body are evident.

Case 32. ×30,000

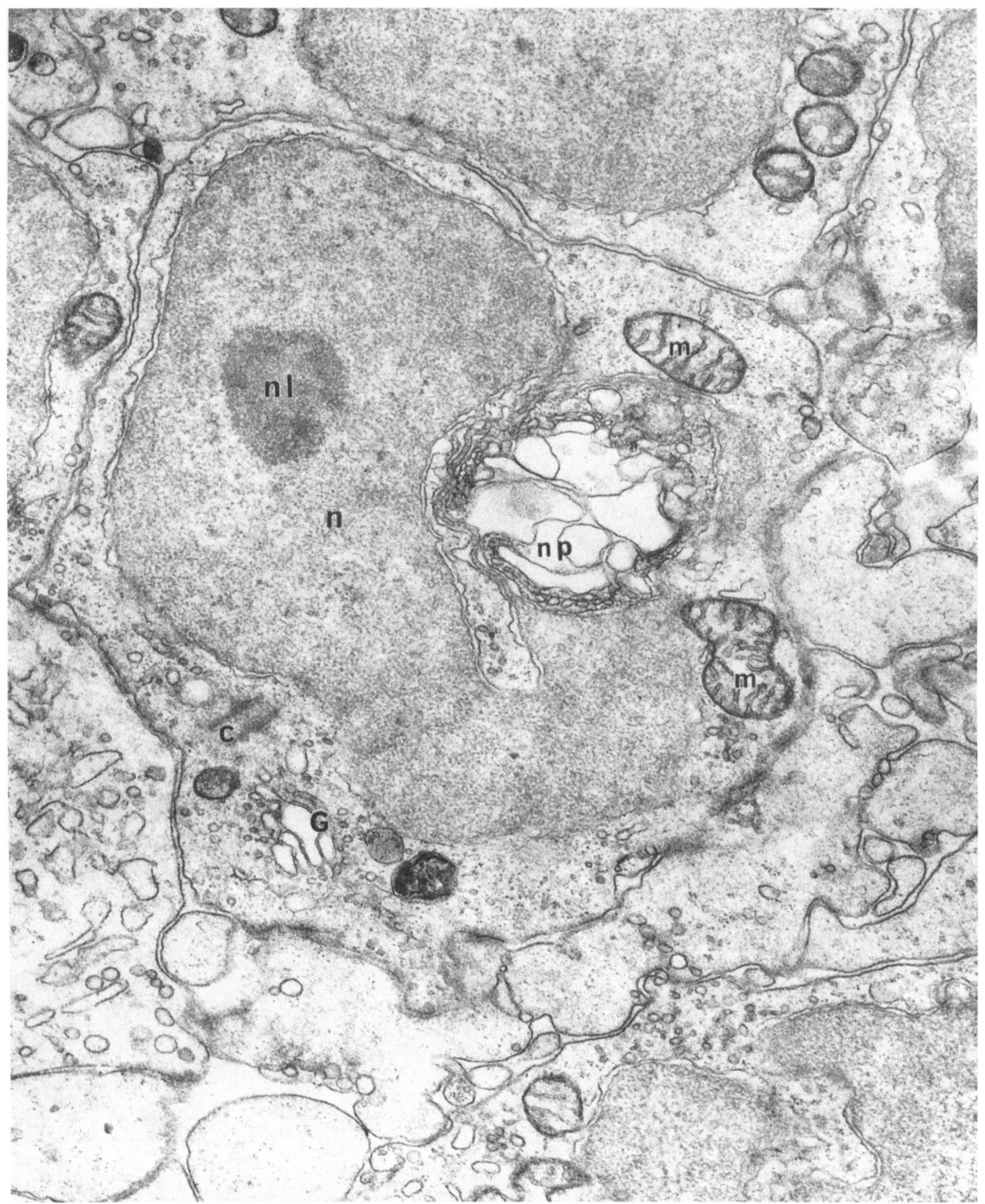

Fig. 88   Follicular lymphoma (germinoblastoma) demonstrating long-branching reticulum cells with desmosomal connections

Two slender cytoplasmic branches of reticulum cells exhibiting filaments are seen extending between two germinoblasts. Desmosome indicates point of anchorage.

Photomicrograph contributed by H. R. NIEDORF
Case 32. ×38,000

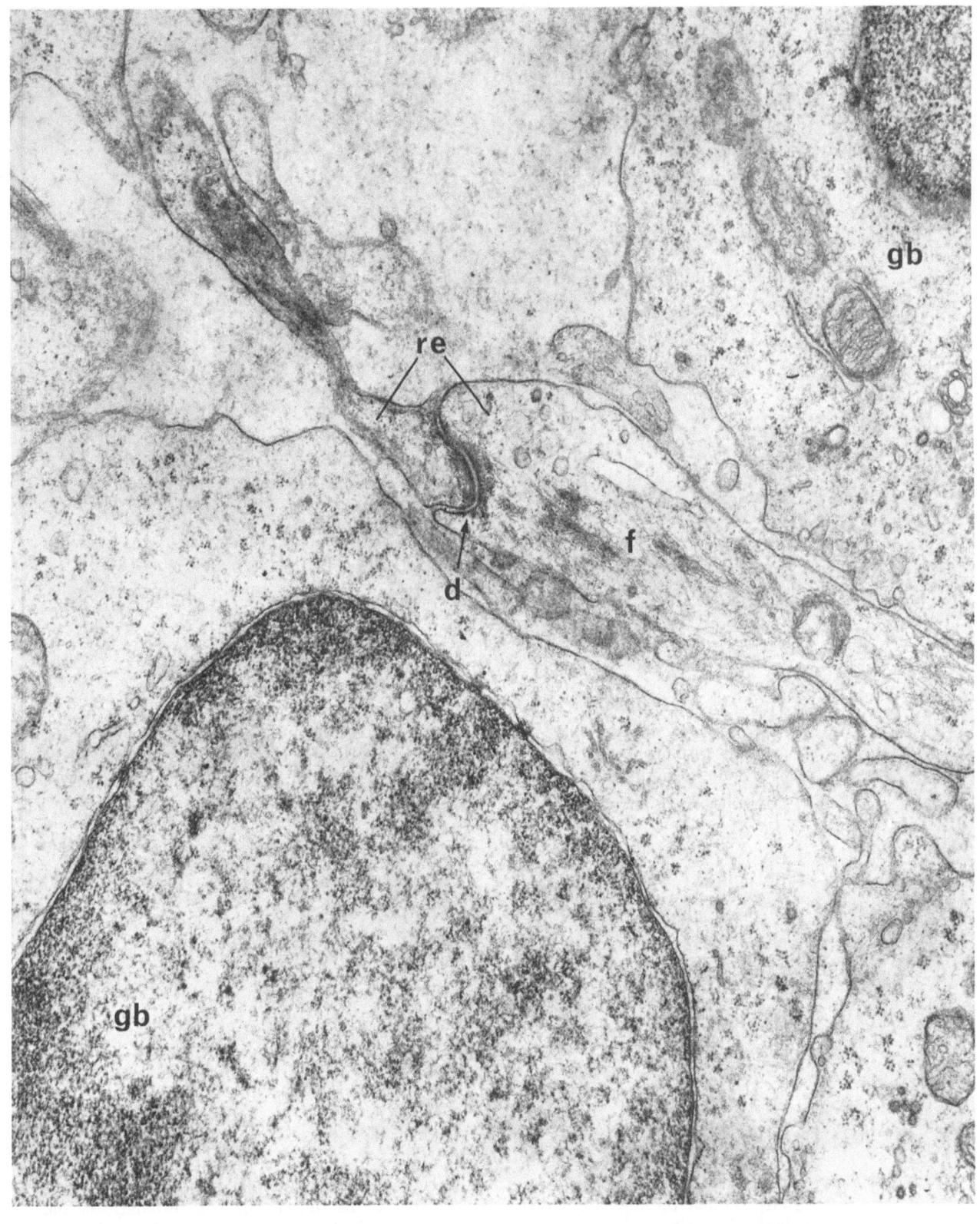

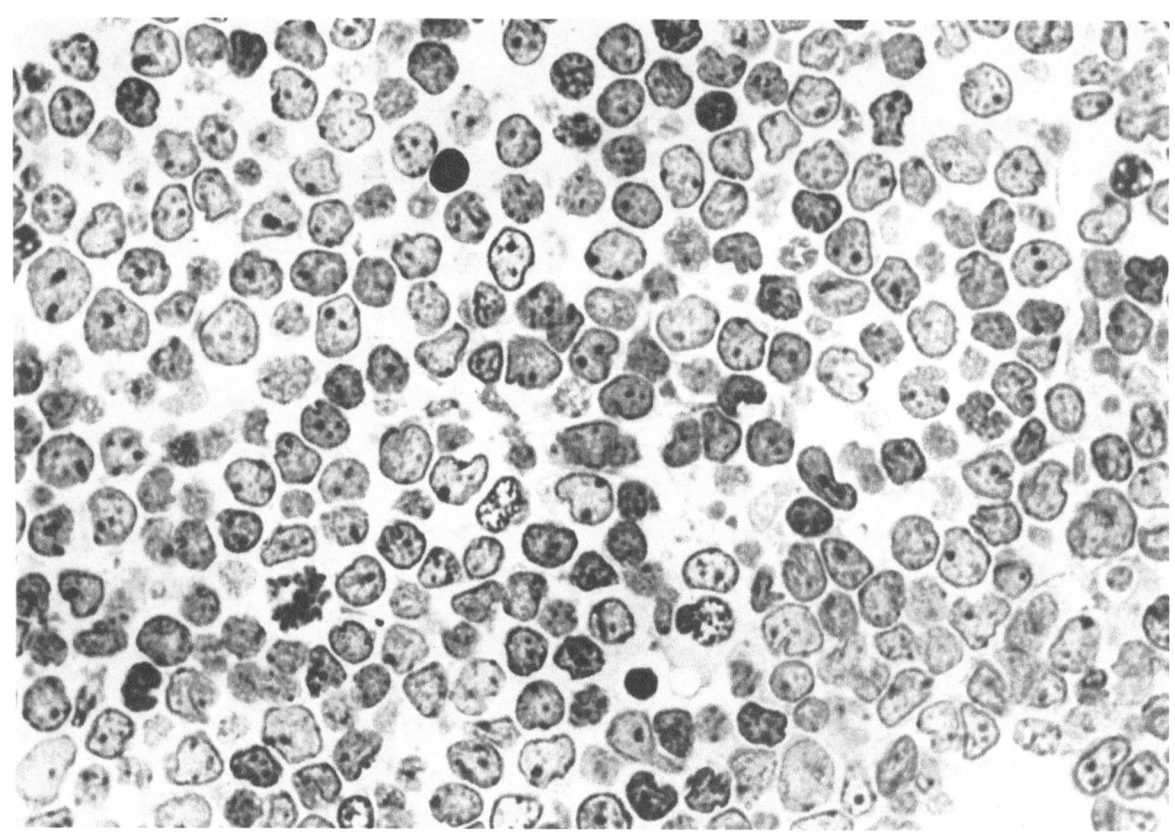

Fig. 89   Lymphoblastic lymphosarcoma of child

Moderately pleomorphic lymphoid tumor cells without true precursors but with striking variations of morphologic features are evident. All cells represent variants of the same cell type. Some medium-sized nucleoli are seen.

Case 53. Osmium tetroxide fixation. Semithin section. Methylene blue-Azure II. ×1,230

Fig. 90   Lymphoblastic lymphosarcoma of child

Tumor cells with striking pleomorphism. Tightly packed rounded tumor cells displaying sharp cellular outlines are shown. Nuclei vary in size and shape. The chromatin is marginated and clumped near the nuclear periphery. Nucleoli are of medium size and for the most part in apposition to the nuclear membrane. Cytoplasmic organelles are scanty.

Case 53. ×8,750

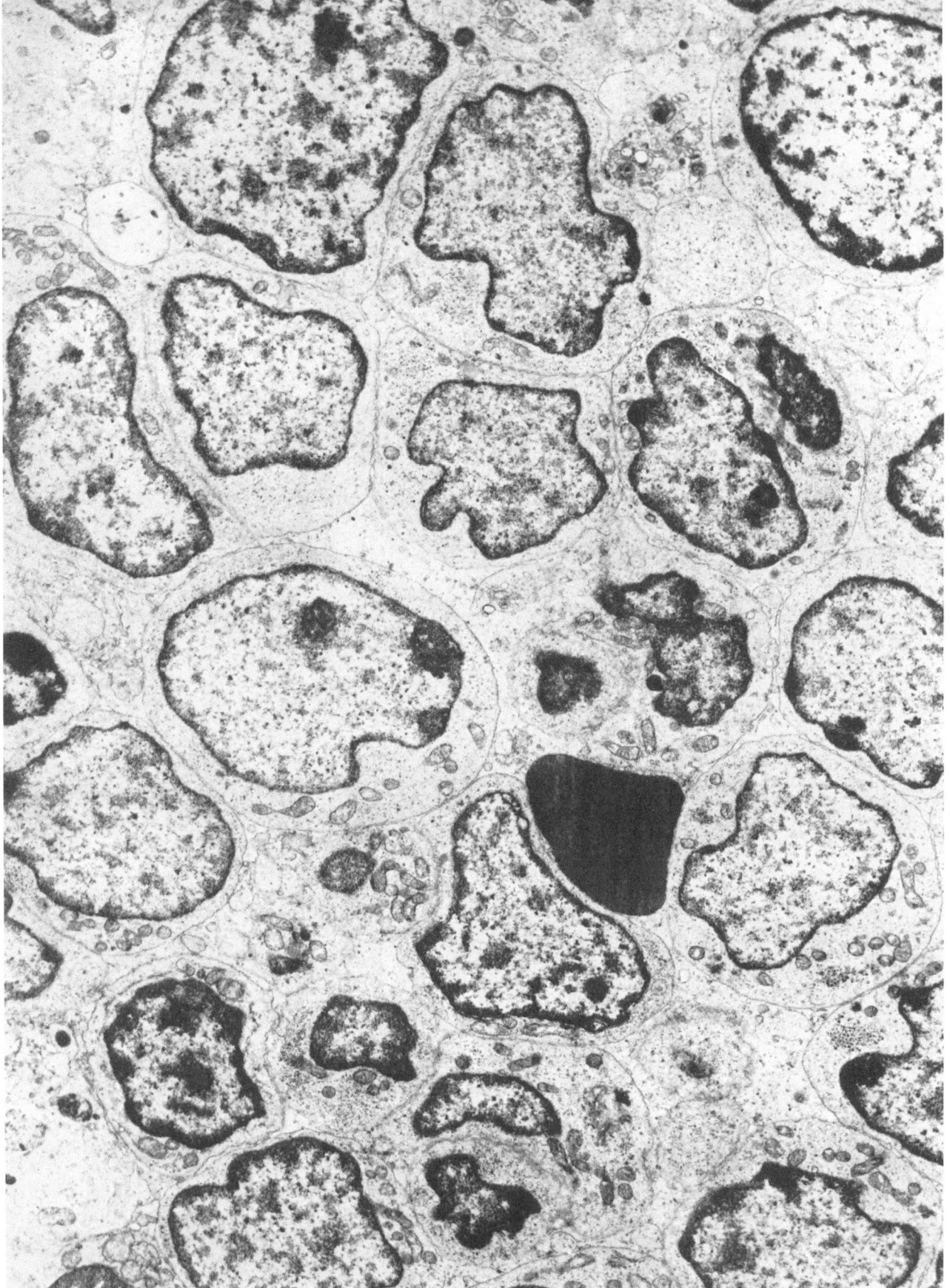

Fig. 91   Lymphoblastic lymphosarcoma of child

Overall view of two lymphosarcoma cells in center of picture. There is a cluster of glycogen deposits and a distinct centriole in the upper cell, and lipid droplets are seen in the lower one. Moderate numbers of polysomes and ribosomes are present in all cells depicted in this micrograph.

Case 53. ×17,500

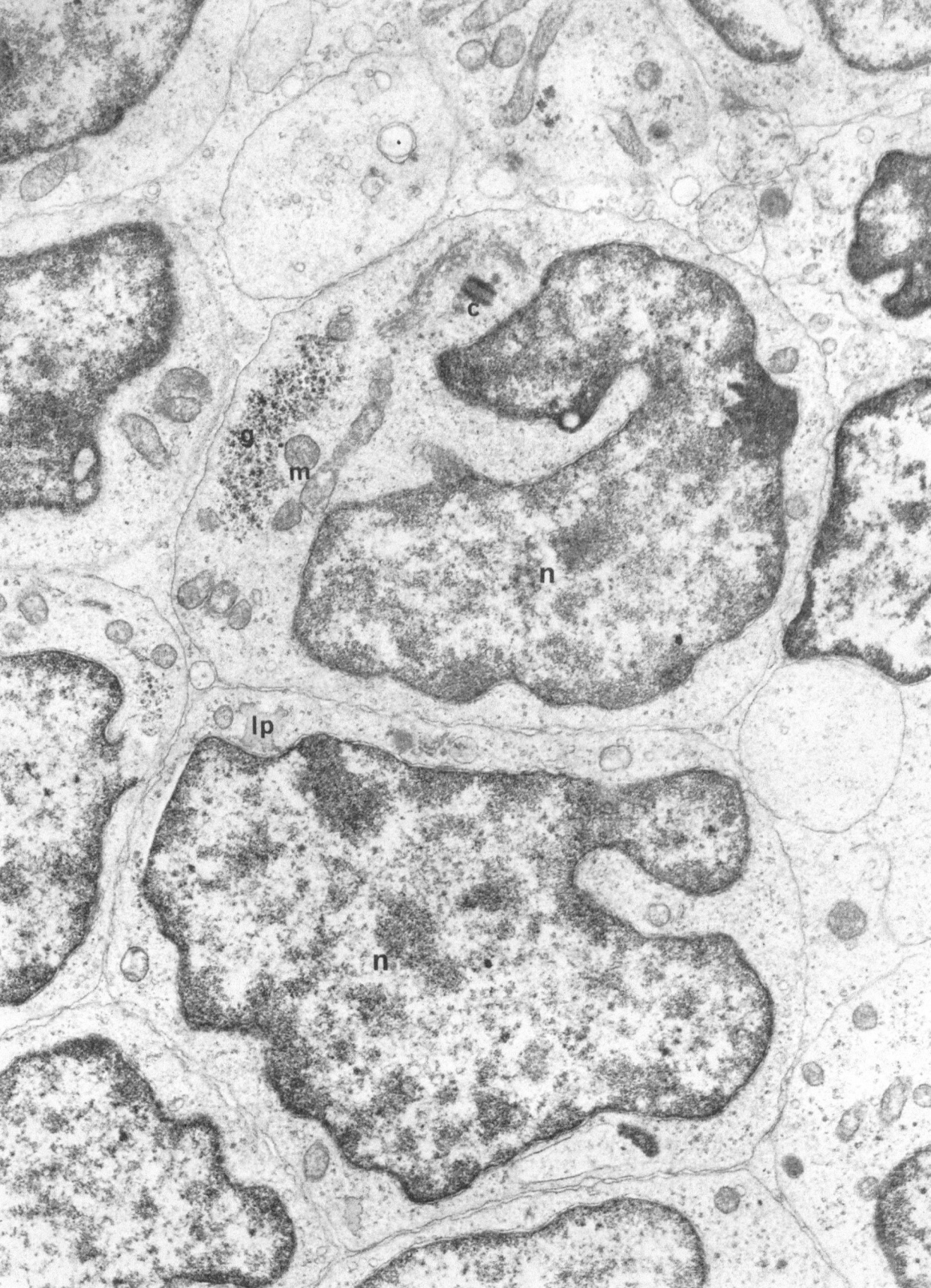

Fig. 92   Lymphoblastic lymphosarcoma of child
The cells show two centrioles (bottom), poorly developed Golgi apparatuses and a few small mitochondria. The cytoplasm is almost completely devoid of endoplasmic reticulum. Some cells contain lysosome-like bodies.

Case 53. ×17,500

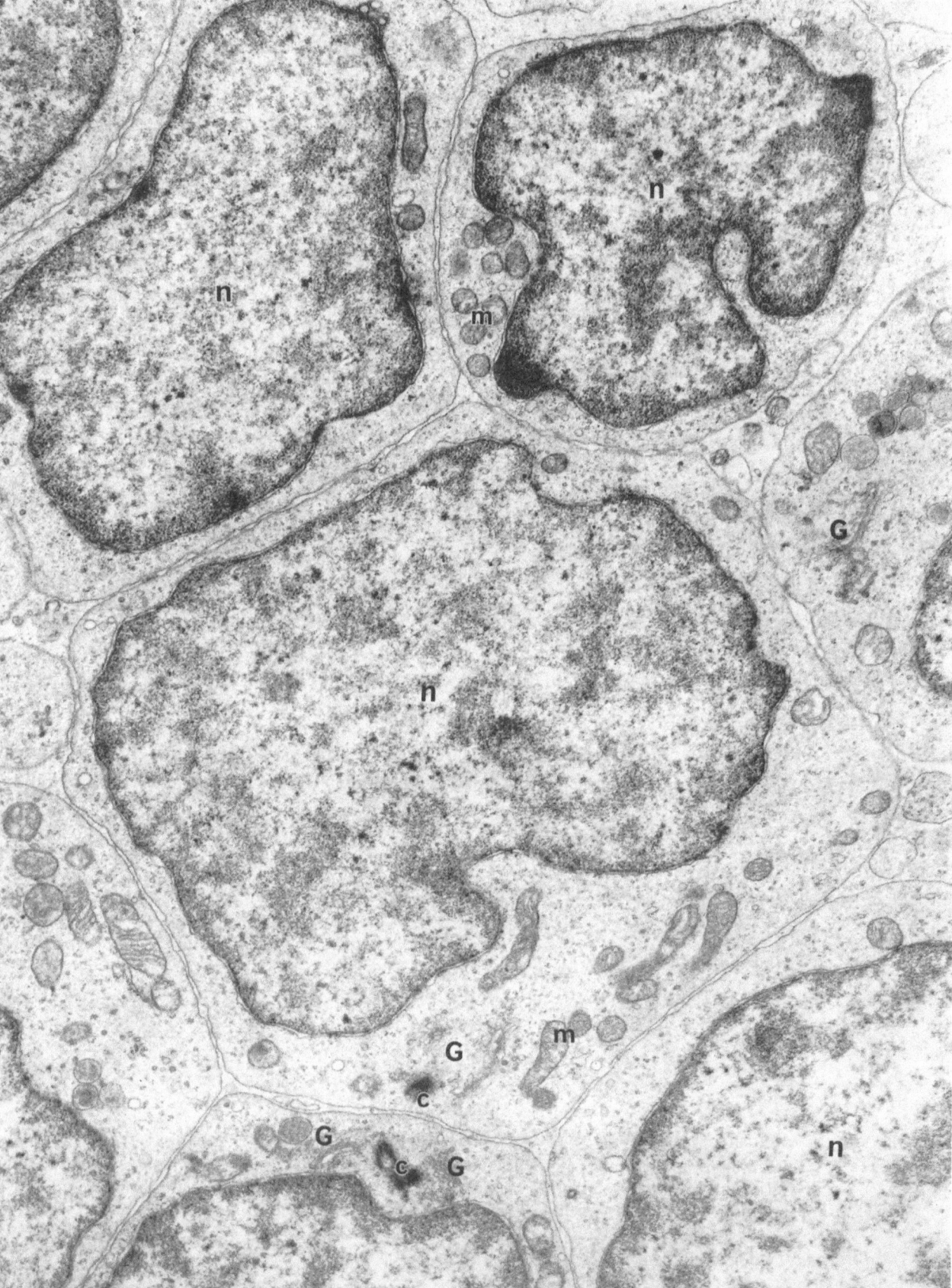

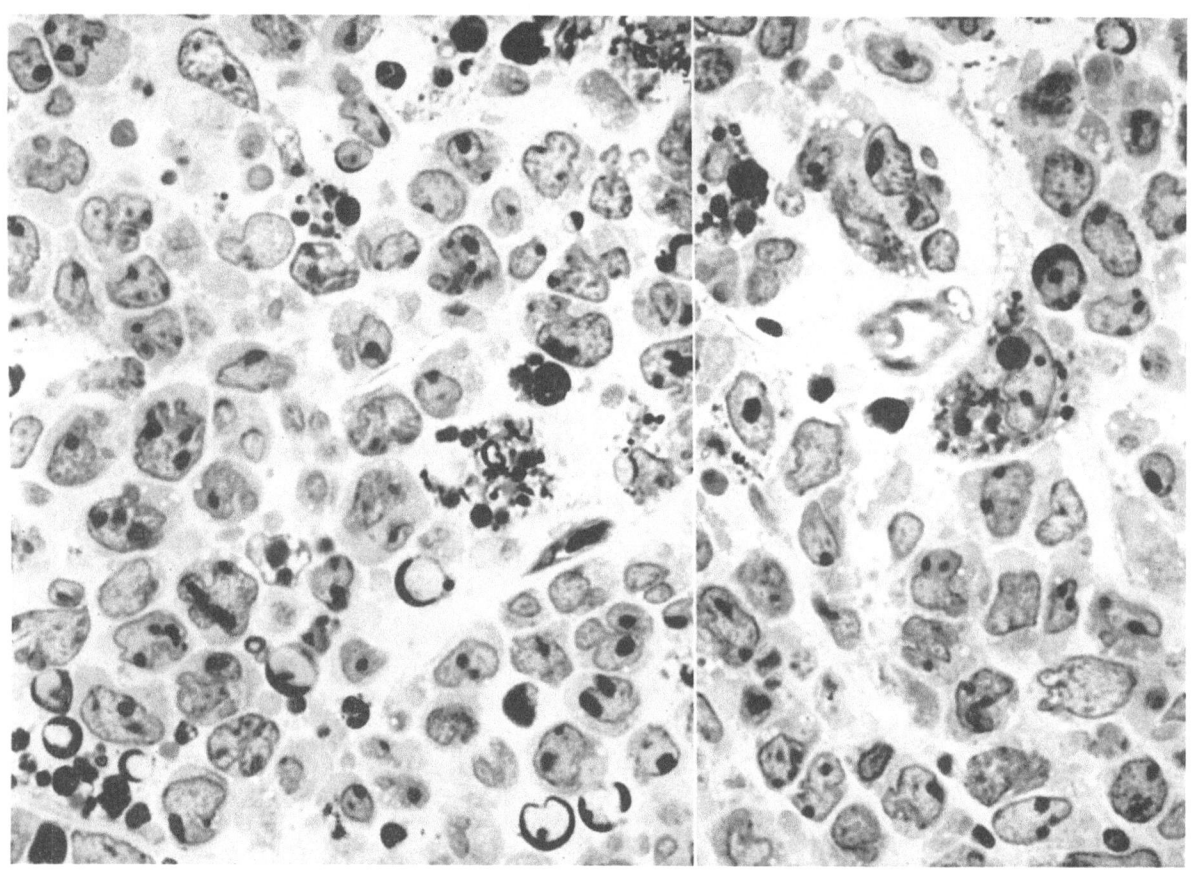

Fig. 93   Reticulum cell sarcoma, type I

The cells have pleomorphic large nuclei and many vacuoles in the abundant cytoplasm. There are numerous degenerating nuclei with margination of condensed chromatin and macrophages containing phagocytized cellular debris.

Case 19. Osmium tetroxide fixation. Semithin section. Methylene blue-Azure II. ×1,230

Fig. 94   Reticulum cell sarcoma, type I

Undifferentiated tumor cells with numerous ribosomes and polyribosomes. Endoplasmic reticulum is almost completely lacking and there are few irregularly shaped mitochondria and occasional small vesicles. A centriole and small vesicular Golgi apparatuses are seen in the center.

Tumor of epipharynx
Case 19. ×17,500

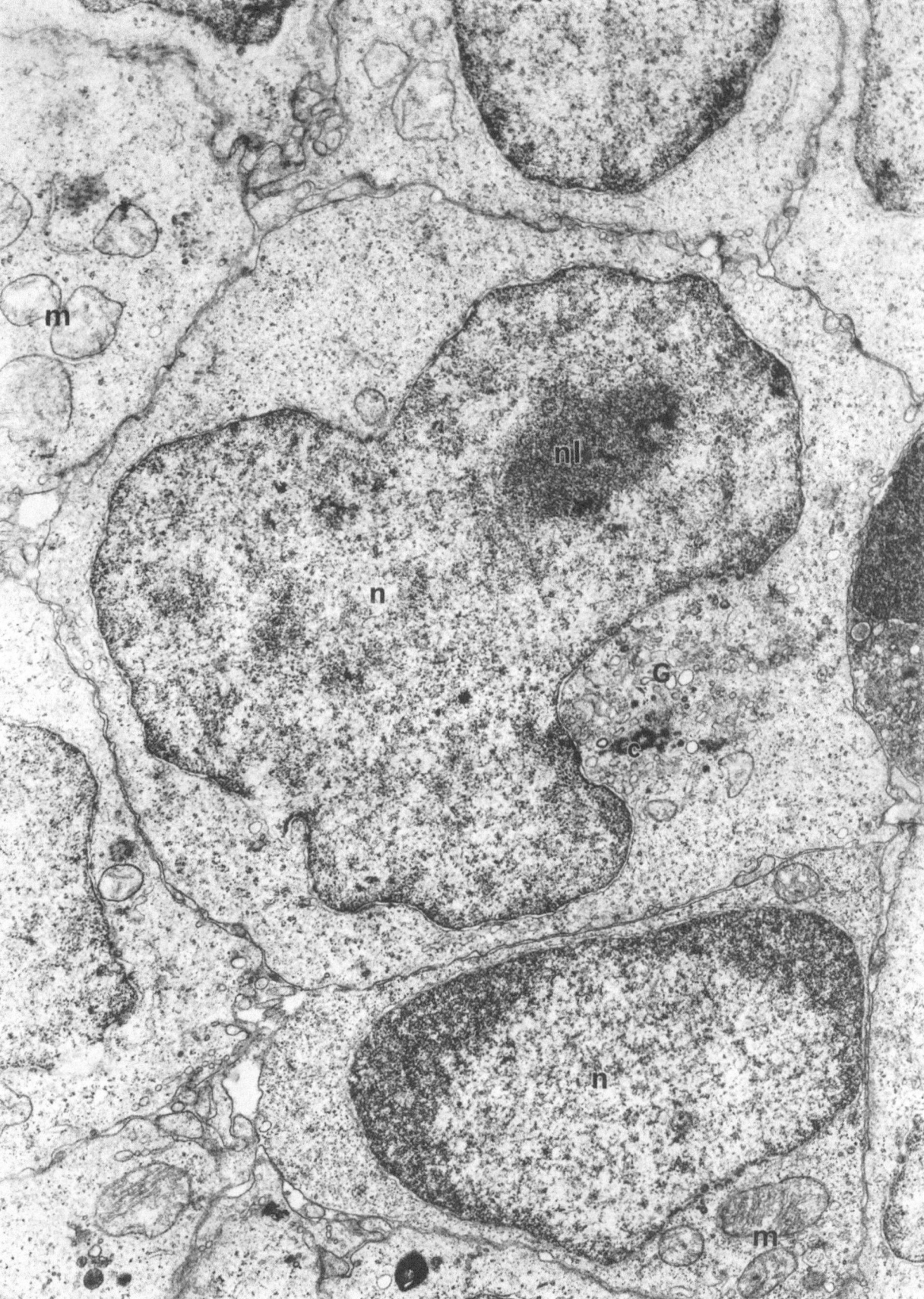

Fig. 95   Reticulum cell sarcoma, type I

A somewhat better differentiated tumor cell with many ribosomes and polyribosomes is illustrated. There are a few isolated vesicles and strands of endoplasmic reticulum.

Case 19. ×17,500

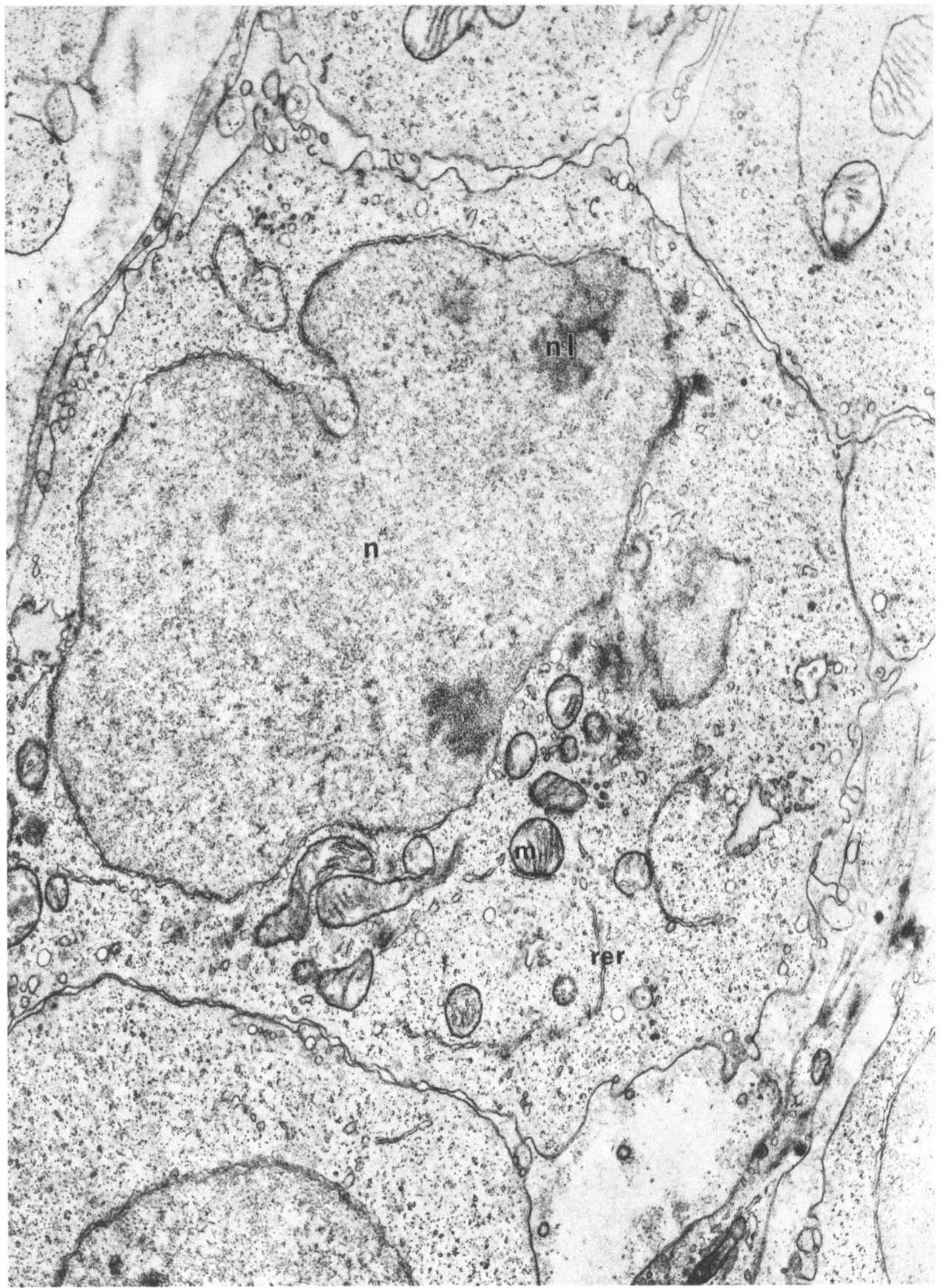

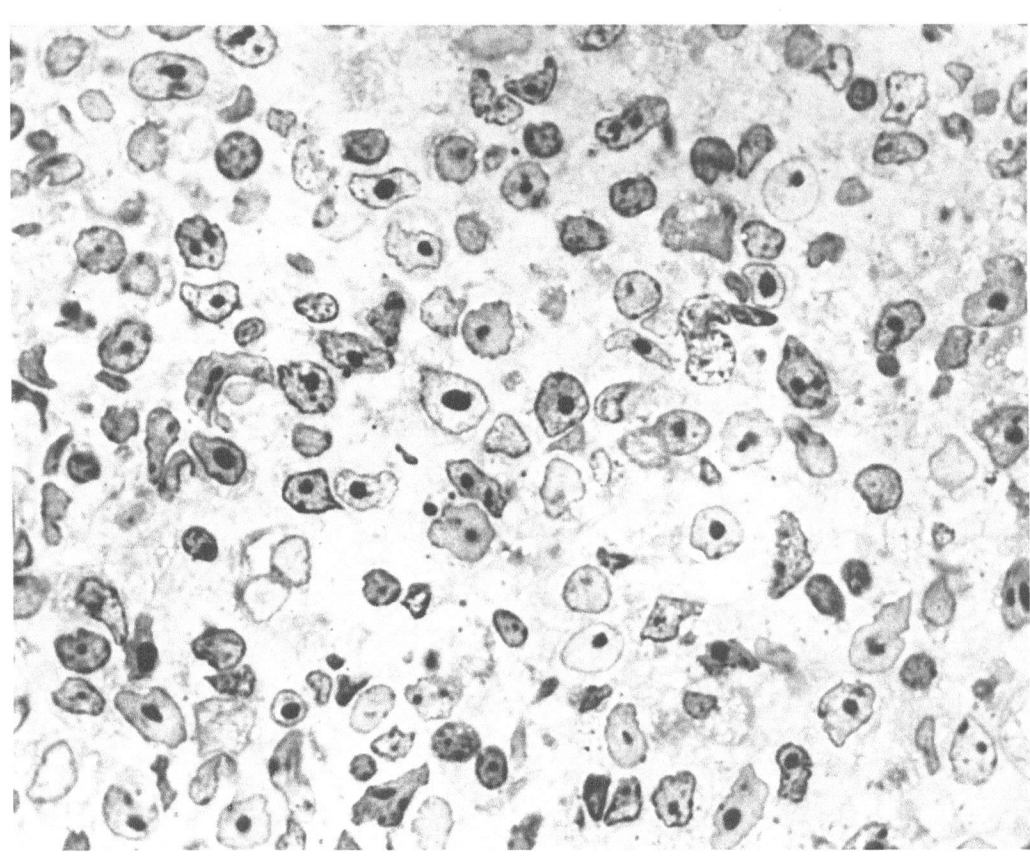

Fig. 96   Reticulum cell sarcoma, type II
Pleomorphic reticulum cells with large prominent nucleoli.

Case 2. Osmium tetroxide fixation. Semithin section. Methylene blue-Azure II. ×1,230

Fig. 97   Reticulum cell sarcoma, type II
The cells contain a moderate amount of smooth endoplasmic reticulum. The arrow points to a long cytoplasmic extension of a reticulum cell.

Case 2. ×8,750

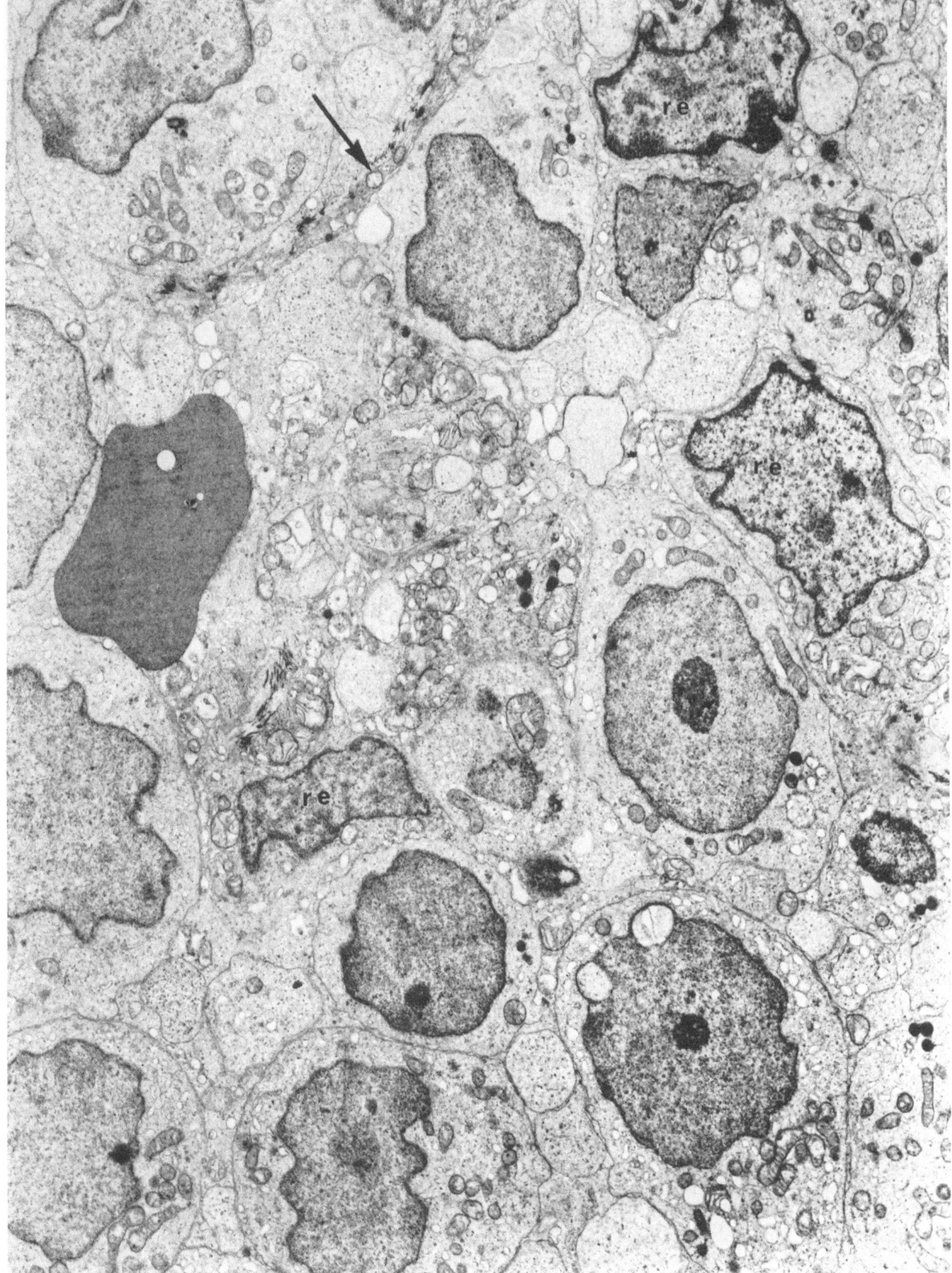

Fig. 98   Reticulum cell sarcoma, type II
Fairly well developed vacuolar structures, some with a small number of ribosomes, are seen. Two nuclear pockets containing cytoplasmic components and occasional lysosome-like bodies in most of the cells are illustrated in this micrograph.

Case 2. ×17,500

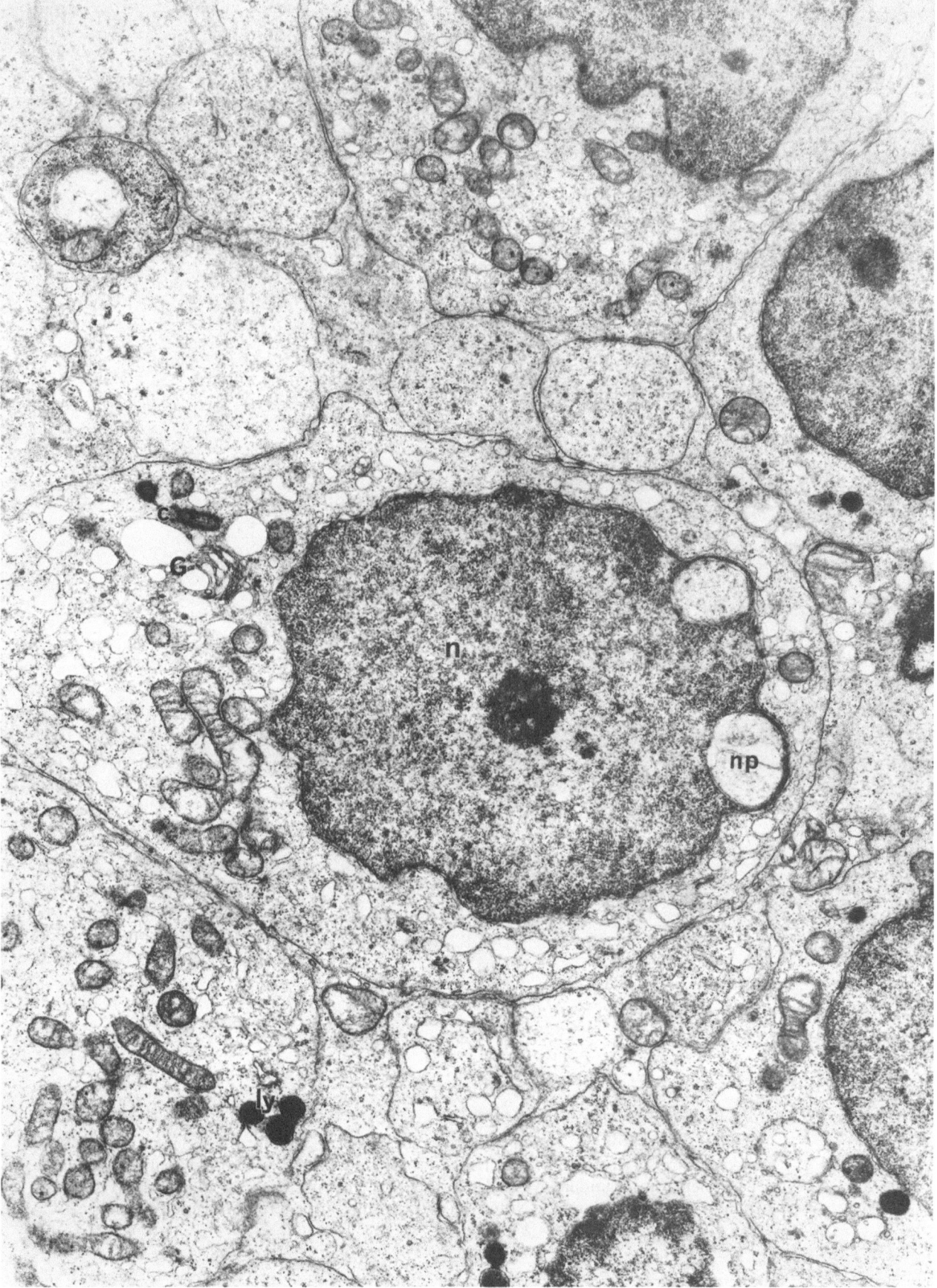

Fig. 99   Reticulum cell sarcoma, type II

Portions of nuclei of two tumor cells with surrounding cytoplasm are demonstrated. Free ribosomes are scattered throughout the cytoplasmic matrix. Only a few polysomes can be seen. There are abundant vesicular profiles of endoplasmic reticulum with occasional attached ribosomal granules. Note widened perinuclear cisterna above lower nucleus.

Case 2. ×40,000

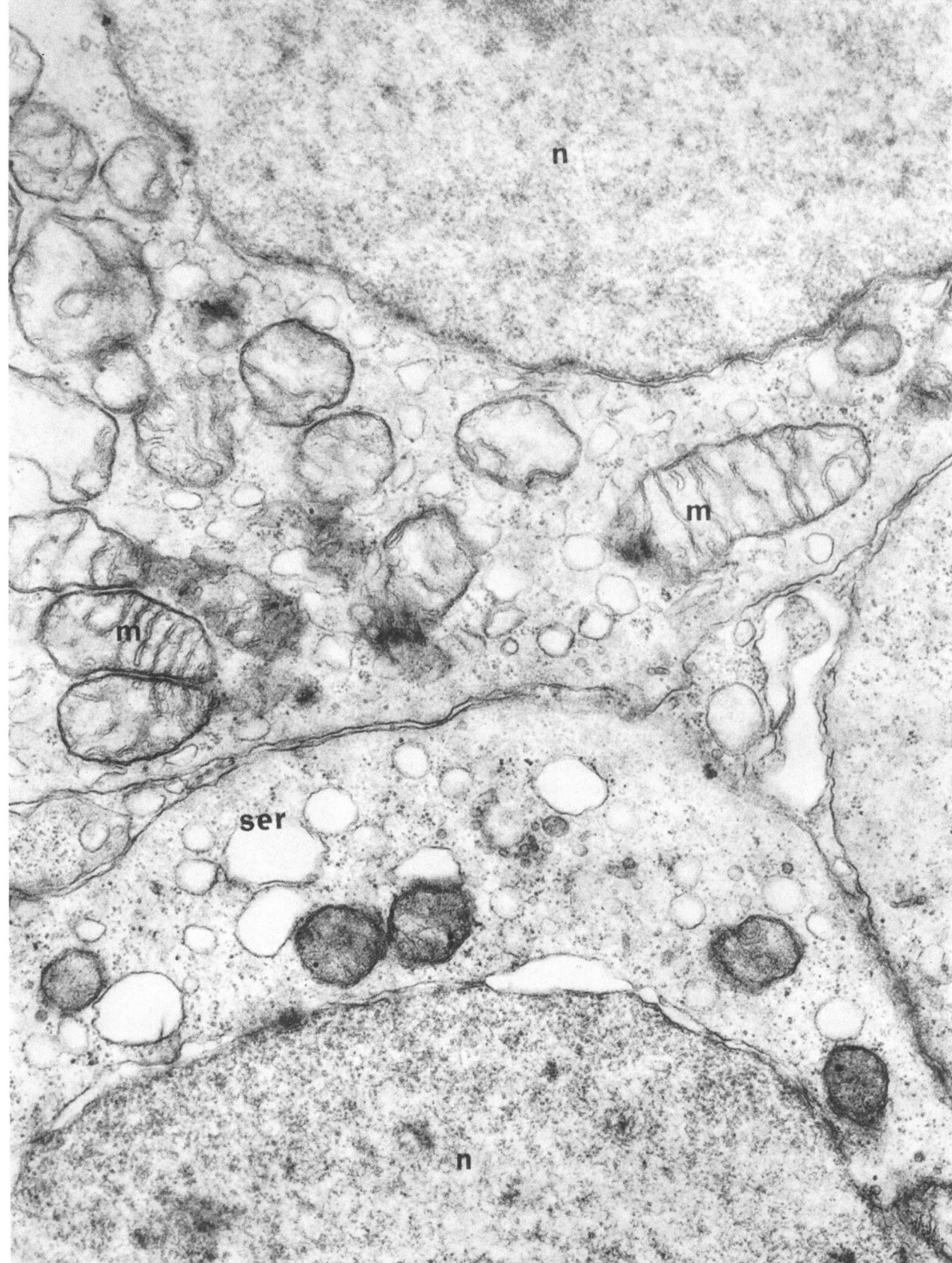

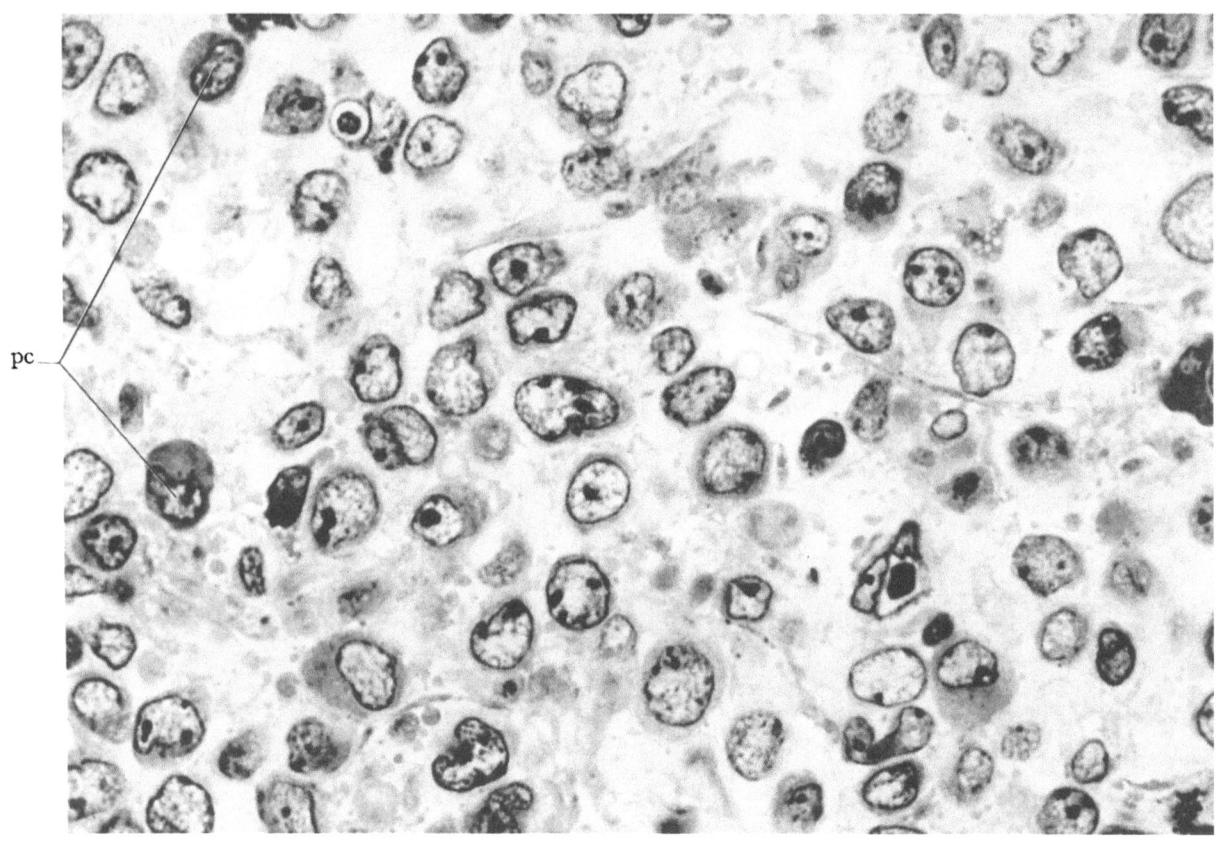

Fig. 100   Reticulum cell sarcoma, type II

Due to prefixation with formalin nuclear and cytoplasmic outlines are more distinct than after osmium tetroxide fixation alone. Note plasma cells (*pc*).

Case 64. Fixed in buffered neutral formalin, postfixed in osmium tetroxide. Semithin section. Methylene blue-Azure II. ×1,230

Fig. 101   Reticulum cell sarcoma, type II

The moderately well differentiated tumor cells contain a medium amount of endoplasmic reticulum. Note the plasma cell on the right which is crowded with endoplasmic reticulum. Macrophages are seen above and below.

Case 64. ×8,750

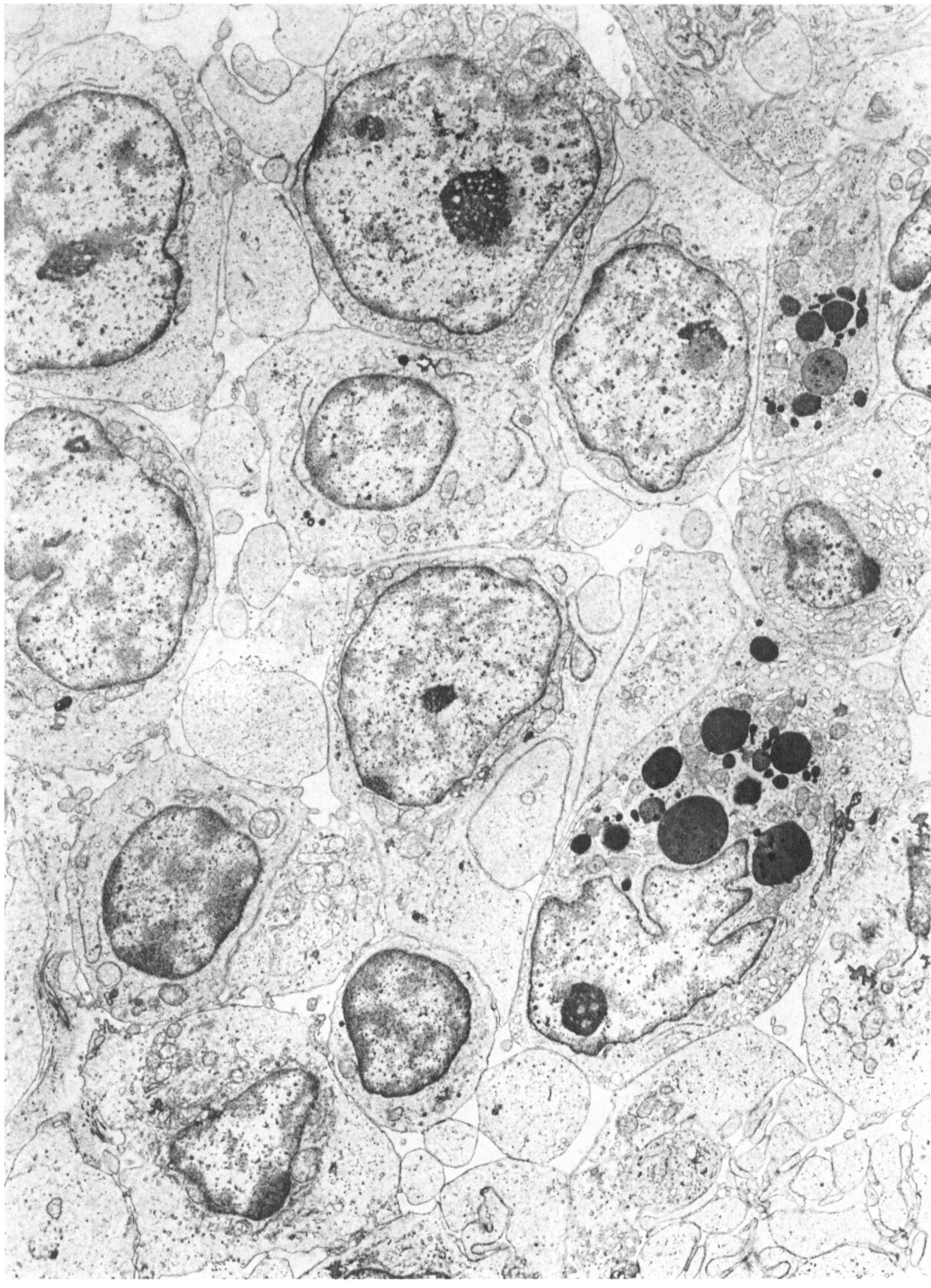

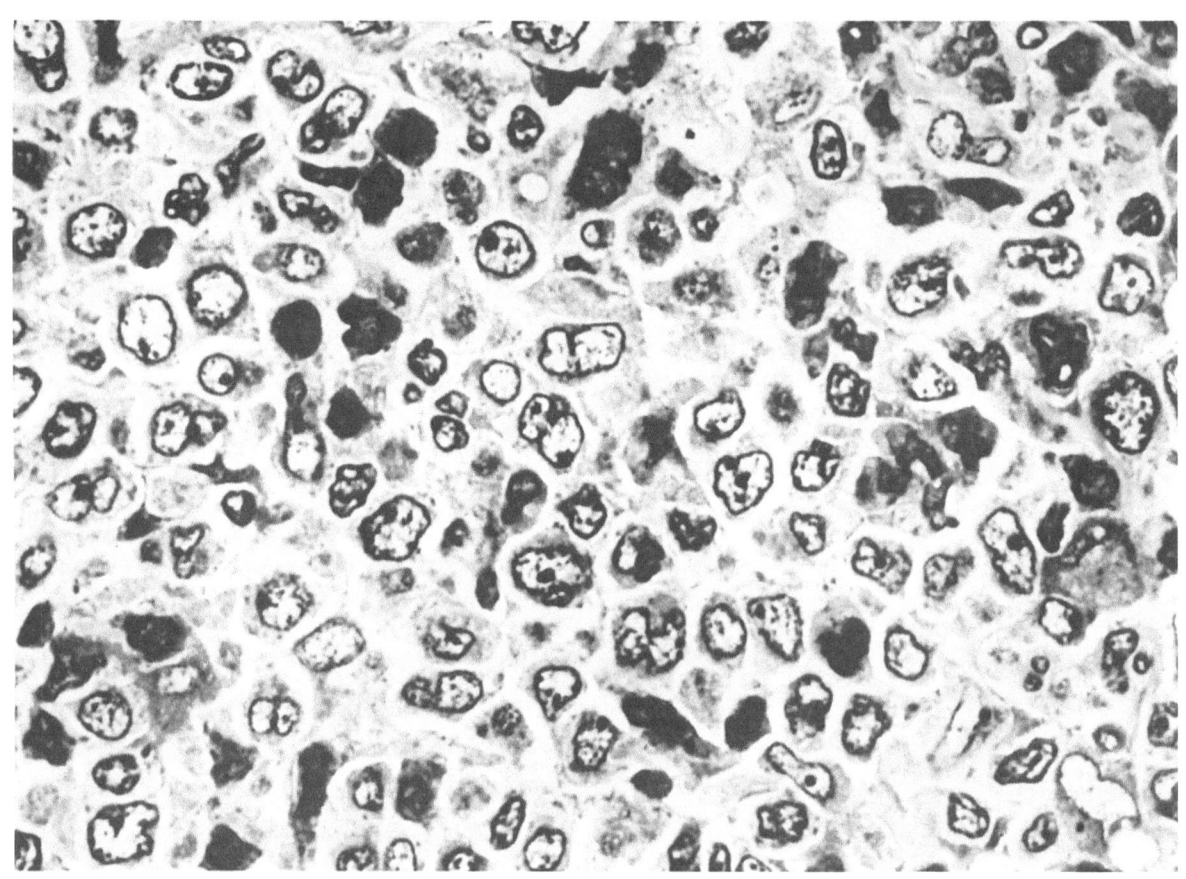

Fig. 102   Reticulum cell sarcoma, type III
Numerous pyknotic and hyperchromatic cells are illustrated.

Case 4. Osmium tetroxide fixation, preceded by fixation in unbuffered formalin (routine surgical material). Semithin section. Methylene blue-Azure II. ×1,230

Fig. 103   Reticulum cell sarcoma, type III
Parallel lamellae of rough endoplasmic reticulum are in apposition to the lower pole of the nuclear membrane.

Case 4. Prefixation in unbuffered formalin (routine surgical material). ×21,000

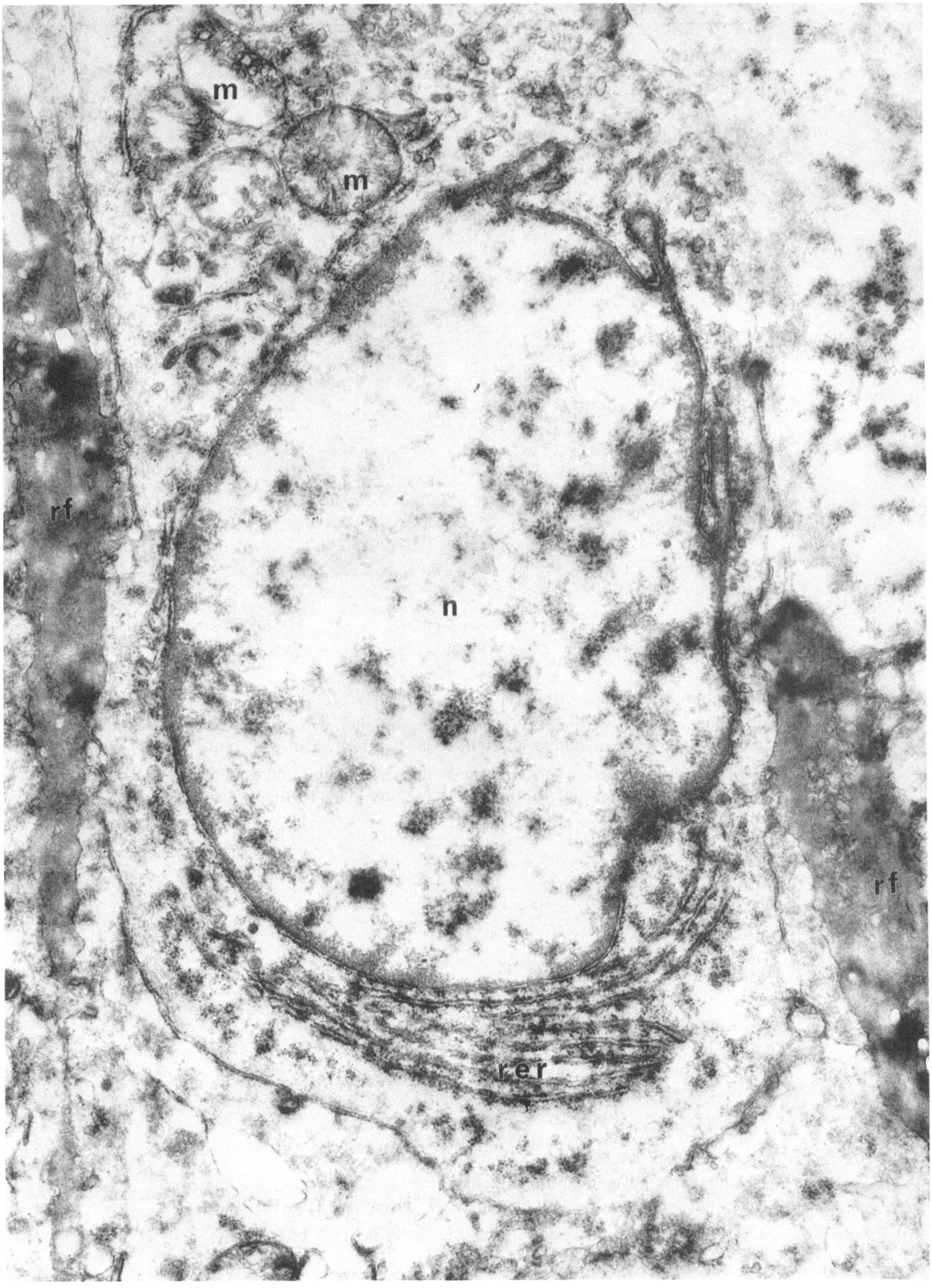
m
m
rf
n
rf
rer

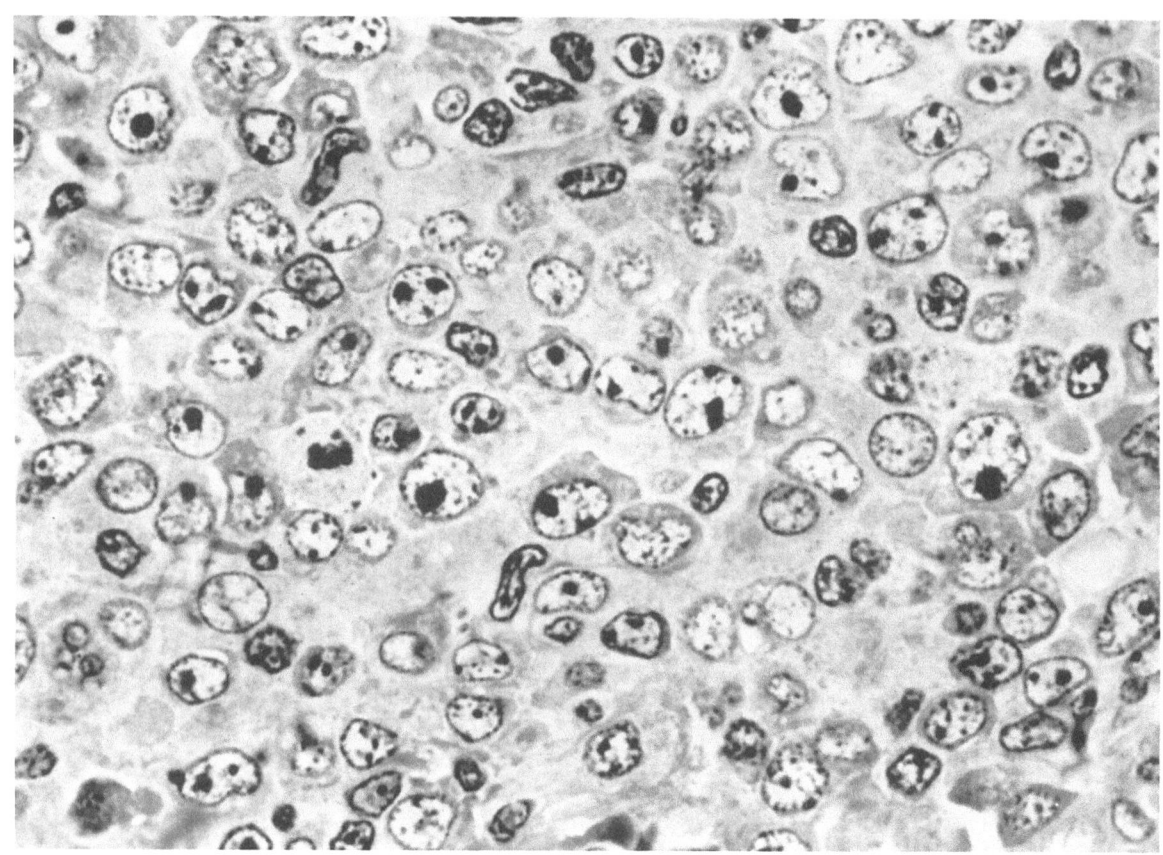

Fig. 104   Reticulum cell sarcoma, type III

Pleomorphic cells with deeply stained (basophilic) cytoplasm are evident, many with morphologic features similar to stem cells. Large nucleoli are common.

Case 62. Short initial fixation in unbuffered formalin, followed by osmium tetroxide fixation. Semithin section. Methylene blue-Azure II. ×1,230

Fig. 105   Reticulum cell sarcoma, type III

Channels of rough endoplasmic reticulum may be seen in moderate numbers. Free ribosomes and polyribosomes are common. There is a prominent large nucleolus.

Case 62. Short initial fixation in unbuffered formalin. ×17,500

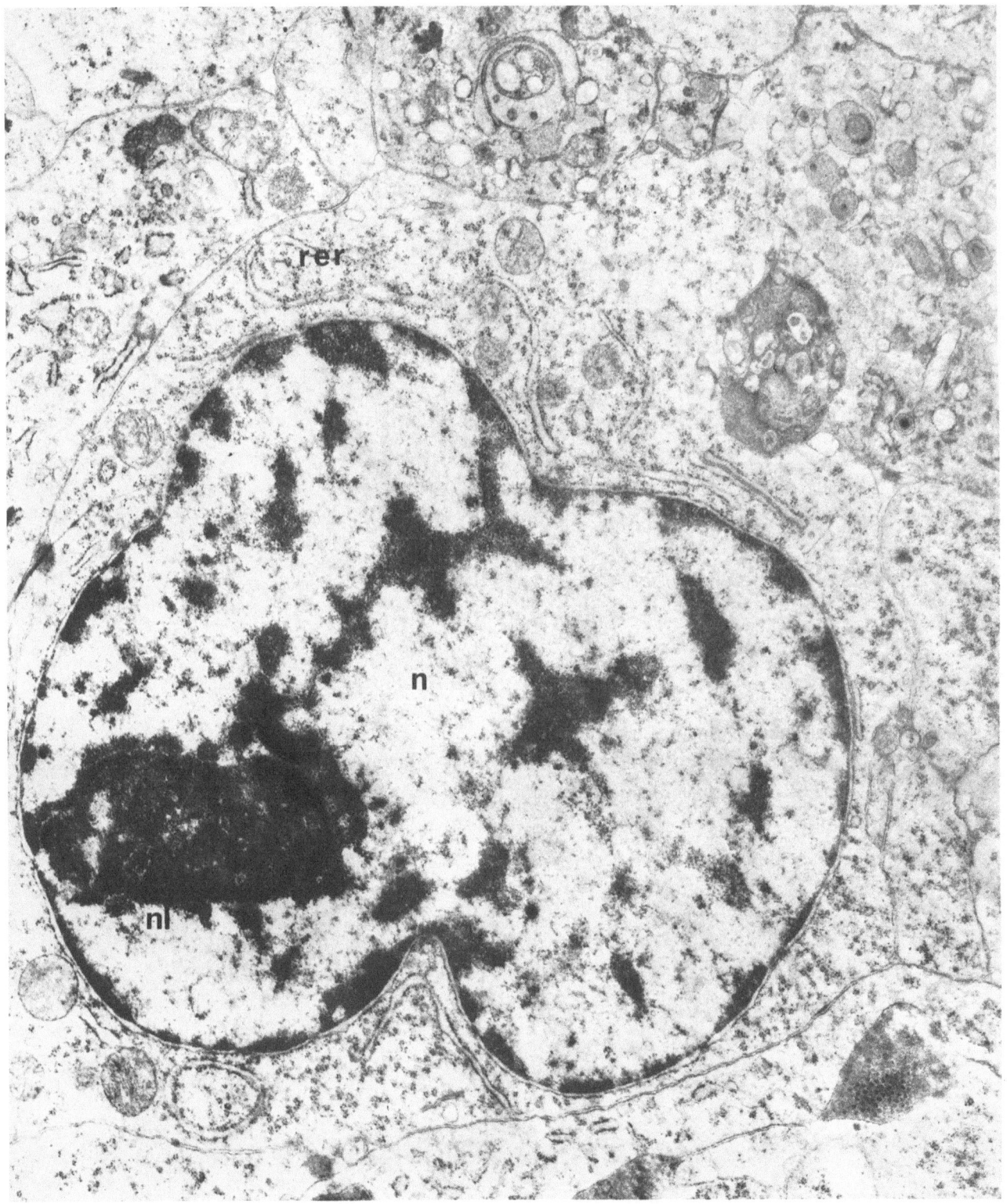

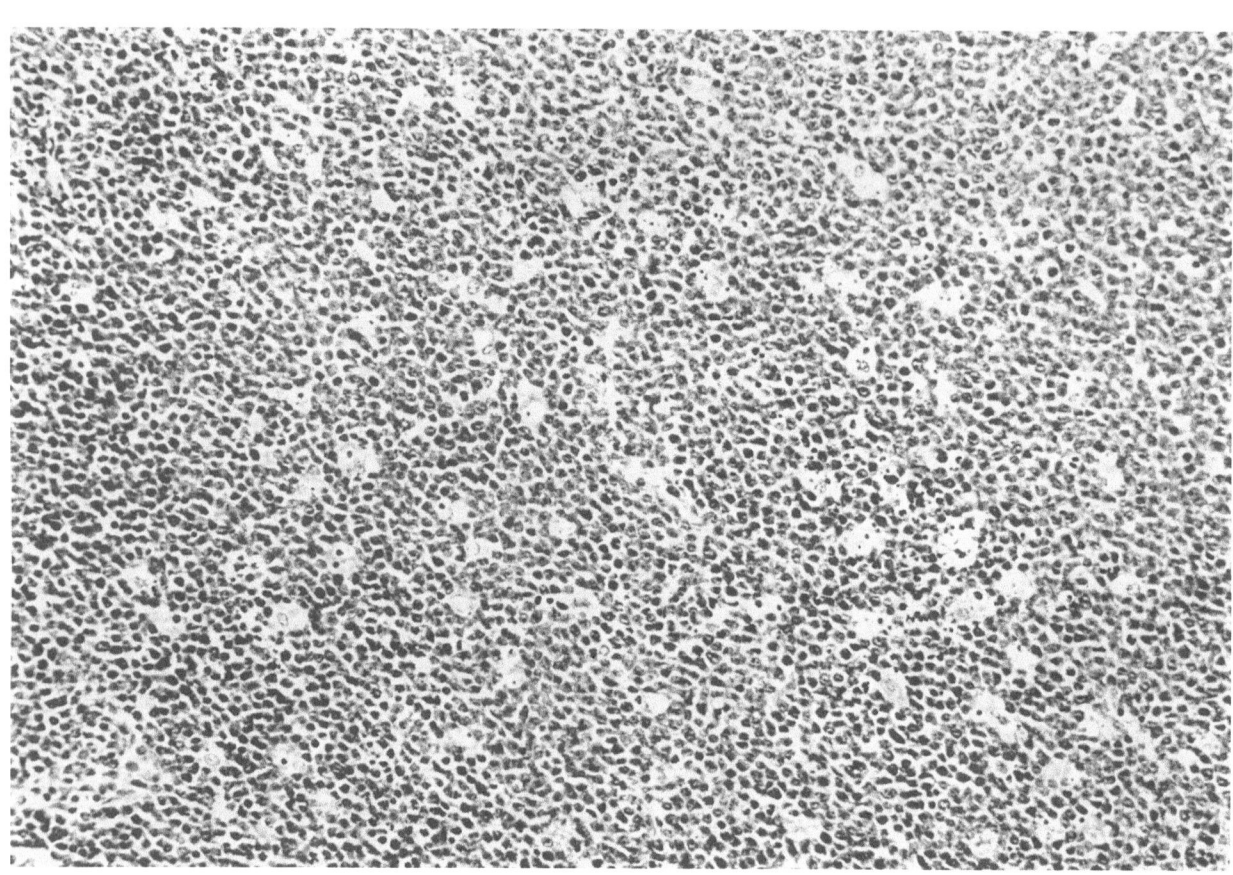

Fig. 106   Burkitt's tumor, European case
Representative overall view with numerous "starry sky cells".
Case 46. Formalin fixation. Paraffin section. Hematoxylin and eosin stain. ×200

Fig. 107   Burkitt's tumor, European case
Large pleomorphic tumor cells with abundant ribosomes and polyribosomes and poorly
developed cytoplasmic organelles are seen. Note occasional lipid droplets in the cytoplasm.
Case 46. ×8,750

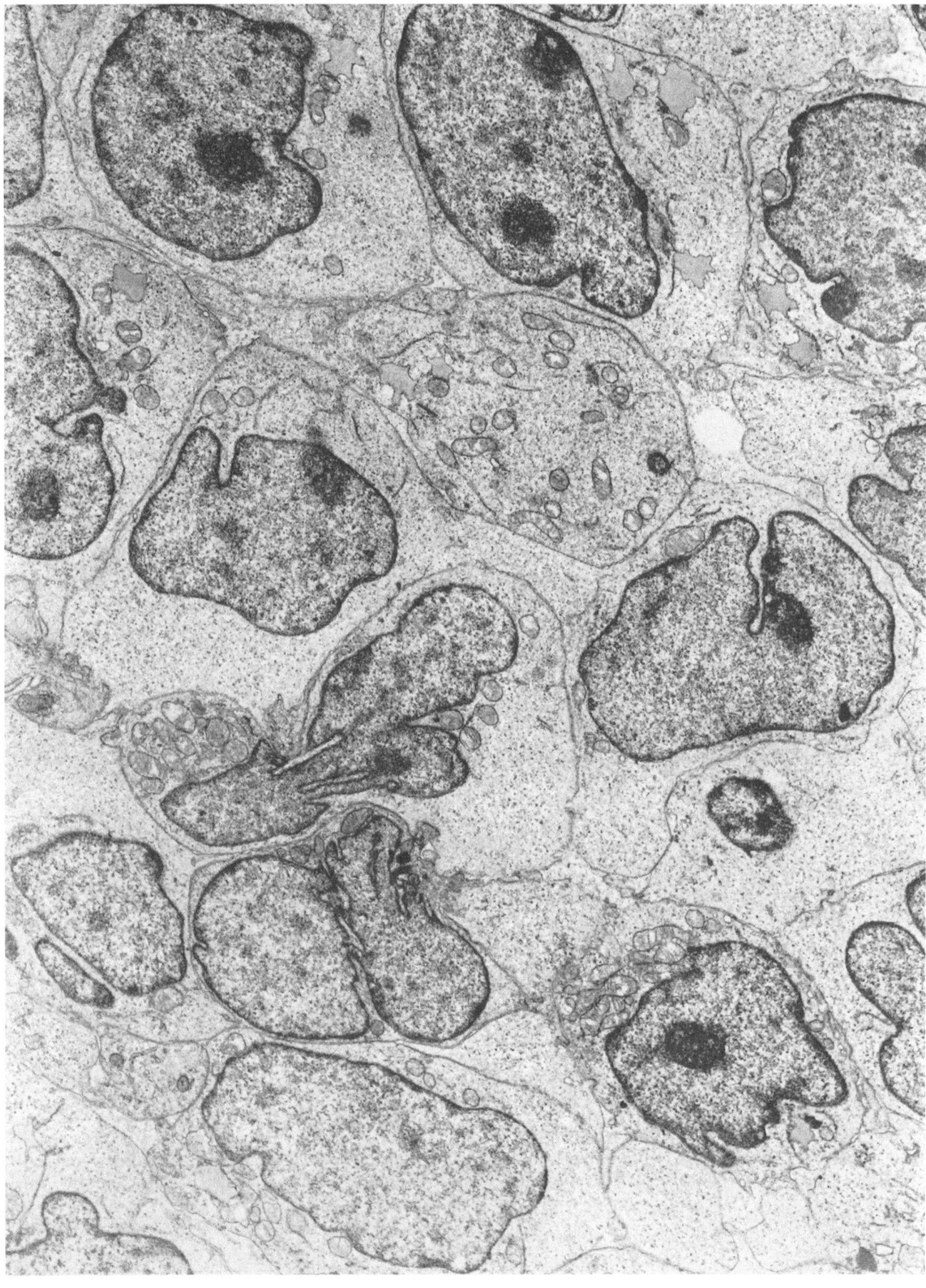

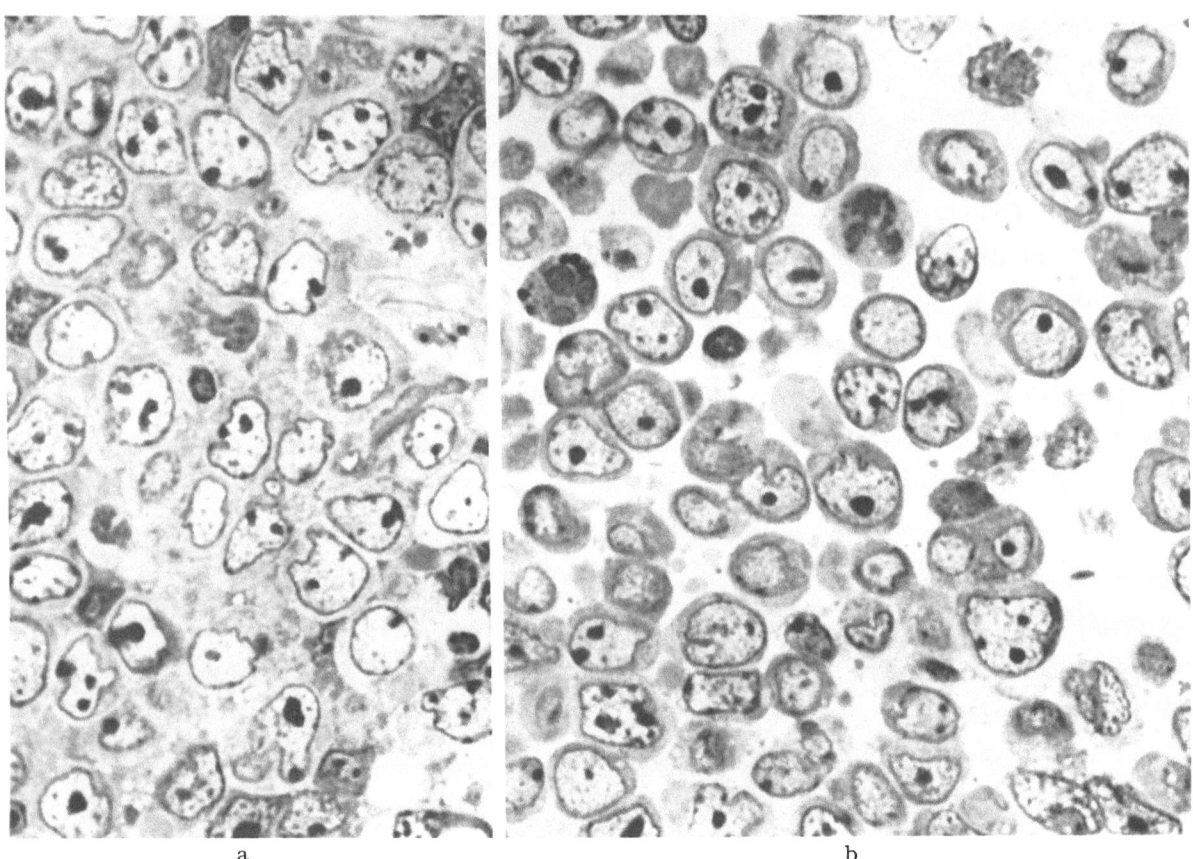

a   b

Fig. 108a and b   Burkitt's tumor, European case

Large cells with basophilic cytoplasm and prominent nucleoli in pleomorphic, often rounded nuclei in a) central and b) peripheral part of the section. Note the similarity to basophilic stem cells. These cells may be demonstrated much better in imprints, where the numerous little vacuoles (Sudan red positive) are clearly visible.

Case 46. Osmium tetroxide fixation. Semithin section. Methylene blue-Azure II. ×1,230

Fig. 109   Burkitt's tumor, European case

The cells contain abundant ribosomes and polysomes, few large mitochondria, a paucity of endoplasmic reticulum, few lipid droplets and an inconspicuous Golgi apparatus (the lower cell). Large nucleoli are striking.

Case 46. ×17,500

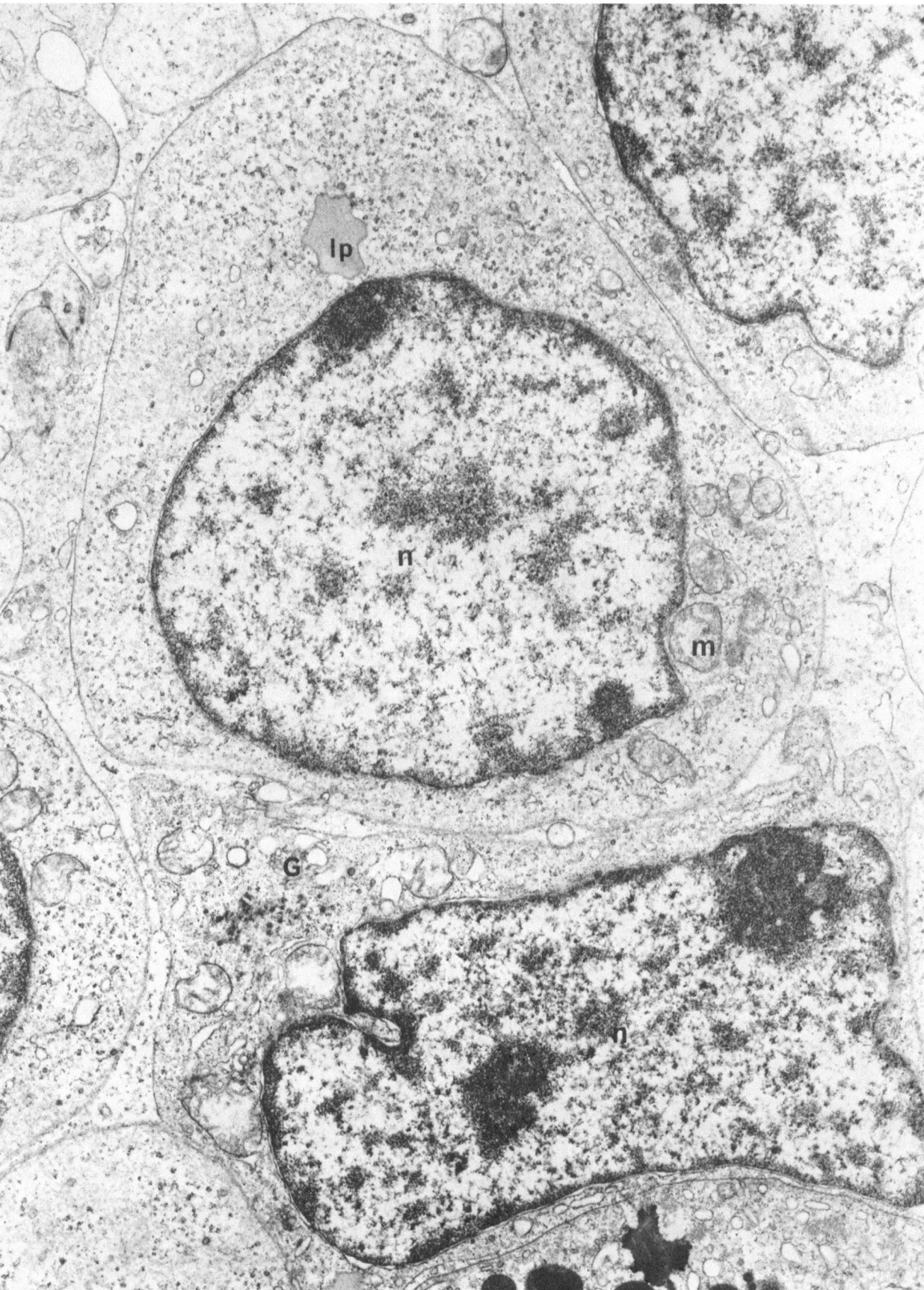

Fig. 110   Burkitt's tumor, European case
There is scanty rough endoplasmic reticulum, but evenly distributed free ribosomes and
polysomes are numerous. An eosinophilic leukocyte is seen in the lower left corner.
Case 46. ×17,500

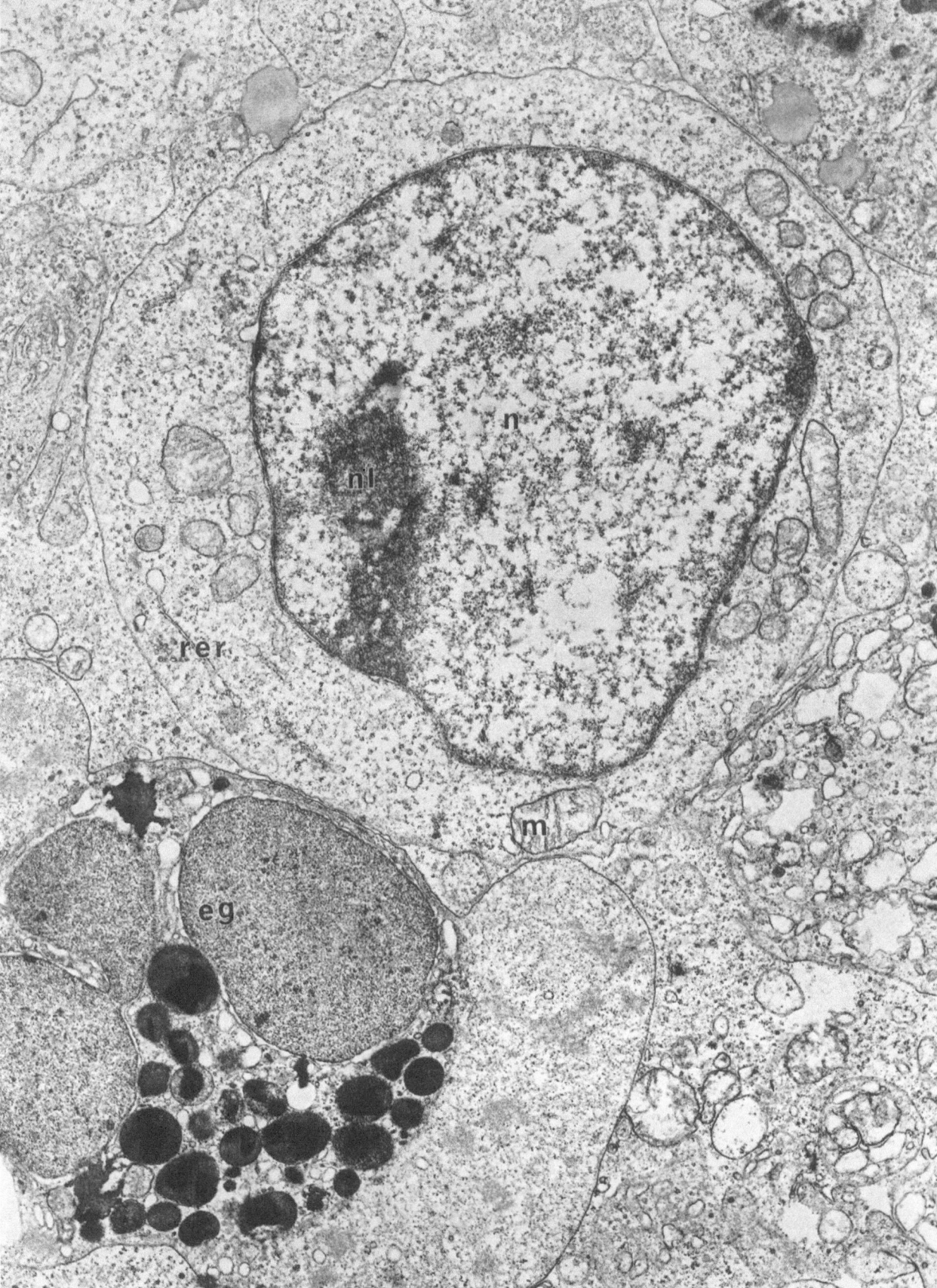

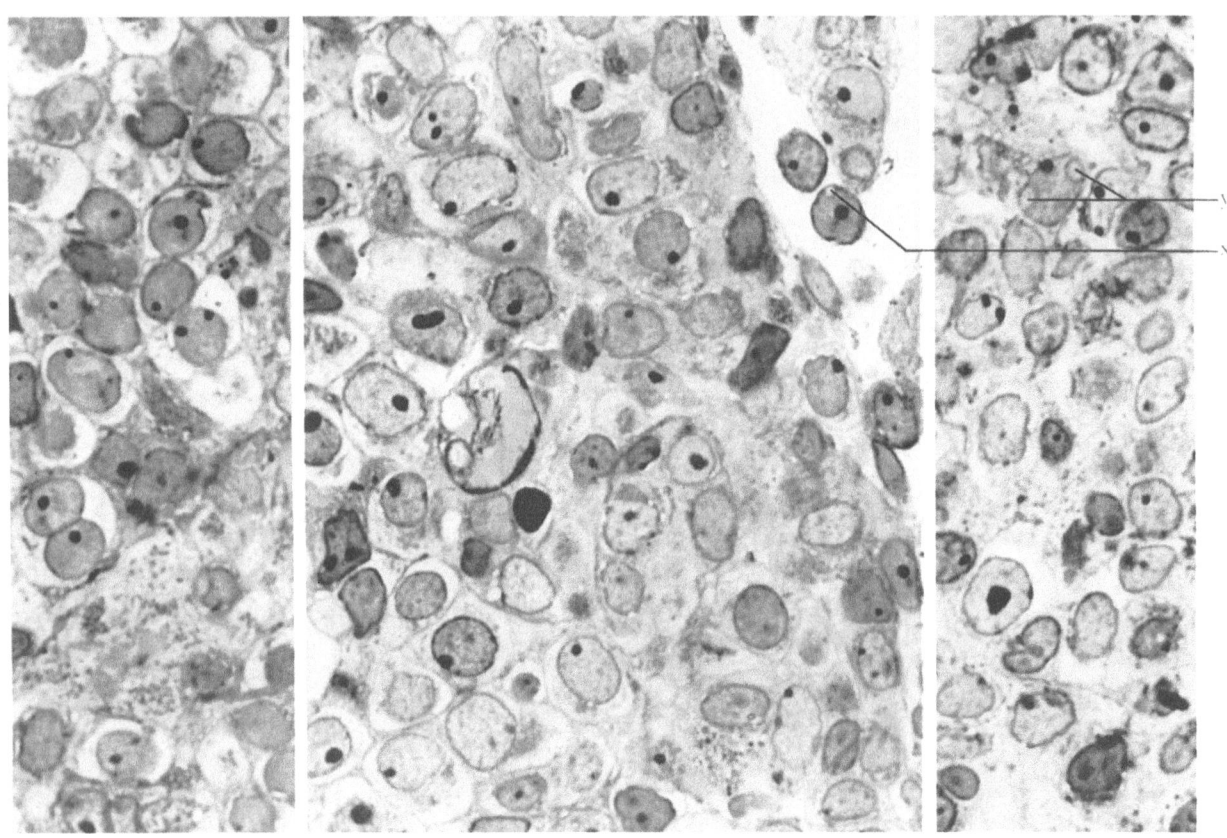

Fig. 111   Macroglobulinemia of WALDENSTRÖM

Pleomorphic tumor cells show considerable variation in size of nucleoli. Occasionally numerous vacuoles, lysosomes or "granules" are seen in the cytoplasm. There are nuclear inclusion bodies in an intrasinusoidal tumor-cell (x) and in the "pulp" (y).

Case 3. Osmium tetroxide fixation. Semithin section. Methylene blue-Azure II. ×1,230

Fig. 112   WALDENSTRÖM's macroglobulinemia

The basic structure of the tightly packed tumor cells approximates that of lymphocytes. Note, however, the distinct Golgi complex and a variable amount of rough endoplasmic reticulum. Widened perinuclear cisternae are discernible in most of the cells.

Case 3. ×17,500

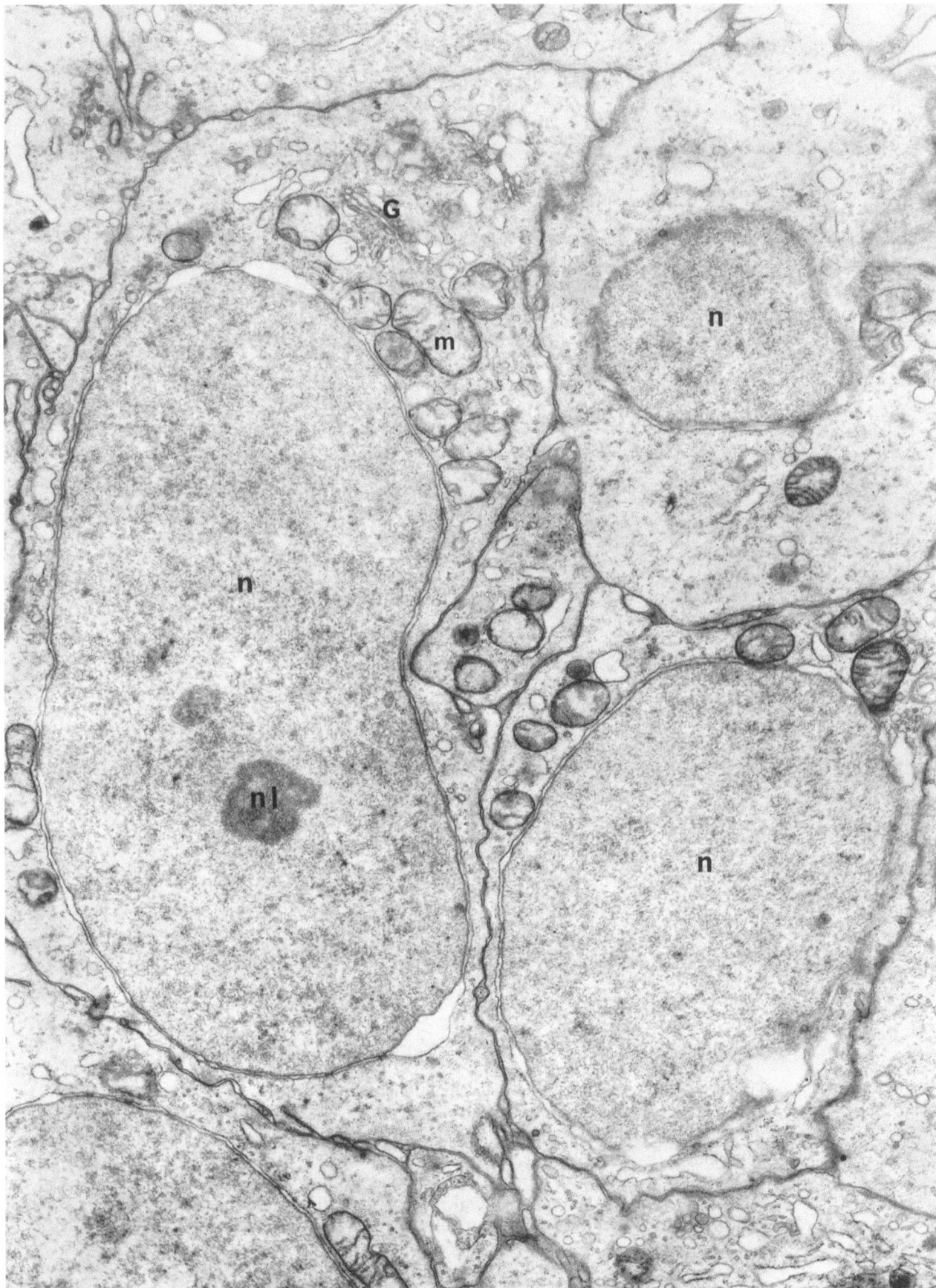

Fig. 113   Tumor cell of WALDENSTRÖM's macroglobulinemia

This cell bears no similarity to lymphocytes, but has morphologic features of plasma cells. Note large vesicular and tubular cavities of ribosome-associated endoplasmic reticulum containing a flocculent granular precipitate believed to be macroglobulin. The Golgi apparatus is prominent with a number of small dark-staining and a few extremely large pale vesicles. Two dark-staining lysosome-like bodies are easily distinguished from other vesicular elements. Free ribosomes and a few polysomes are scattered throughout the cytoplasm.

Case 3. ×35,000

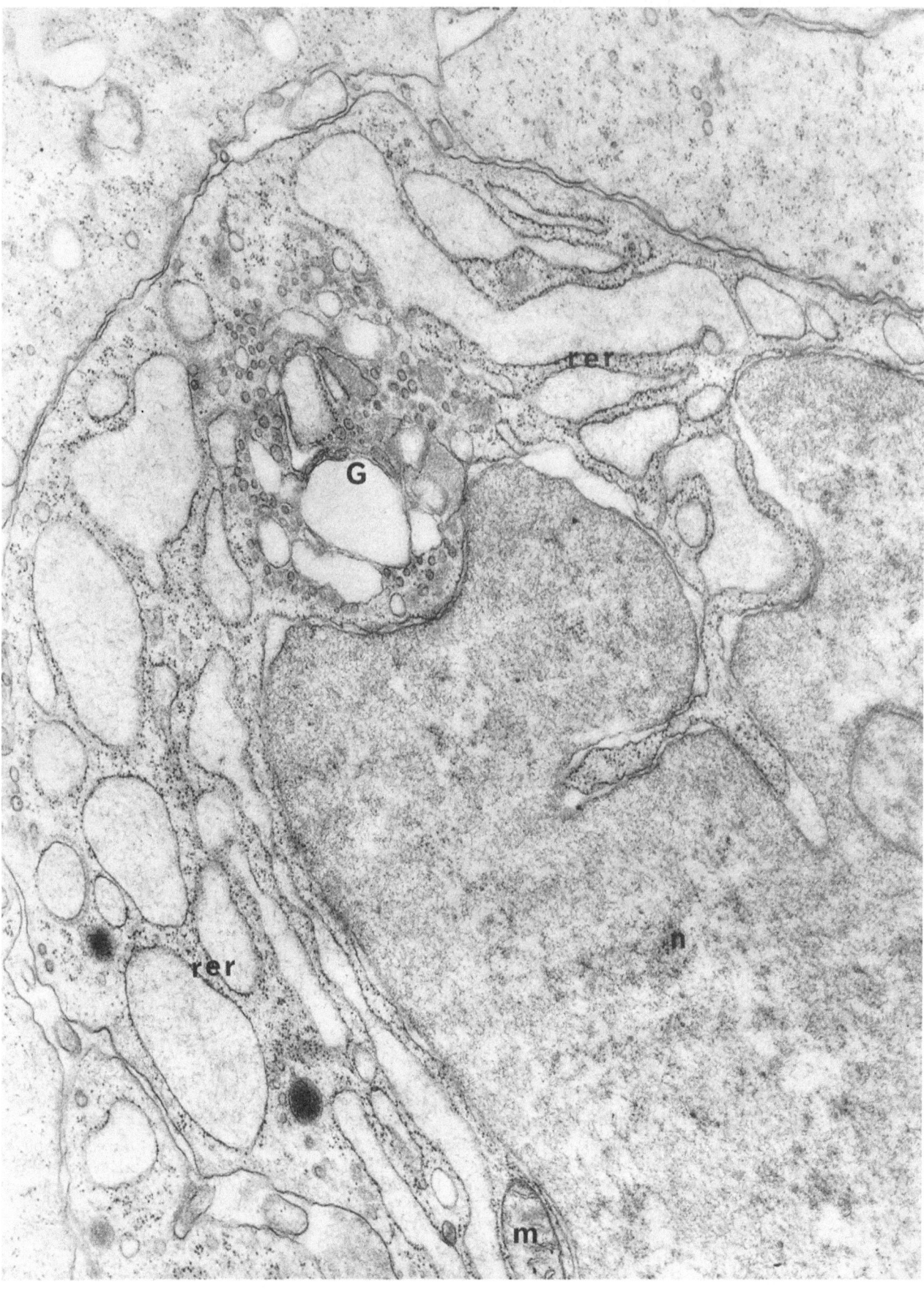

Fig. 114   Intranuclear inclusion of tumor cell in WALDENSTRÖM's macroglobulinemia

A large intranuclear protein inclusion with few vesicular elements occupies a considerable part of the nuclear area. It cannot be excluded that the protein material has taken its origin from the perinuclear cisterna. It may have been secondarily incorporated into the nucleus by invagination of the inner nuclear membrane.

Case 3. ×35,000

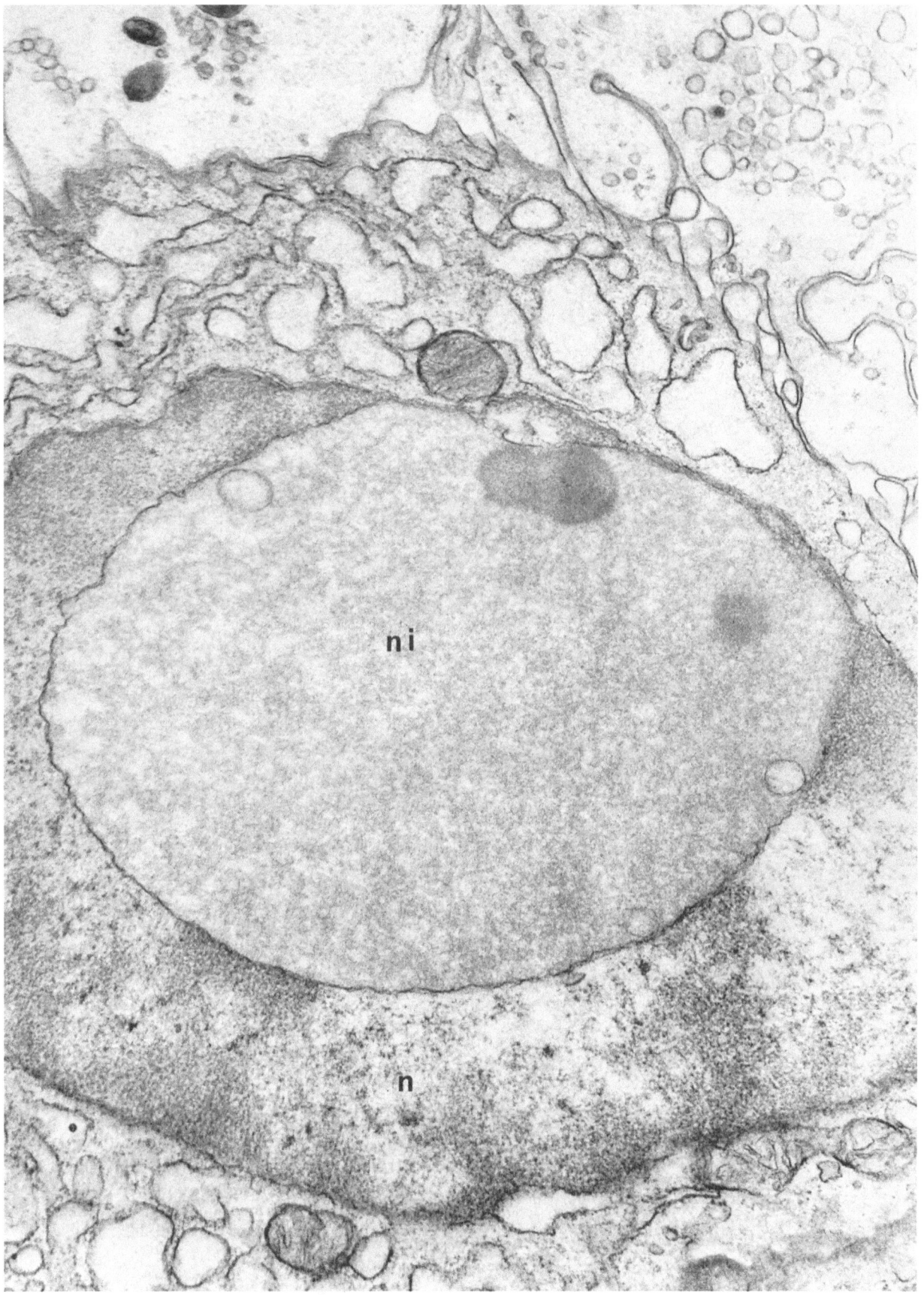

Fig. 115   Macroglobulinemia of WALDENSTRÖM. Protein deposits in intercellular spaces

There are contiguous tumor cells with intercellular protein deposits (*pr*). Note the irregularly widened perinuclear cisterna of the lower nucleus. The Golgi complex is moderately well developed. Elongated somewhat swollen mitochondria are prominently displayed. Scattered vesicular profiles of rough endoplasmic reticulum are discernible.

Case 3. ×28,000

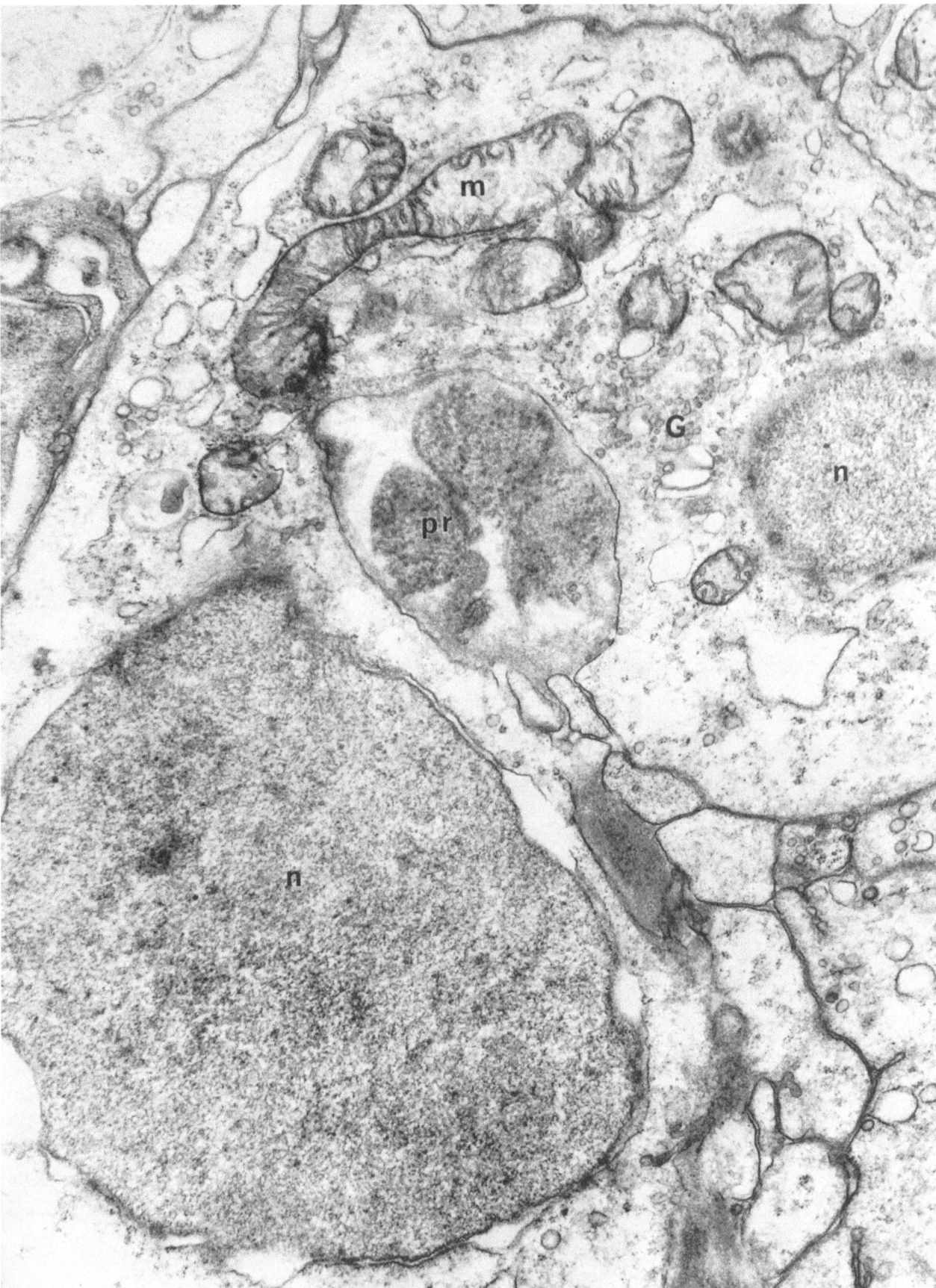

Chronic Lymphocytic Leukemia

Poorly Differentiated Leukemias and
Erythremias
  Lymphoblastic Type
  Granulocytic (Paramyeloblastic) Type
  Monocytic Type
  Undifferentiated Type
  Acute Erythremia (Di Guglielmo's
  Disease)

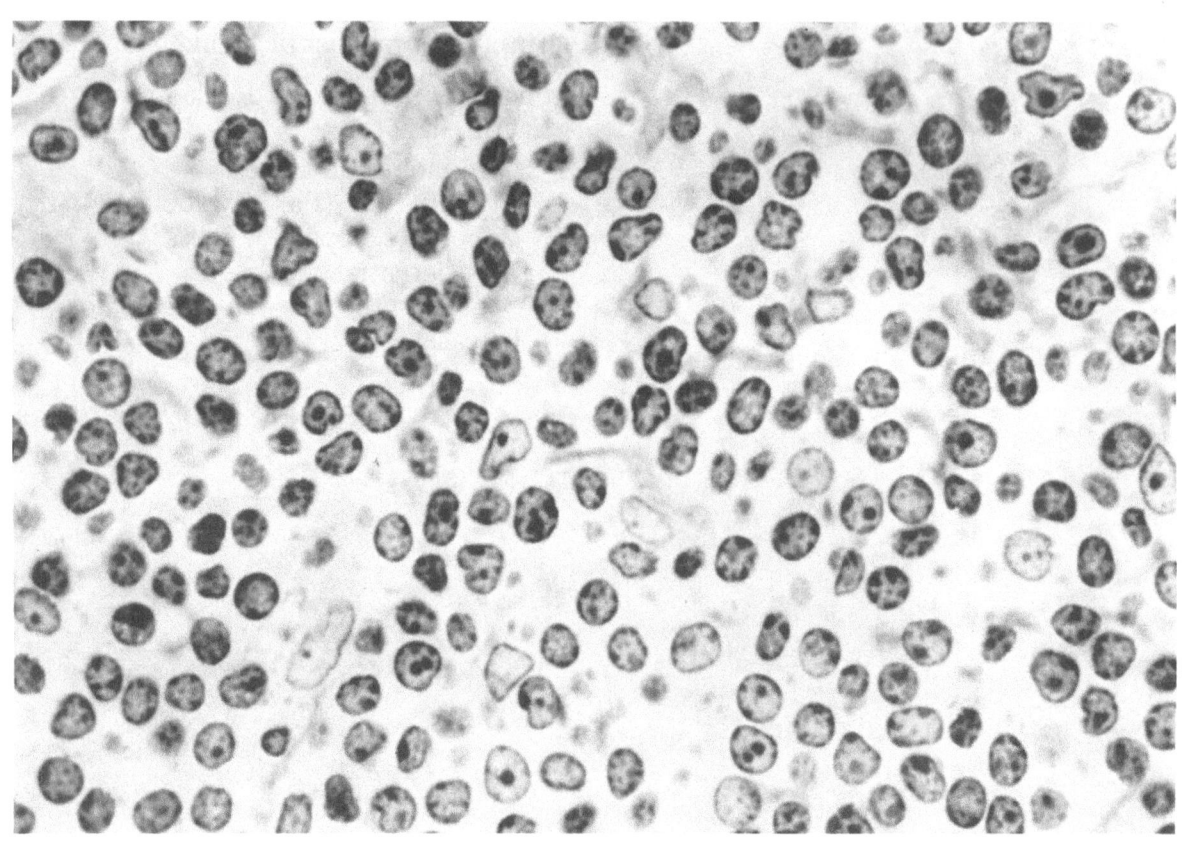

Fig. 116   Chronic lymphocytic leukemia
There is predominance of lymphocytes with coarse chromatin clumps, and occasional lymphoblasts and reticulum cells are seen.

Case 13. Osmium tetroxide fixation. Methylene blue-Azure II. ×1,230

Fig. 117   Chronic lymphocytic leukemia
Cell showing an odd-shaped nucleus with multiple indentations. There are a relatively large nucleolus with a central zone of lesser electron density and a slightly widened perinuclear cisterna. The cytoplasm is relatively broad with ovoid large mitochondria and a finely vesicular Golgi apparatus. Sparse endoplasmic reticulum, predominantly of the rough variety and displaying vesicular profiles, and numerous free ribosomes and polysomes are scattered throughout the cytoplasm. One centriole is recognizable in the center.

Case 13. ×20,000

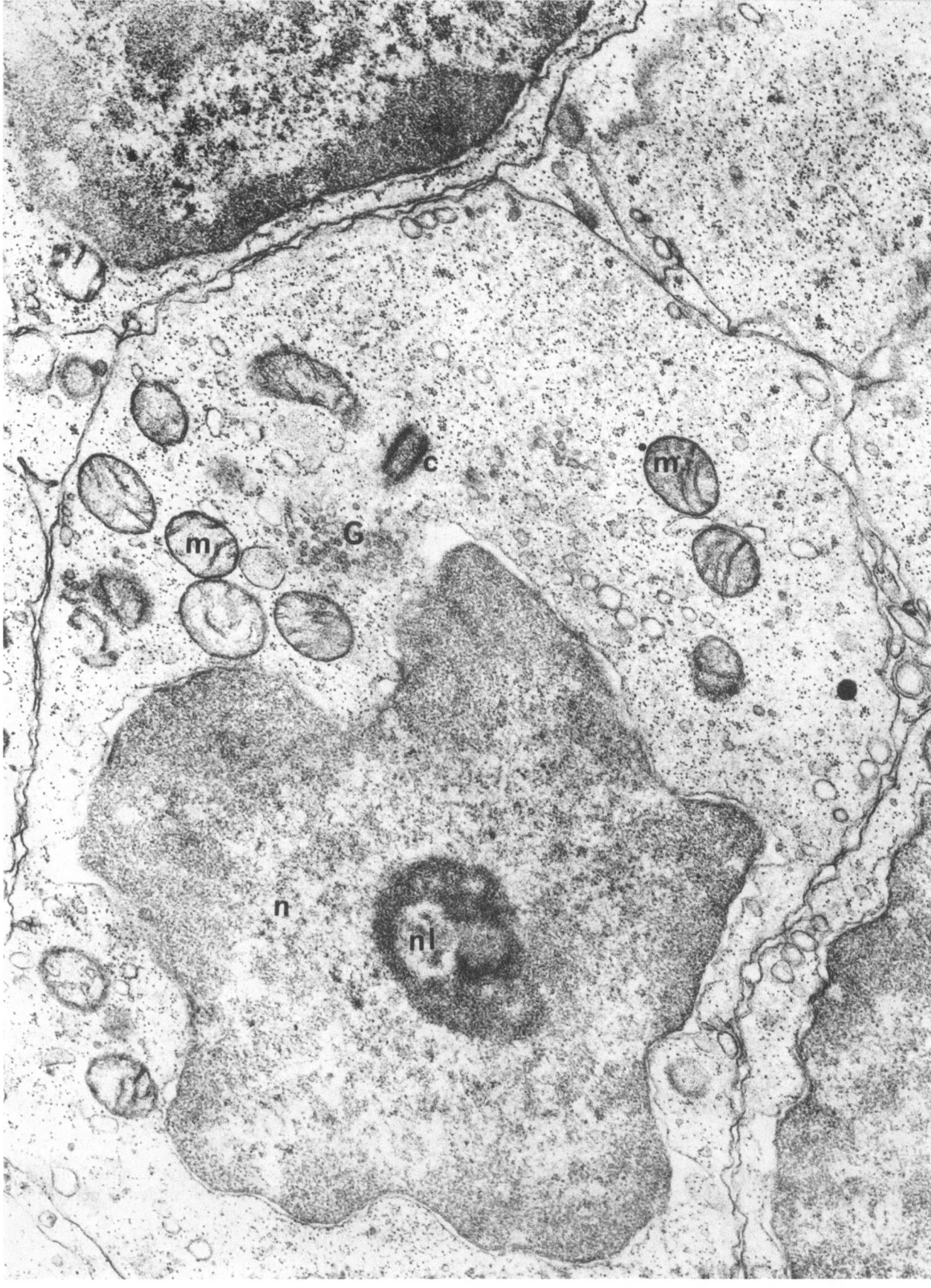

Fig. 118   Chronic lymphocytic leukemia

This leukemic cell is reminiscent of a normal lymphocyte. It has globular mitochondria, a finely vesicular Golgi area, some strands of rough endoplasmic reticulum and many free ribosomes and polysomes.

Case 13. ×17,500

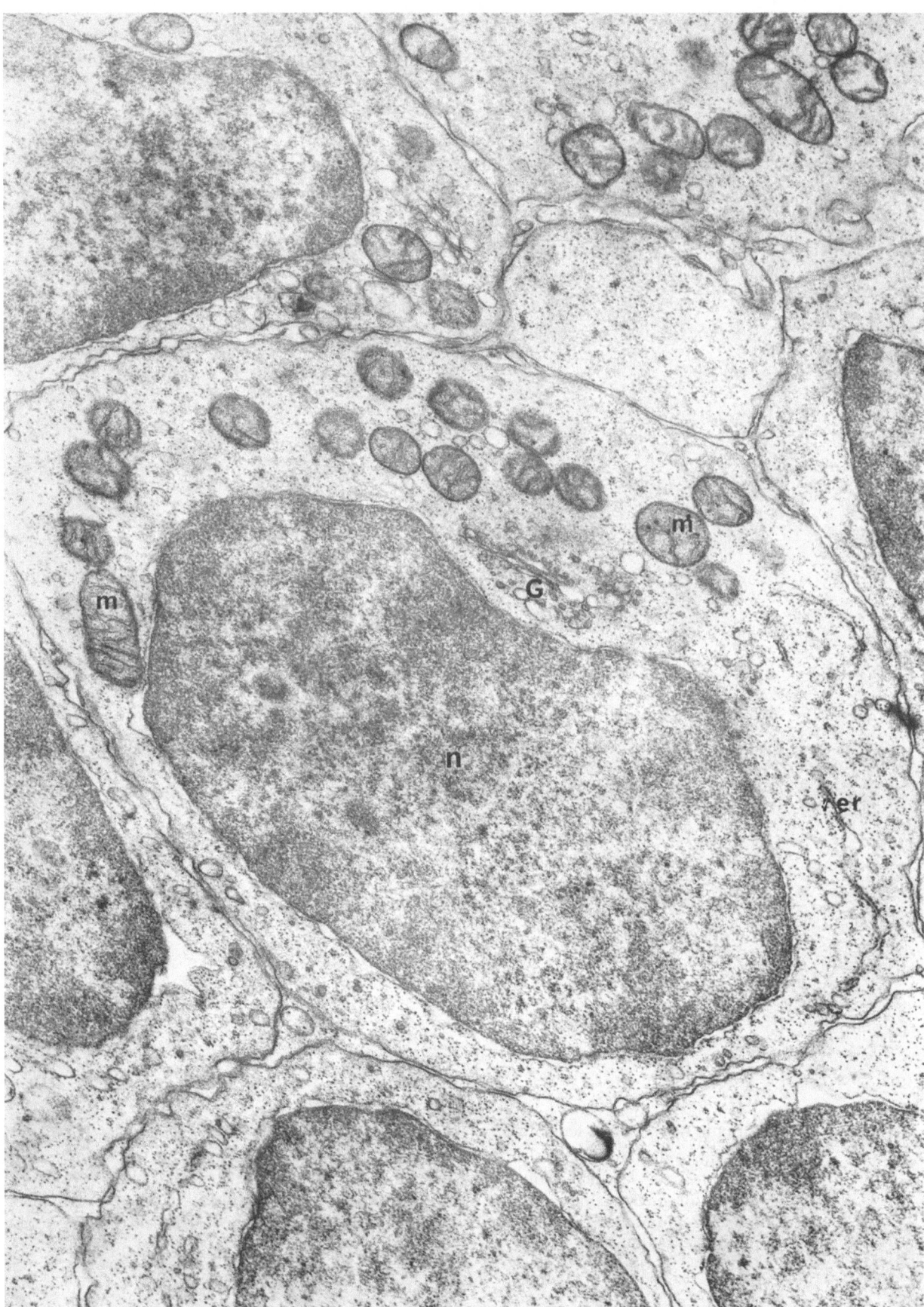

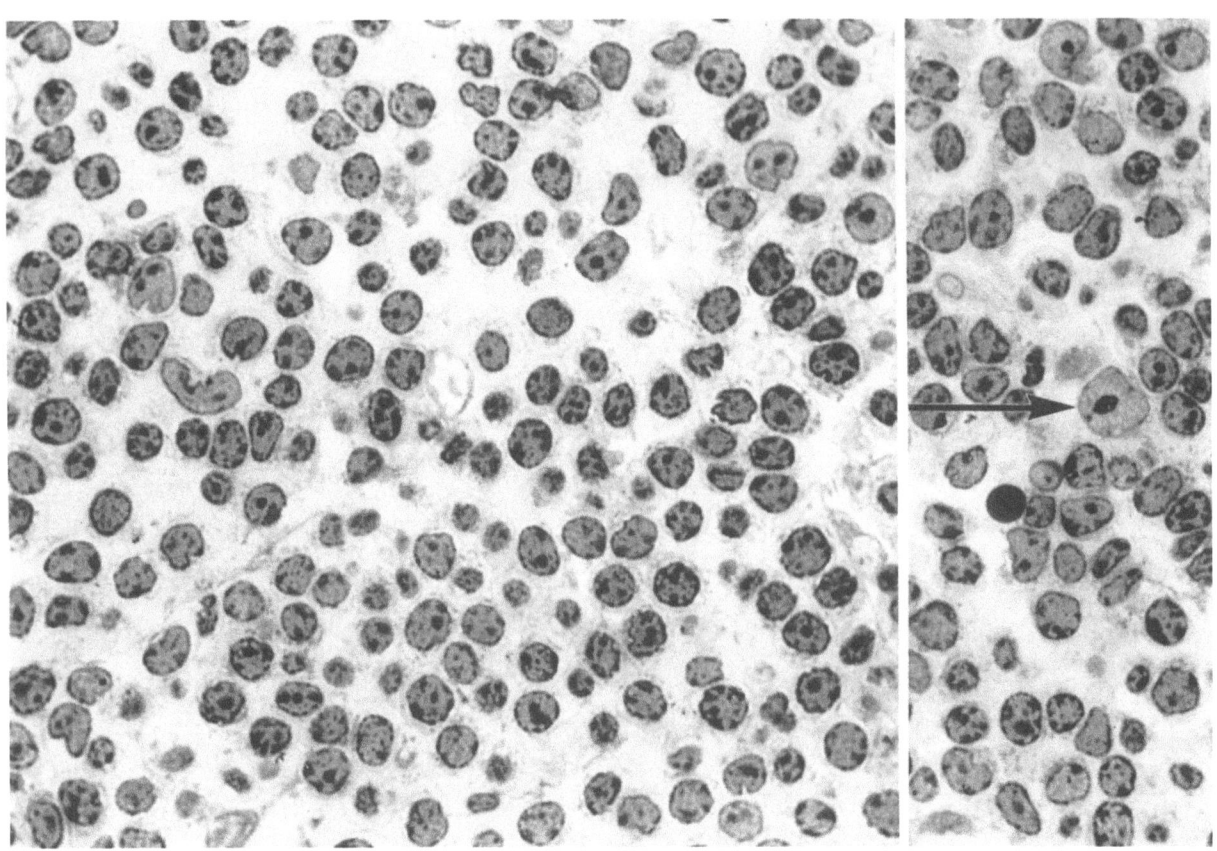

Fig. 119 Chronic aleukemic lymphocytic leukemia
Note clumping of lymphocytic chromatin, one lymphoblast (arrow) and occasional reticulum cells.

Case 9. Osmium tetroxide fixation. Semithin section. Methylene blue-Azure II. ×1,230

Fig. 120 Chronic lymphocytic leukemia
Cellular uniformity of closely contiguous lymphocytes with relatively scanty cytoplasm is evident. Irregular distensions of perinuclear cisternae may be observed. Mostly vesicular profiles of endoplasmic reticulum studded with only a few ribosomes are demonstrated in moderate numbers. The mitochondria are fairly large and rounded. One centriole is seen in the upper center.

Case 9. ×20,000

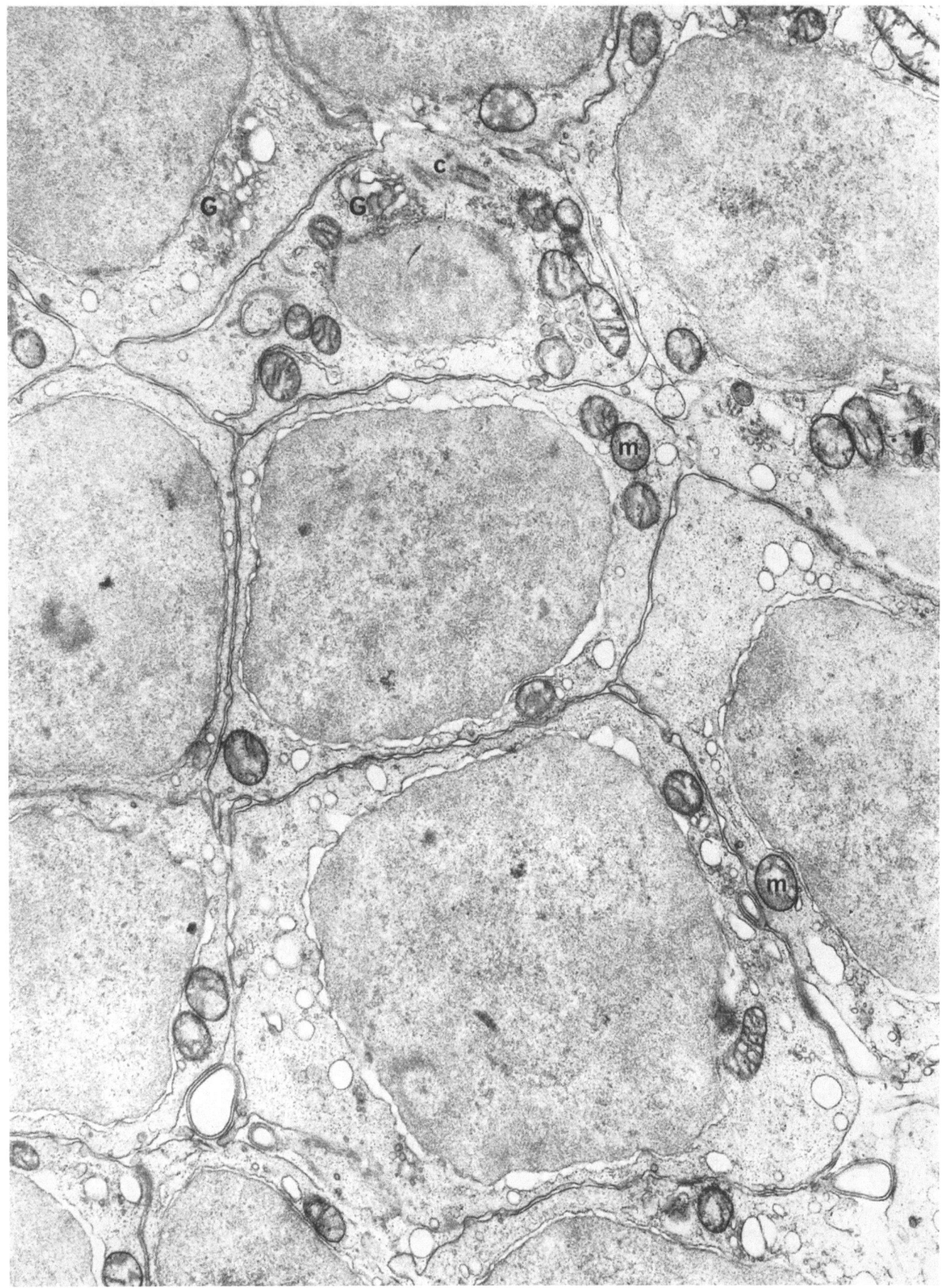

Fig. 121   Chronic lymphocytic leukemia
Relatively large nucleus with widened perinuclear cisterna. Vesicular profiles of rough endoplasmic reticulum associated with a few ribosomal granules and a Golgi zone with some lysosome-like bodies nearby are observed. Mitochondria are large and globular and one centriole is seen at the top. This particular cell bears a certain resemblance to cells found in WALDENSTRÖM'S macroglobulinemia.

Case 9. ×35,000

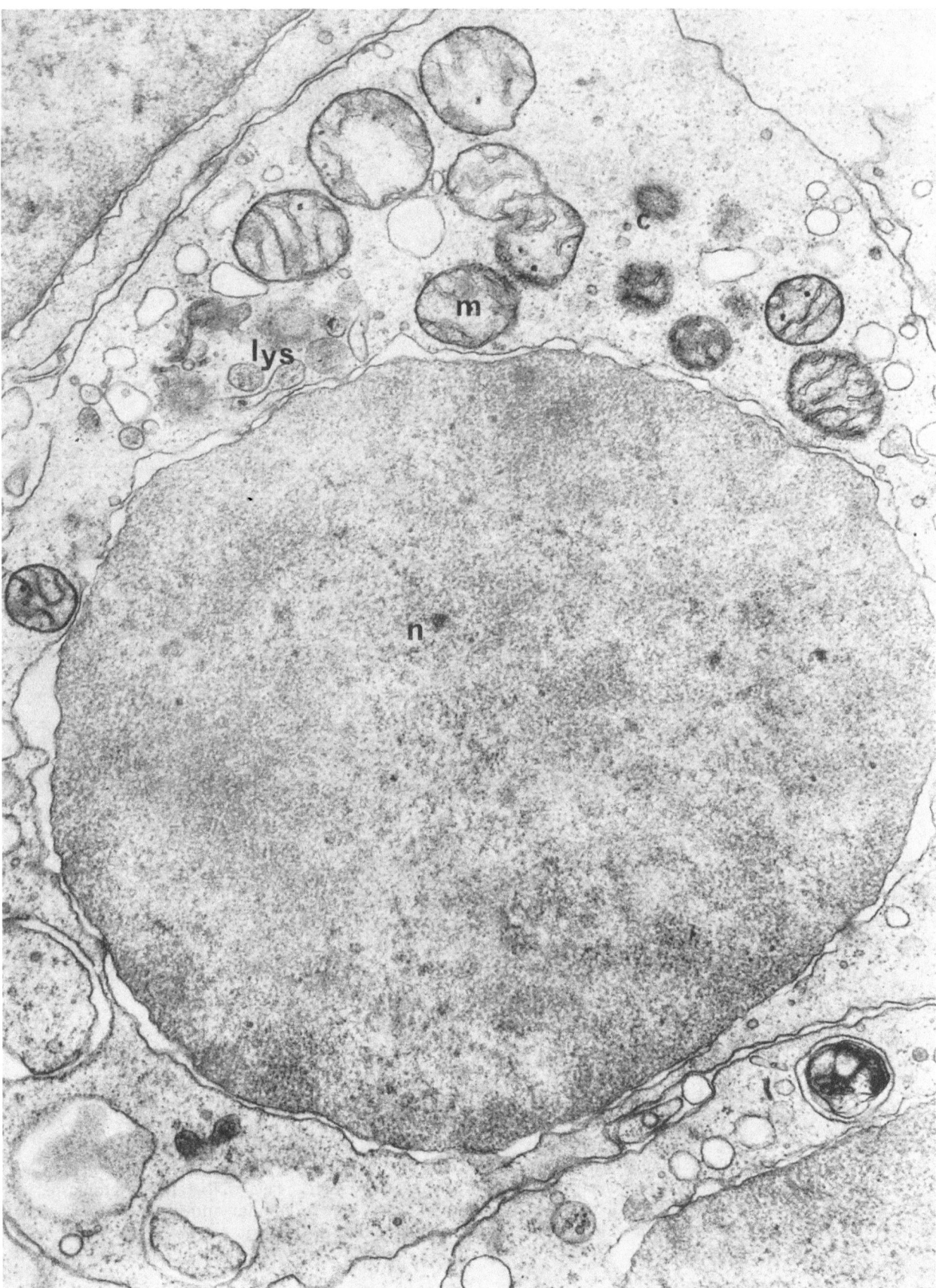

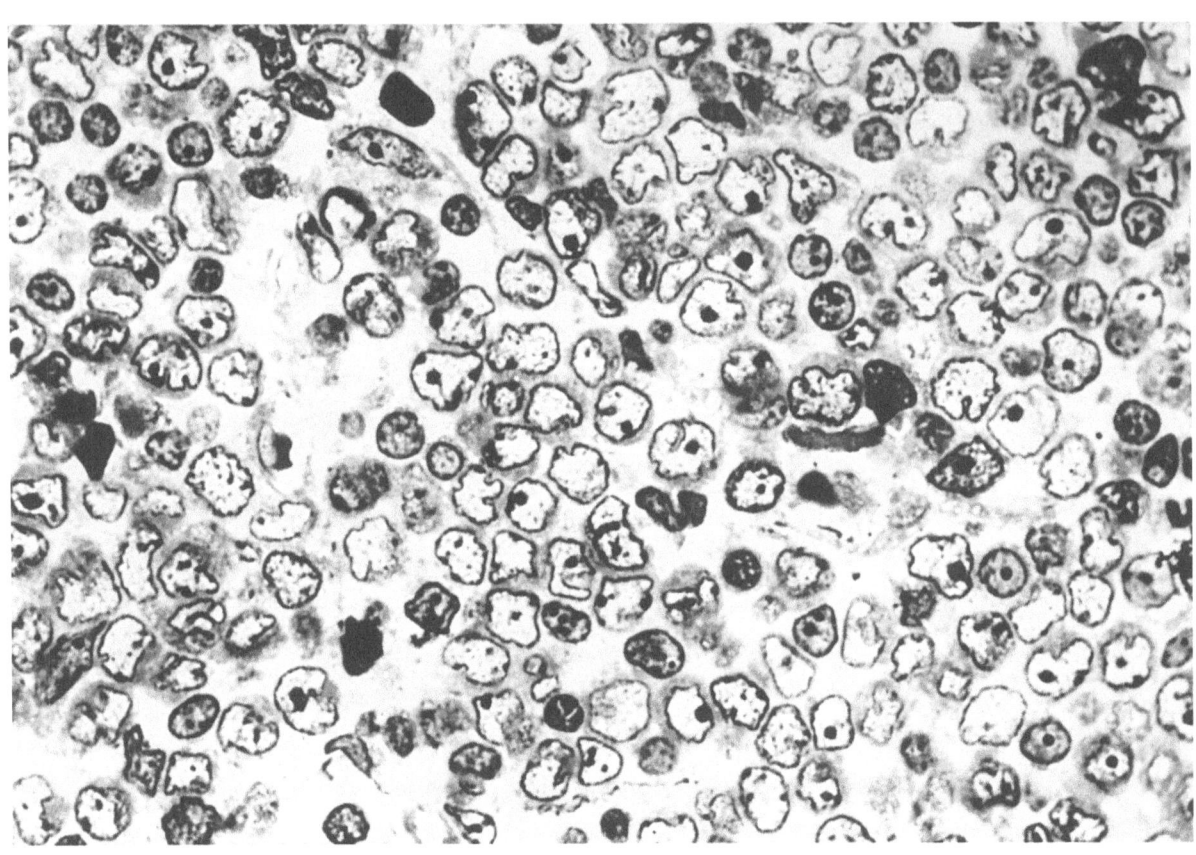

Fig. 122   Poorly differentiated leukemia, lymphoblastic type (child)
The nucleolus is of medium size and frequently placed in a peripheral position subjacent to the nuclear membrane.

Case 31. Leukemic blood picture. Osmium tetroxide fixation. Semithin section. Methylene blue-Azure II. ×1,230

Fig. 123   Poorly differentiated leukemia, lymphoblastic type (child)
The nuclei have an irregular shape with deep indentations. There is margination of chromatin, and a large electron-opaque nucleolus is evident. The well-stained cytoplasm is crowded with small vesicles, ribosomes and polysomes. Tubular and somewhat vesicular profiles of rough endoplasmic reticulum and small mitochondria are also seen.

Case 31. ×17,500

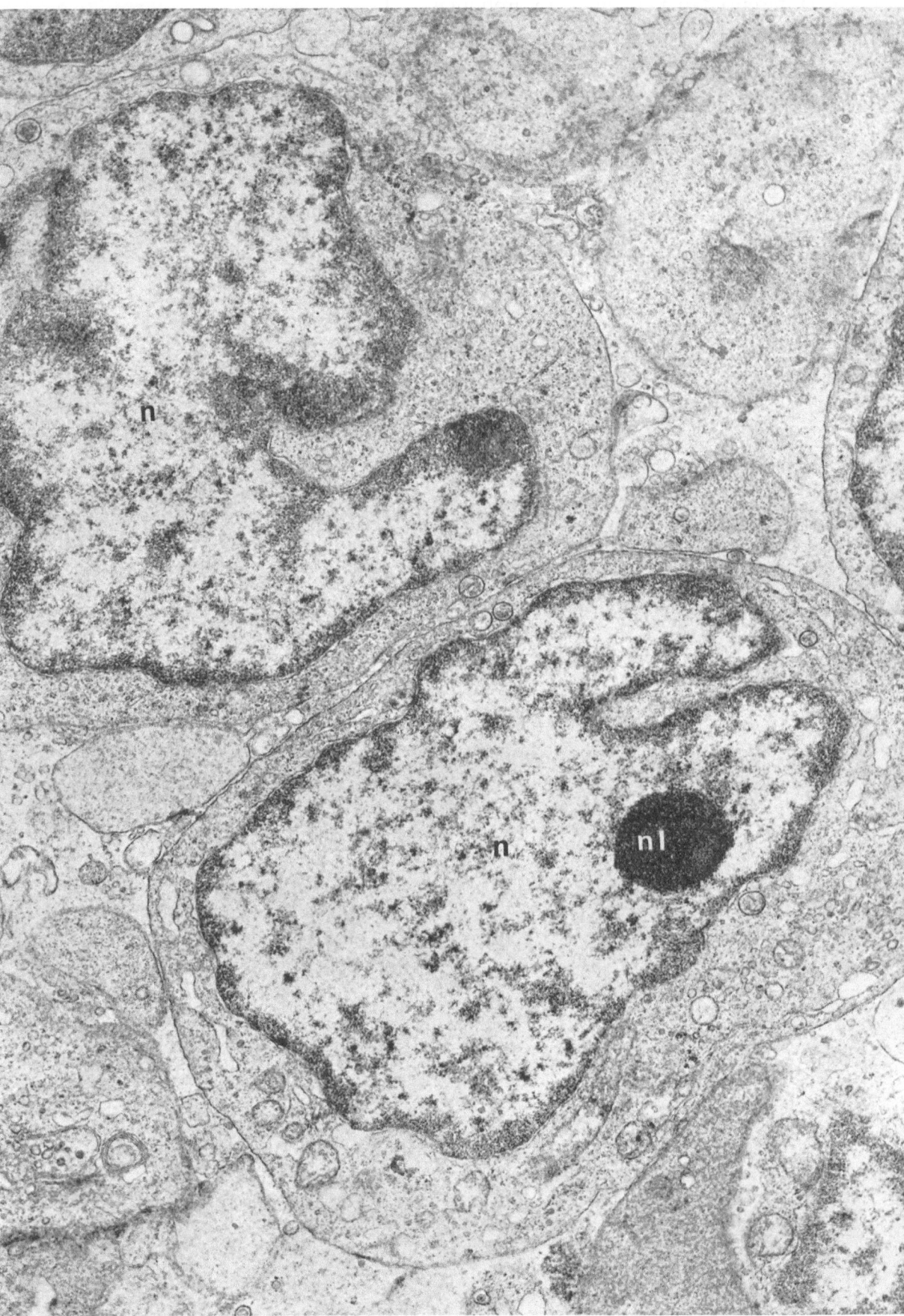

Fig. 124   Poorly differentiated leukemia, lymphoblastic type (child)

Morphologic features are similar to those observed in Fig. 123. In addition prominent glycogen deposits (bottom, *g* and arrow), distinct rod-shaped mitochondria and a few multivesicular bodies are present.

Case 31. ×17,500

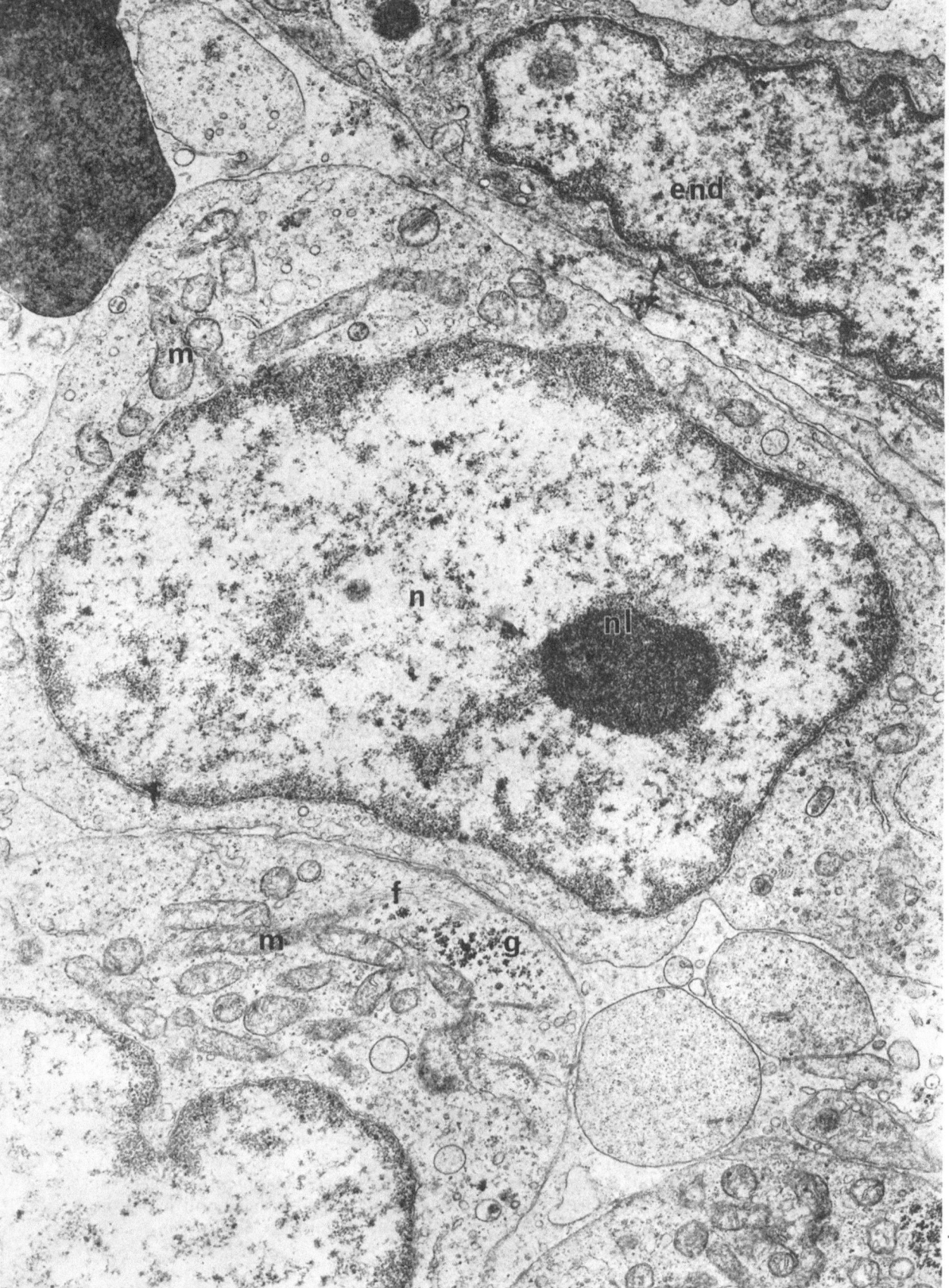

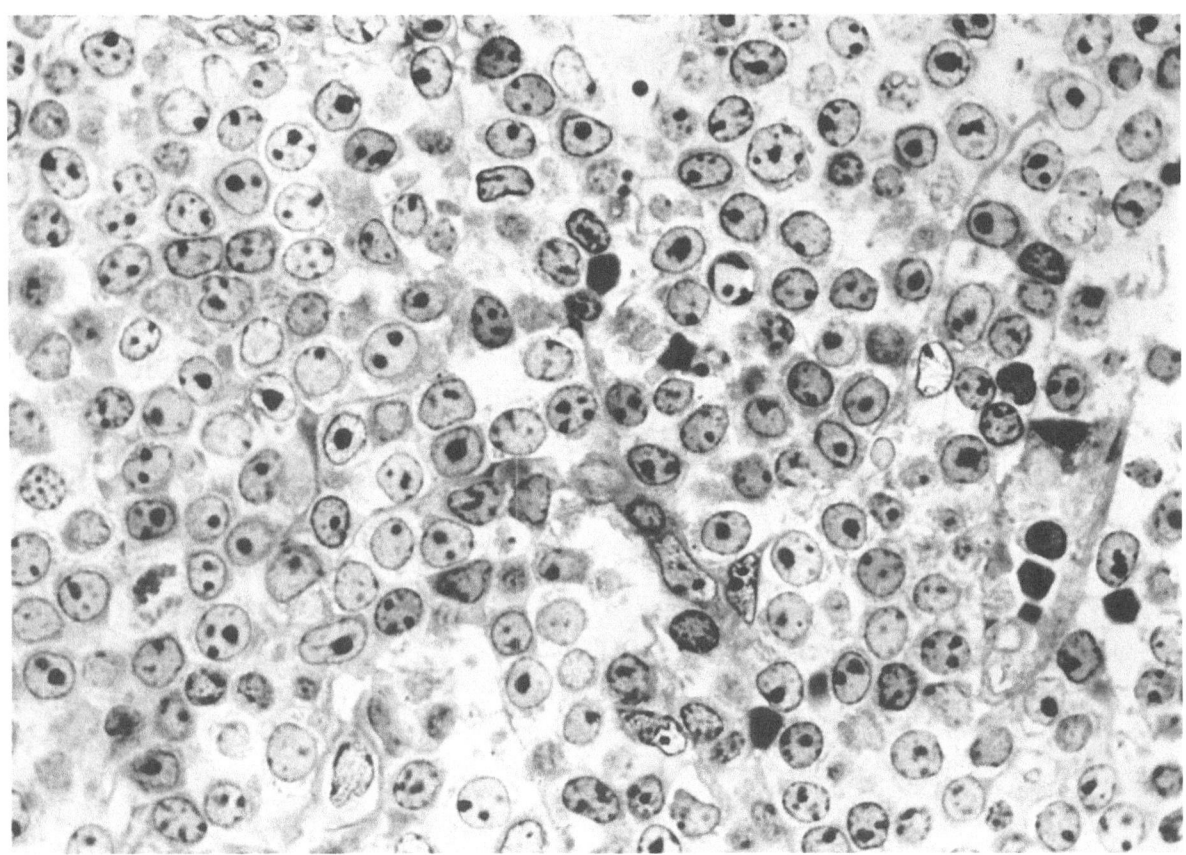

Fig. 125   Poorly differentiated leukemia, granulocytic (paramyeloblastic) type
Note rounded nuclei with prominent nucleoli.

Case 60. Leukemic blood picture. Osmium tetroxide fixation. Semithin section. Methylene blue-Azure II. ×1,230

Fig. 126a and b   Poorly differentiated leukemia, granulocytic (paramyeloblastic) type
The rounded nuclei have an elaborate nuclear membrane and strikingly prominent nucleoli. There are numerous nuclear pores (arrows). The cytoplasm contains few narrow tubules of rough endoplasmic reticulum, scattered free ribosomes, occasional polysomes and a few multivesicular bodies. There is possibly one transversely sectioned centriole (b, near right nuclear margin). Note rounded mitochondria and the absence of glycogen.

Case 60. ×17,500

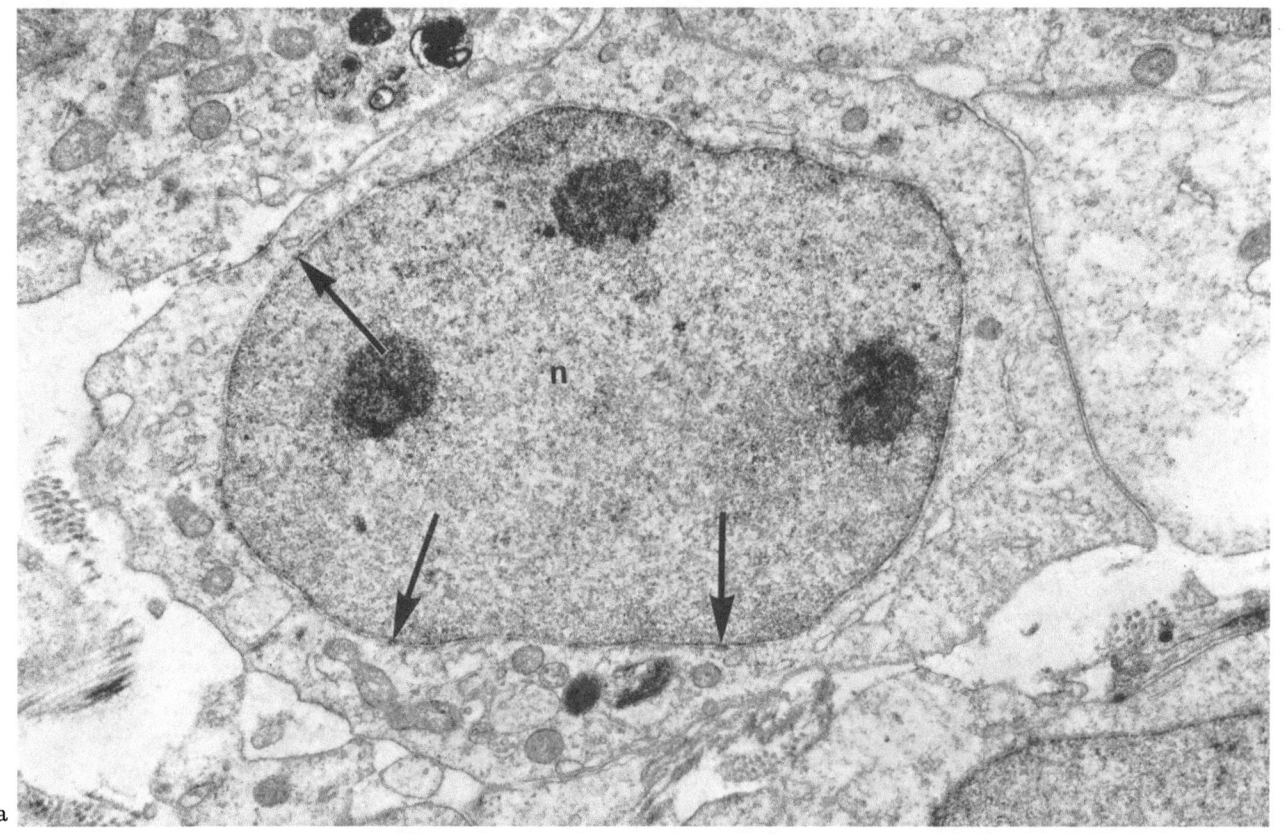

a

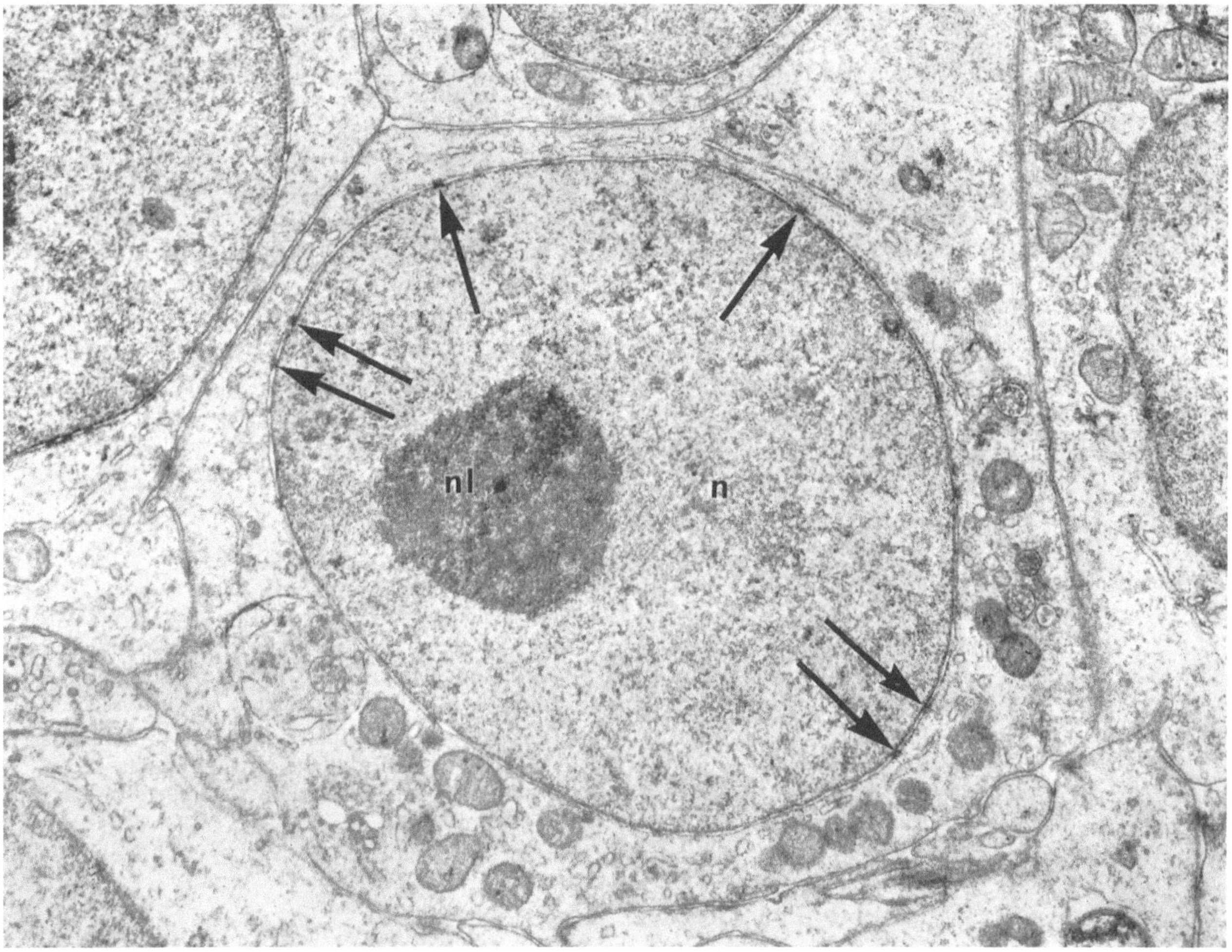

n

b

nl   n

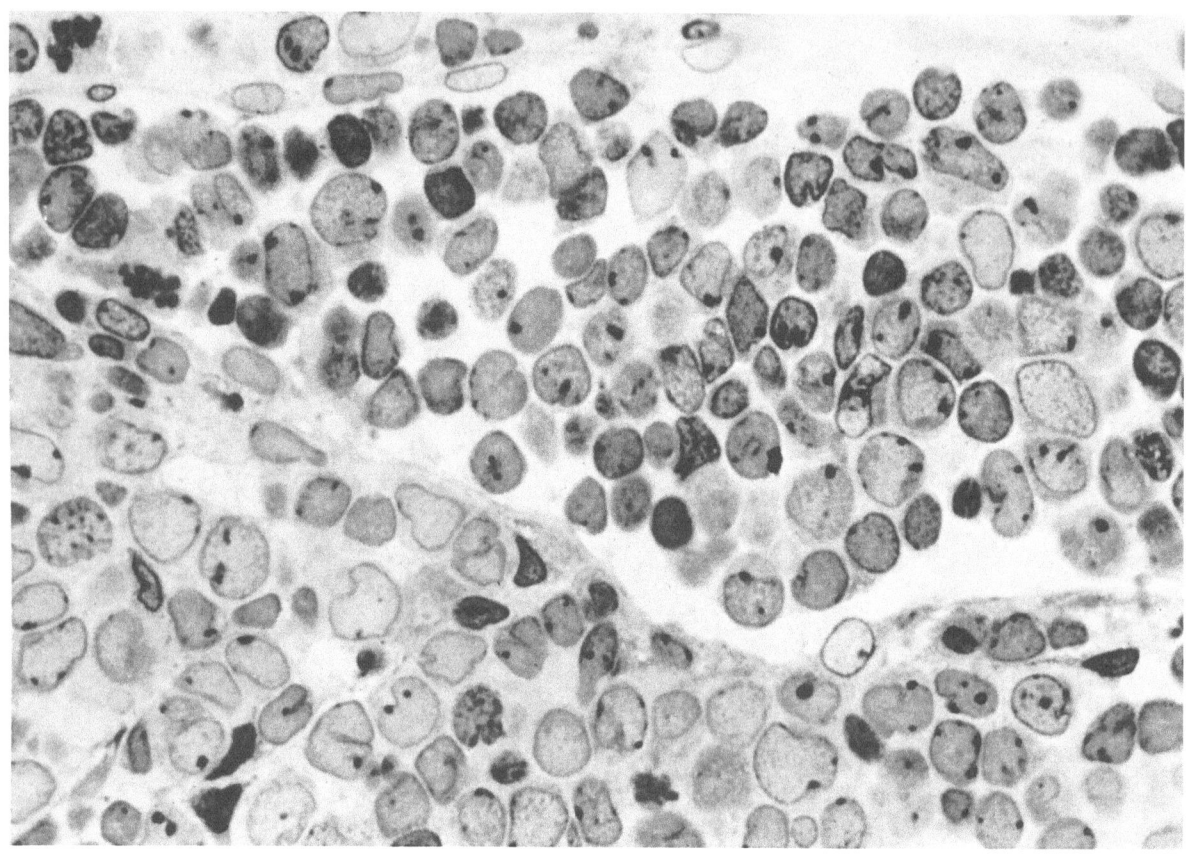

Fig. 127   Poorly differentiated leukemia, granulocytic (paramyeloblastic) type
Granulocytic blasts predominate. There are some monocytes, and numerous leukemic cells are seen in the lumen of a vein.

Case 54. Leukemic blood picture. Osmium tetroxide fixation. Semithin section. Methylene blue-Azure II. ×1,230

Fig. 128a and b   Poorly differentiated leukemia, granulocytic type
a) There is a deeply indented nucleus with two medium-sized nucleoli of intermediate density. Abundant vesicular and occasionally tubular rough endoplasmic reticulum is apparent in the scanty cytoplasm. Polysomes are encountered in great numbers and free ribosomes are present in moderate amounts. Occasional mitochondria can be discerned.
b) Cytoplasmic detail of leukemic cell. Note large lipid droplet, moderately prominent Golgi apparatus with a centriole to the left and swollen mitochondria. There are a considerable amount of ribosomes and polysomes as well as moderate numbers of small vesicles.

Case 54. a) ×17,500   b) ×28,000

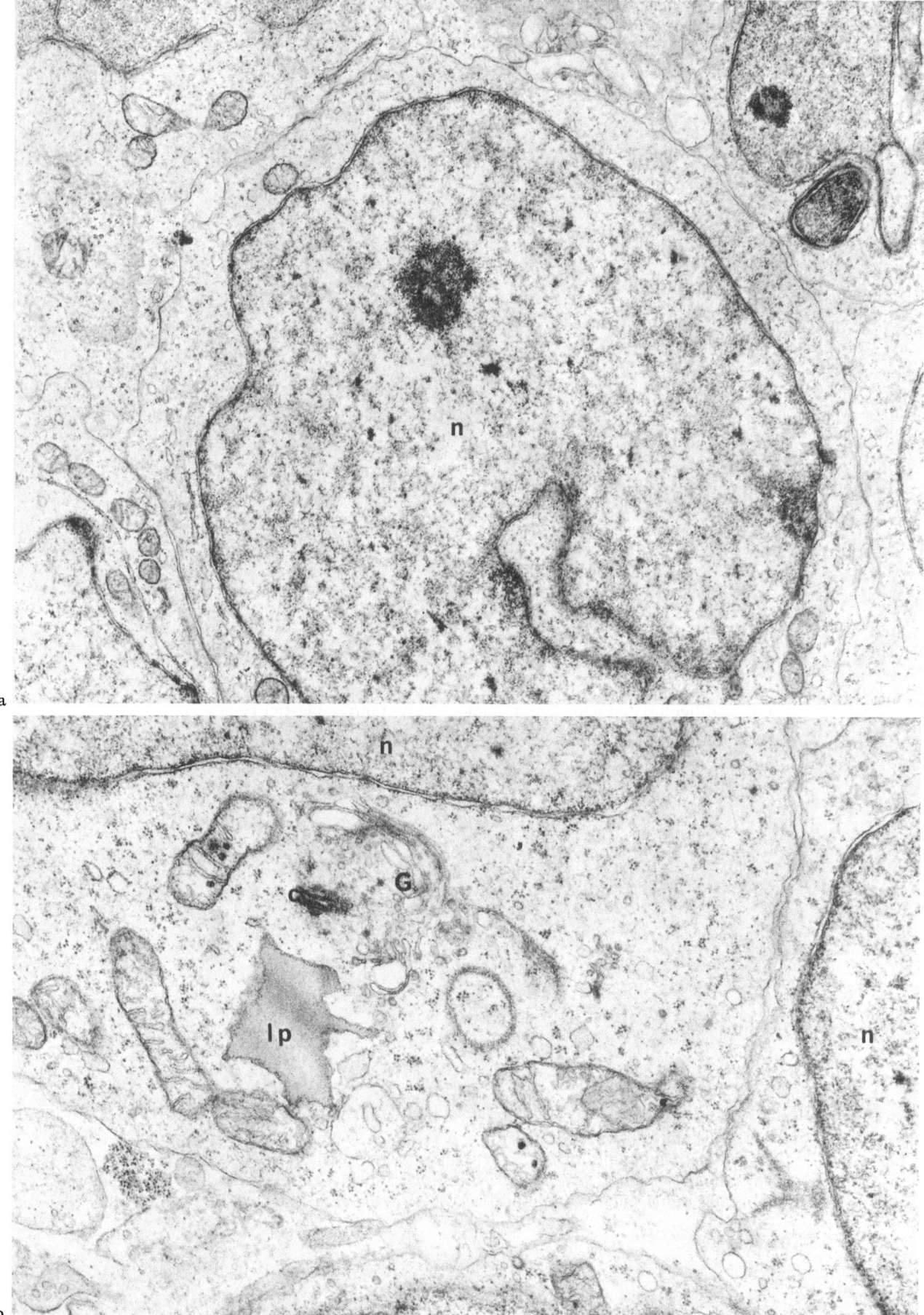

a

b

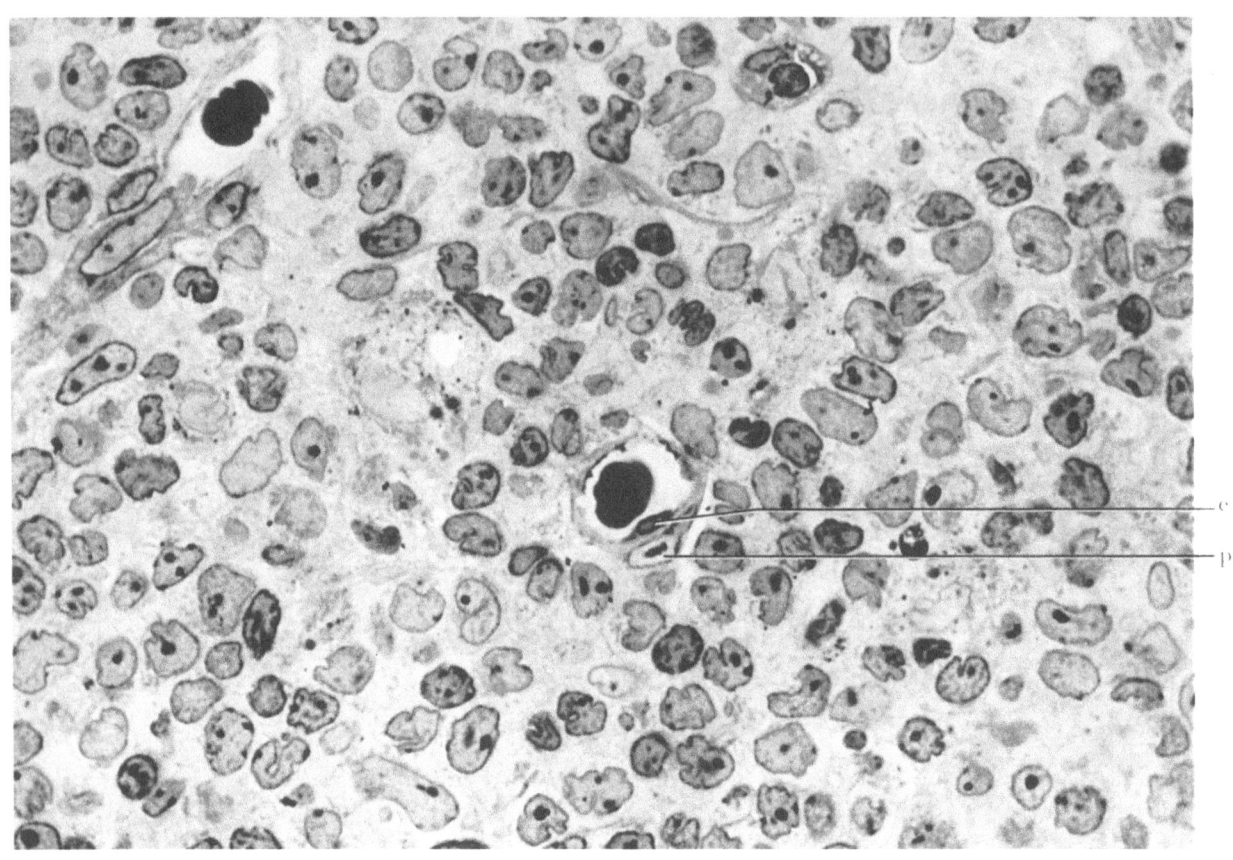

Fig. 129   Poorly differentiated leukemia, granulocytic (paramyeloblastic) type
                  Note arteriole (center) with endothelial cell (e) and pericyte (p).

                  Case 57. Osmium tetroxide fixation. Semithin section. Methylene blue-Azure II. ×1,230

Fig. 130a and b   Poorly differentiated leukemia, granulocytic (paramyeloblastic) type
                  Two leukemic cells with deep narrow indentations. There is a scant amount of rough endo-
                  plasmic reticulum and a small Golgi region (b, bottom). Many ribosomes and polysomes,
                  swollen mitochondria, two probably specific granules with a distinct bounding membrane
                  (b, center right) and a residual body (a, bottom) are seen.

                  Case 57. ×17,500

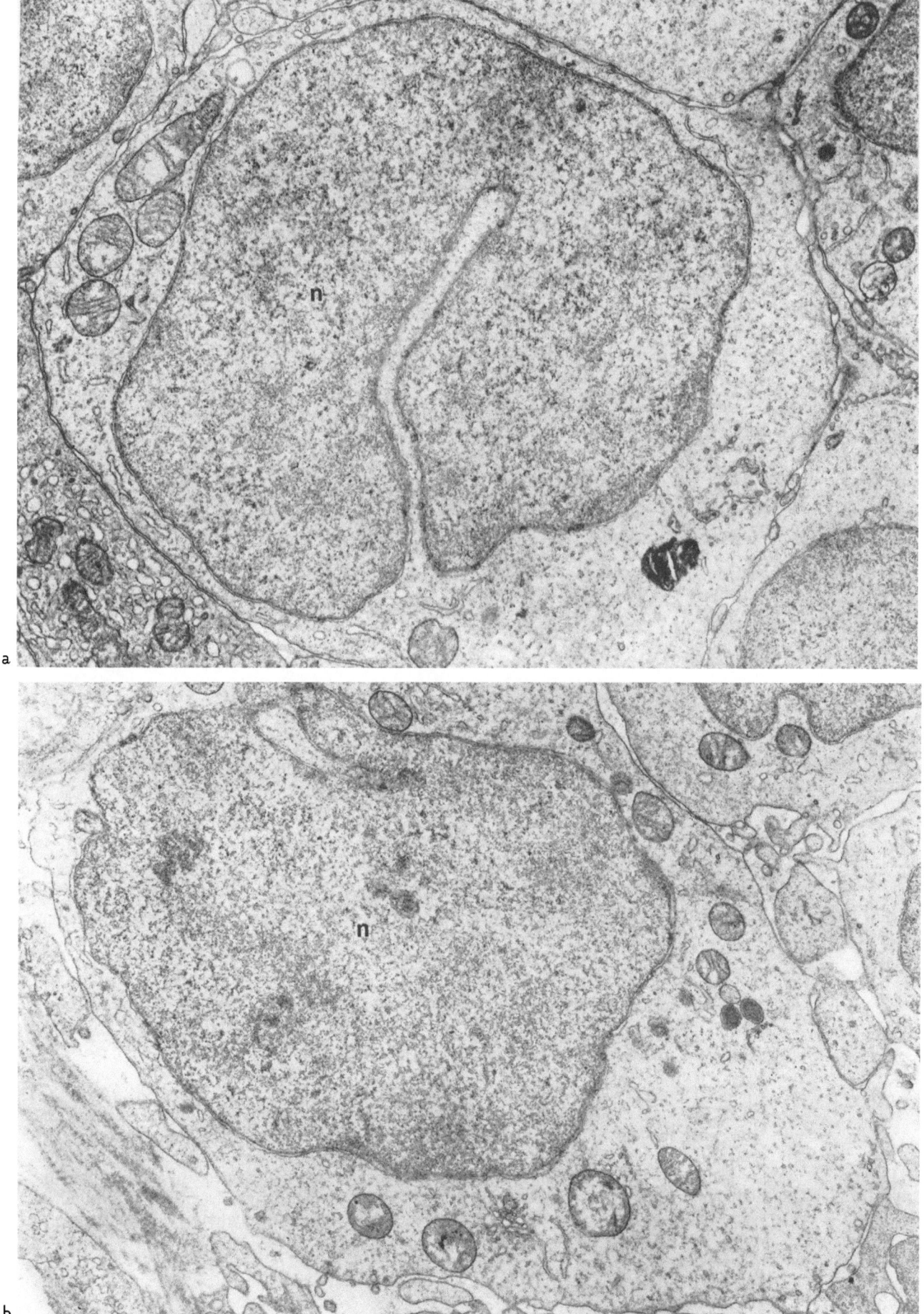

a

b

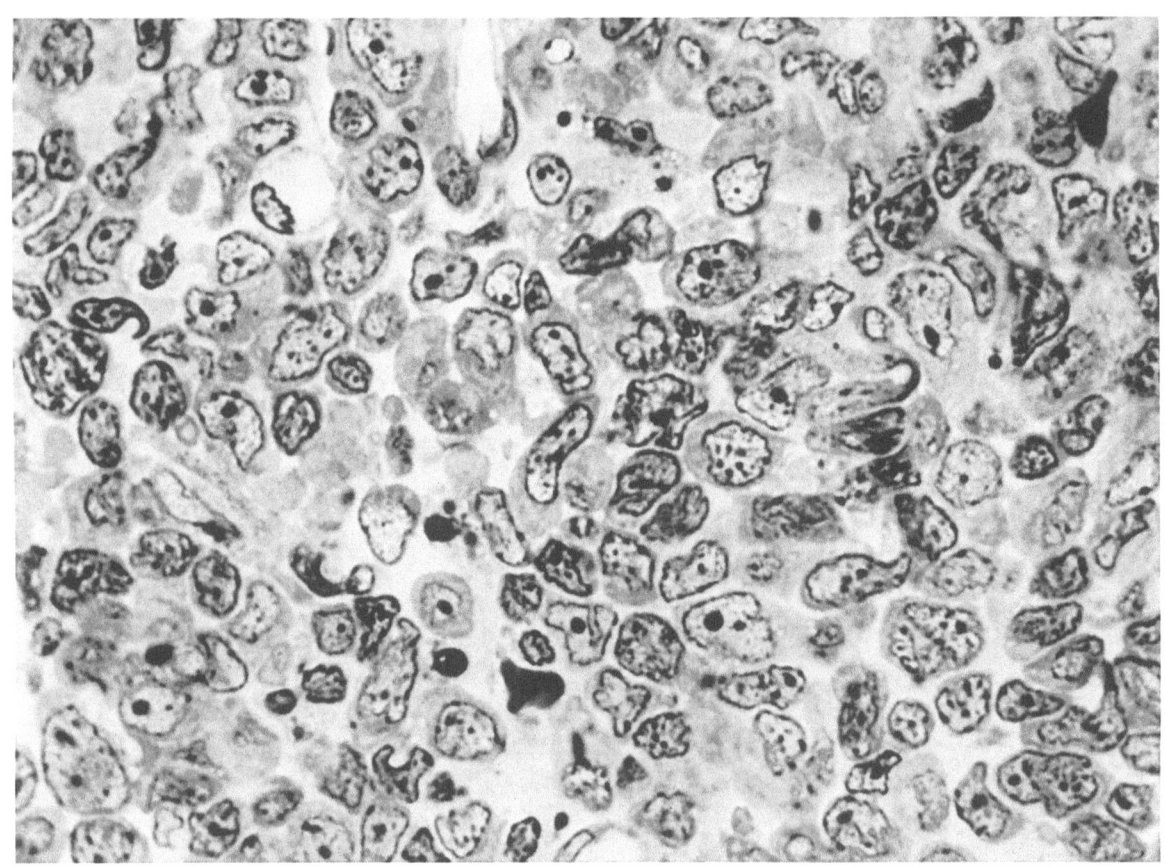

Fig. 131   Monocytic leukemia

There is prominence of pleomorphic monocytes with atypical, often bizarre, folded and indented nuclei. Nucleoli are of small or medium size.

Case 56. Osmium tetroxide fixation. Semithin section. Methylene blue-Azure II. ×1,230

Fig. 132   Monocytic leukemia

These leukemic cells lack the characteristic features of blood monocytes. They have an indented nucleus, a large nucleolus of somewhat reticulated appearance and discrete margination of the chromatin. The cytoplasm shows a considerable accumulation of tubulovesicular rough endoplasmic reticulum. Ribosomes and polysomes are intermediate in amount and an inconspicuous Golgi field with a centriole nearby is seen. Few mitochondria and small accumulations of glycogen deposits are scattered throughout the cytoplasm.

Case 56. ×17,500

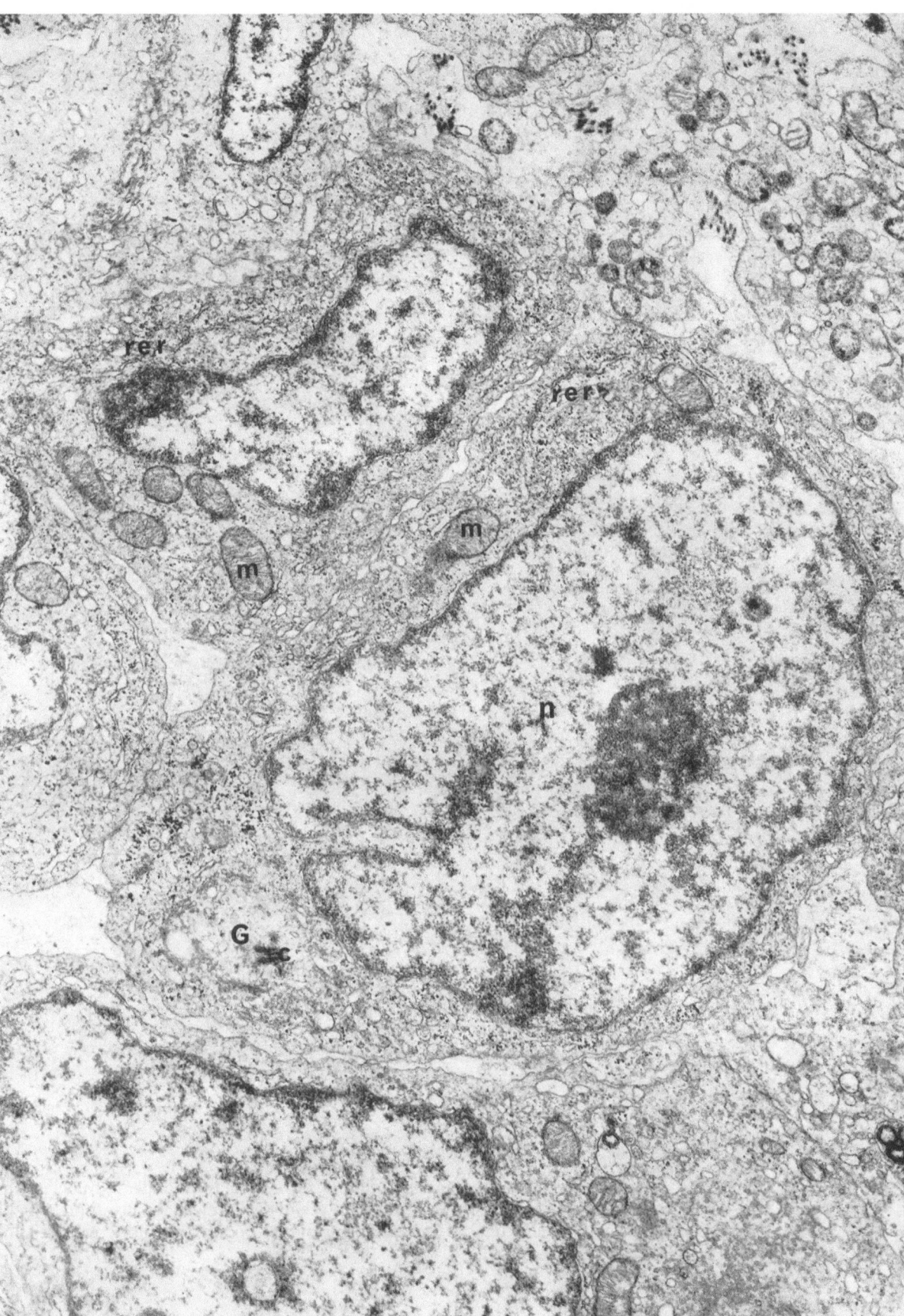

Fig. 133   Monocytic leukemia
Same case as illustrated in previous picture. Note bizarre folded and indented nucleus in the center. Similarity of cytoplasm to that shown in previous picture is readily appreciated. Note, however, a few myelin figures (presumably residual bodies) which are not seen in Fig. 132.

Case 56. ×17,500

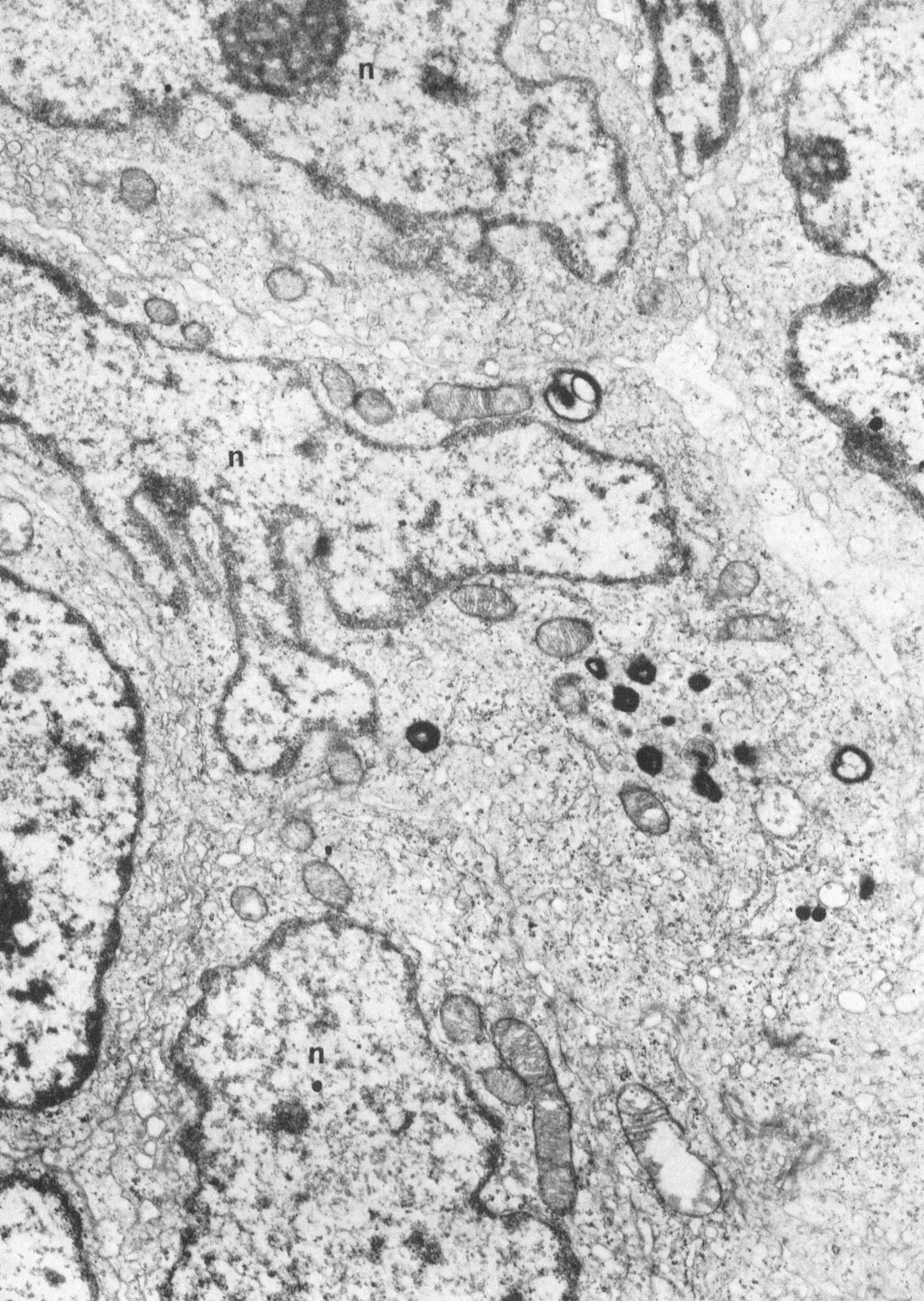

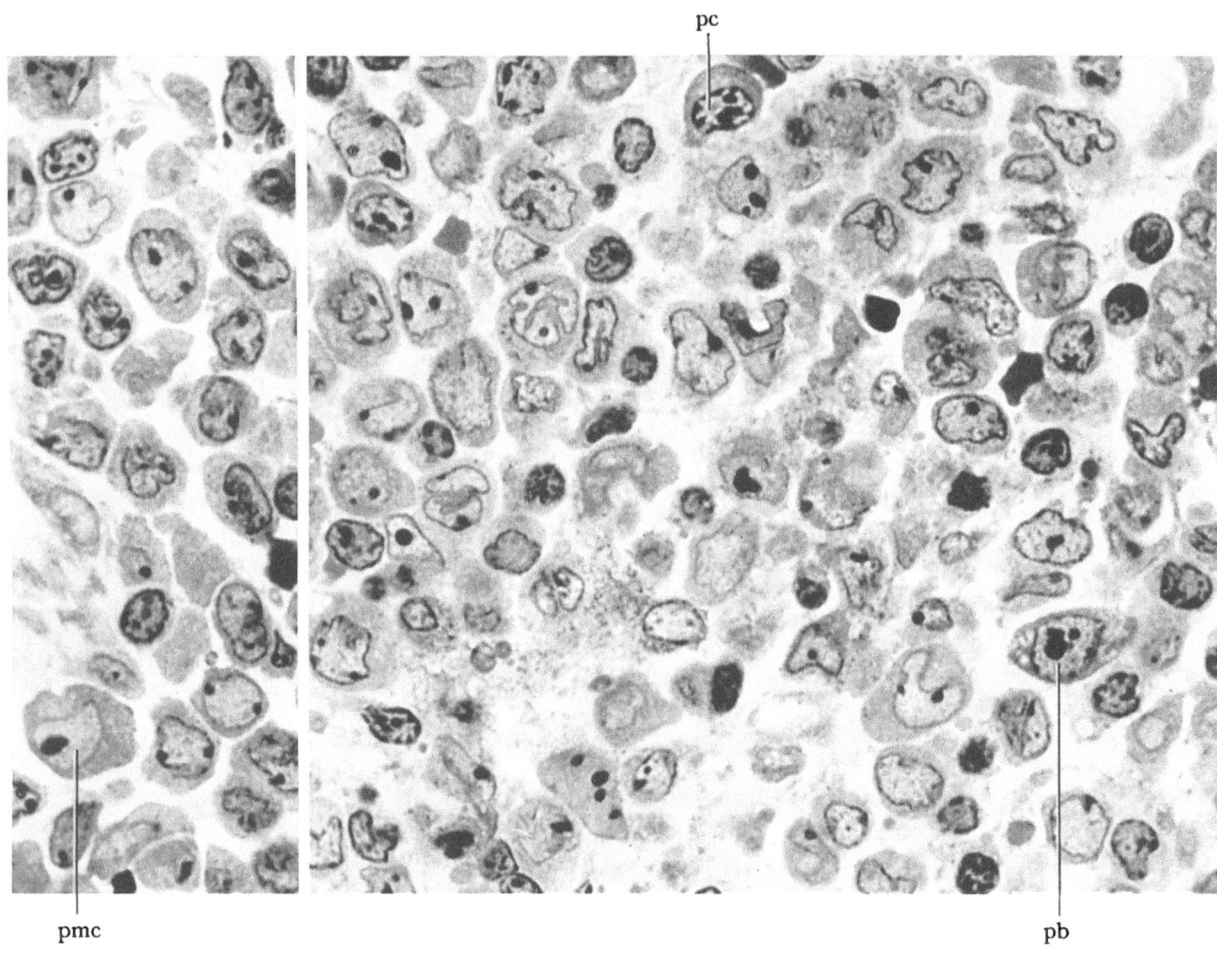

Fig. 134   Acute monocytic leukemia

Numerous pleomorphic monocytes with considerable nuclear atypia. Note wrinkled or deeply indented, sometimes kidney-shaped nuclei. Nucleoli are small or of medium size. Promyelocyte (*pmc*). Plasmablast (*pb*). Plasma cells (*pc*).

Case 56. Osmium tetroxide fixation. Semithin section. Methylene blue-Azure II. ×1,230

Fig. 135   Monocytic leukemia

Leukemic cell reminiscent of blood monocyte containing a lobulate nucleus and a few specific granules. There are moderate development of mitochondria, one centriole and occasional pseudopodal extensions which contrast with smooth cellular outlines of circulating blood monocytes.

Case 55. ×17,500

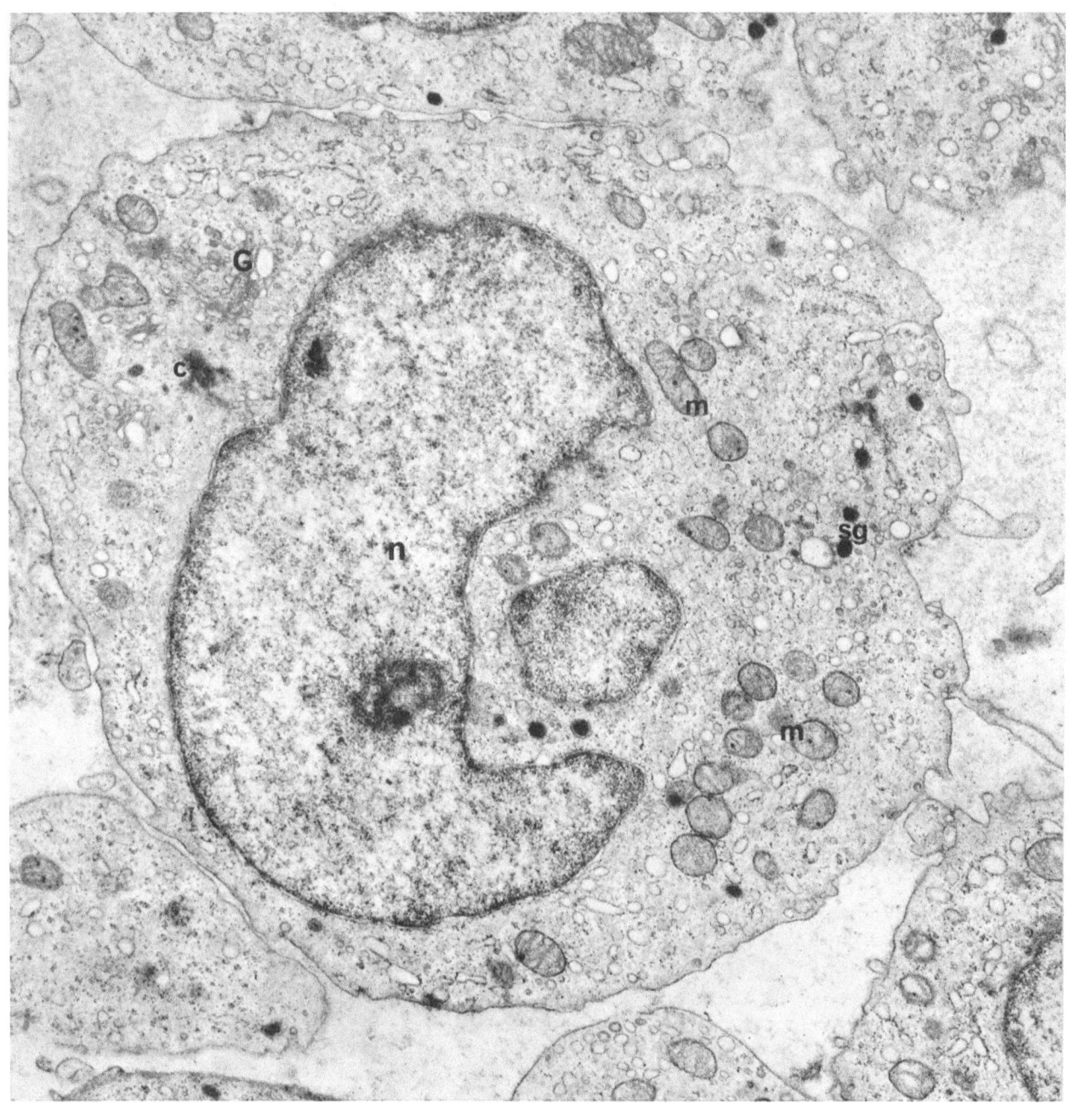

Fig. 136   Monocytic leukemia

Myelocytes with azurophilic and specific (neutrophilic) cytoplasmic granules. The azurophilic granules are delimited by a distinct membrane. The specific granules comprise electron-dense material that in part overshadows the bounding membrane. There is an abundance of polysomes and ribosomes but rough endoplasmic reticulum is inconspicuous. Mitochondria are globular.

Case 55. ×17,500

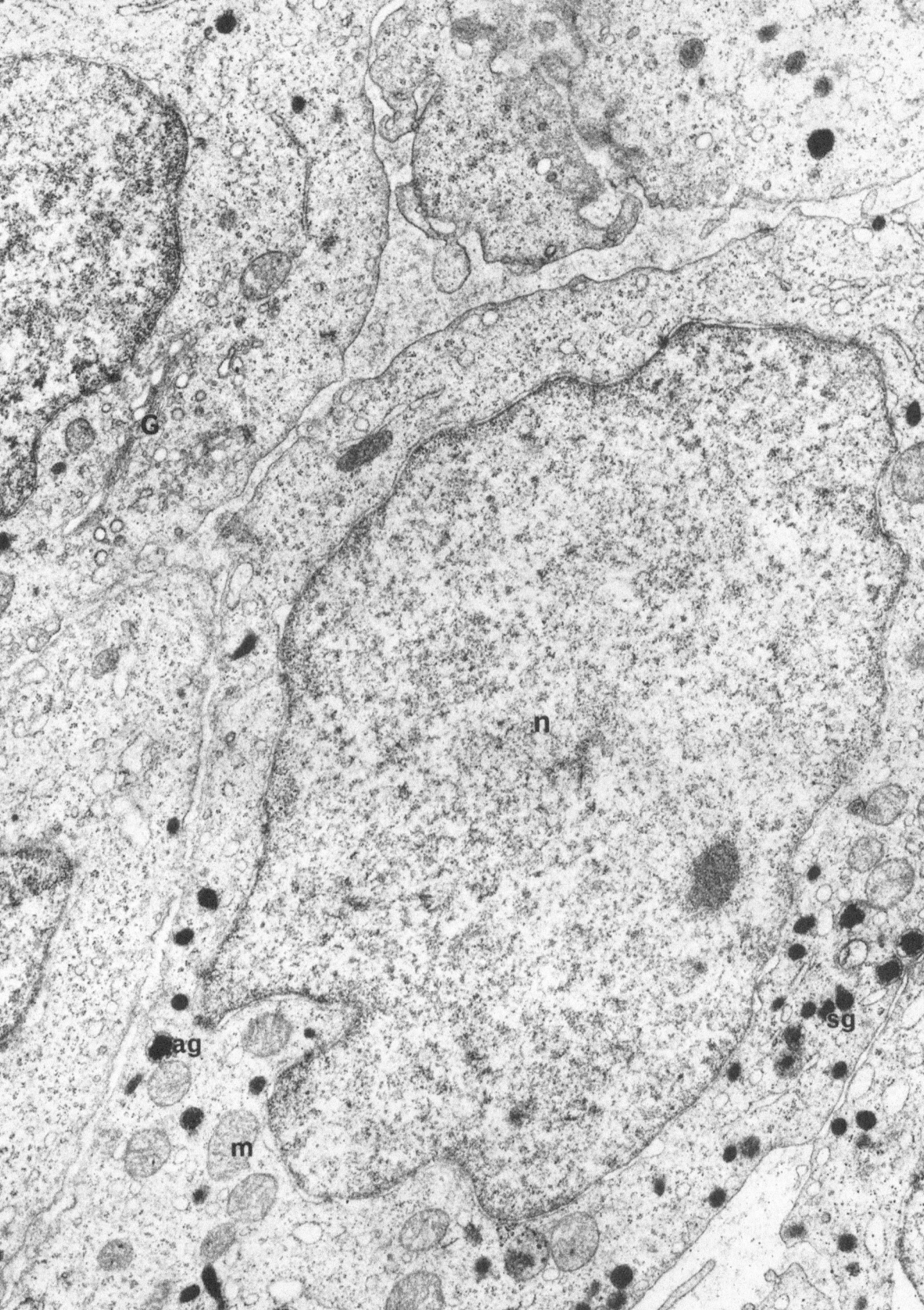

Fig. 137   Promyelocyte in monocytic leukemia
              Ovoid nucleus with large weakly staining nucleolus. The large granules (*ag*) with a dense
              core, sometimes surrounded by an empty halo and provided with a well-defined limiting
              membrane, correspond to the azurophilic granules of promyelocytes. Some small dark-
              staining granules can also be discerned. Rough endoplasmic reticulum is well in evidence.

              Case 55. ×17,500

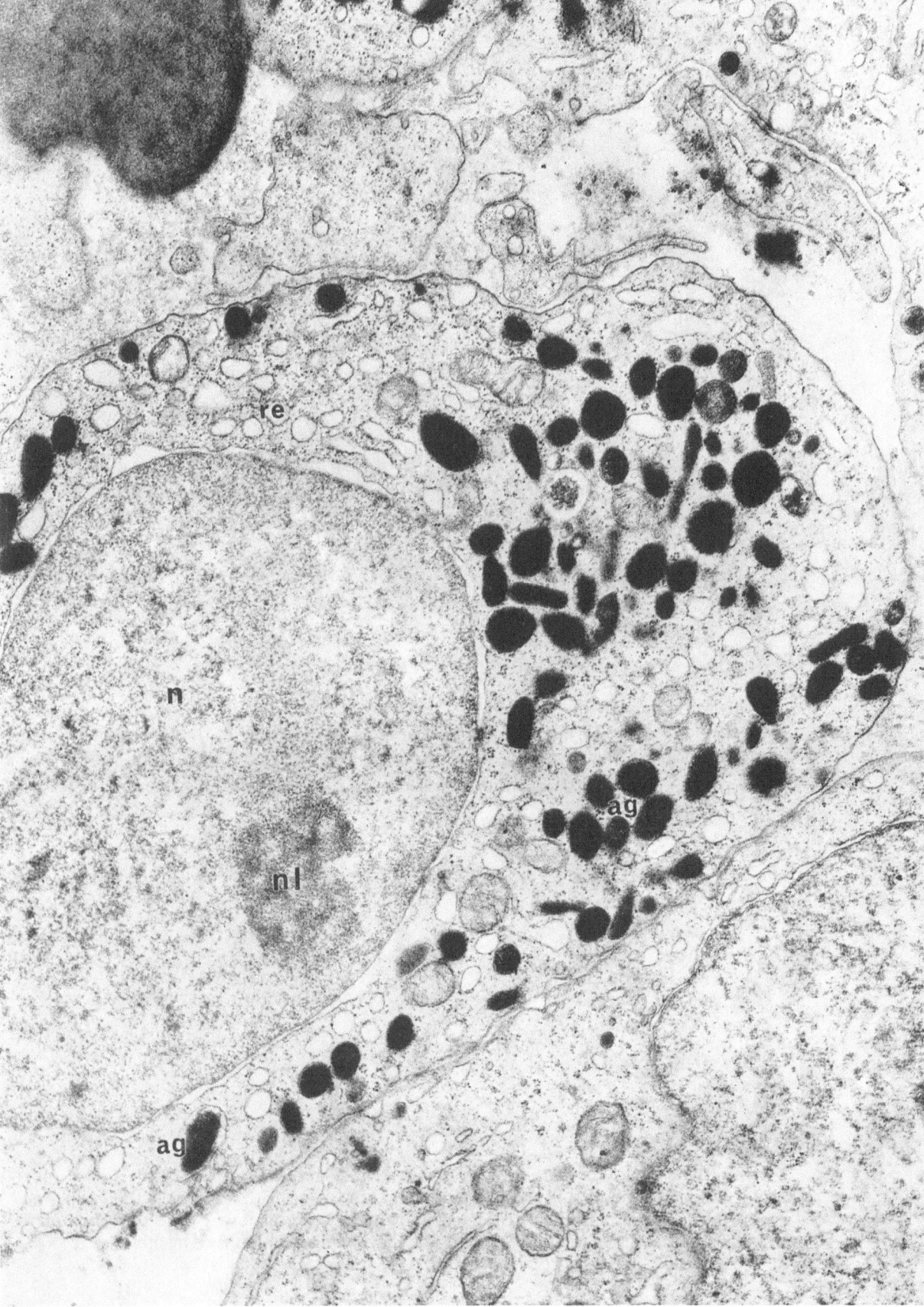

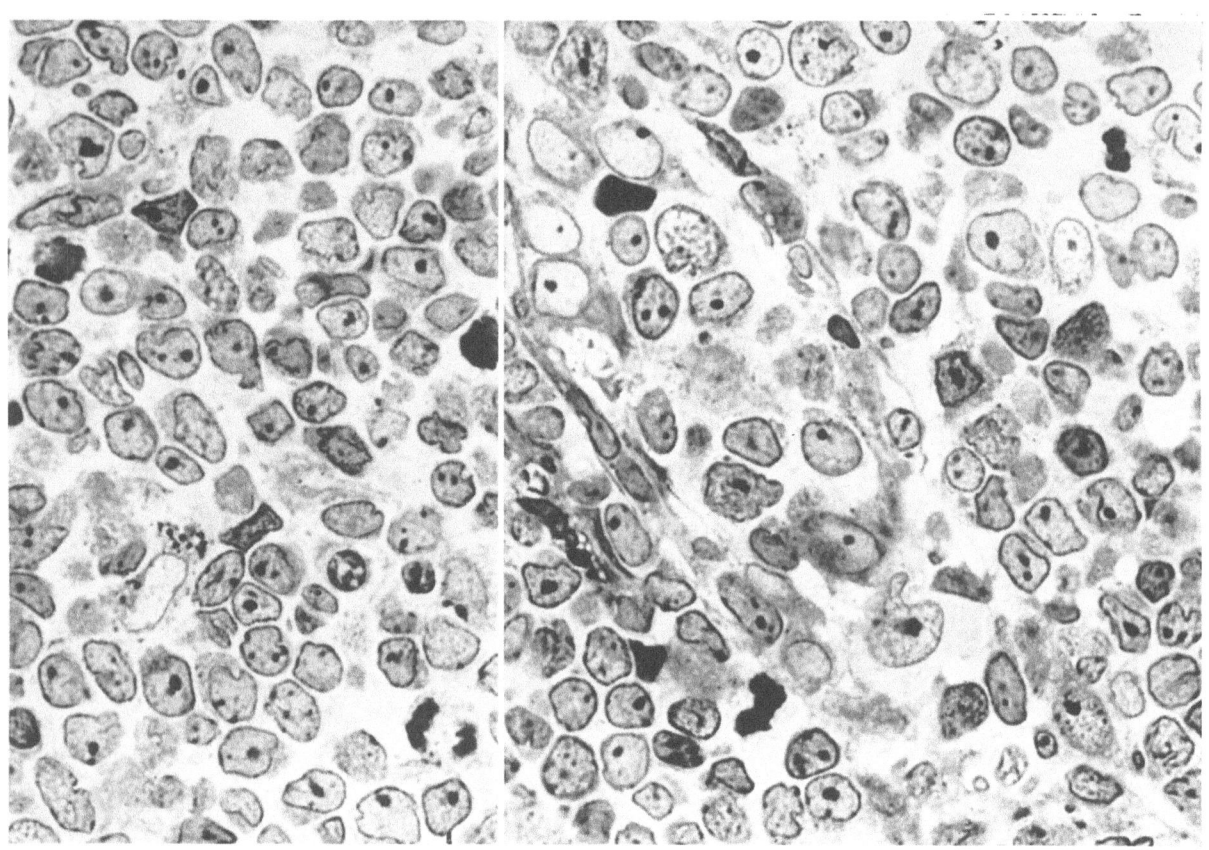

Fig. 138   Acute erythremia of child
A diagnosis cannot be made by conventional histologic technics, but must be based on histochemical reactions.

Case 59. Osmium tetroxide fixation. Semithin section. Methylene blue-Azure II. ×1,230

Fig. 139   Acute erythremia of child
Pleomorphic nuclei contain nucleoli of intermediate size. There are a few strands of rough endoplasmic reticulum and a prominent Golgi area along with a centriole. Polysomes and ribosomes are a prominent feature and mitochondria are globular. Note occasional dark-staining granules, for the most part displaying well-defined limiting membranes.

Case 59. ×17,500

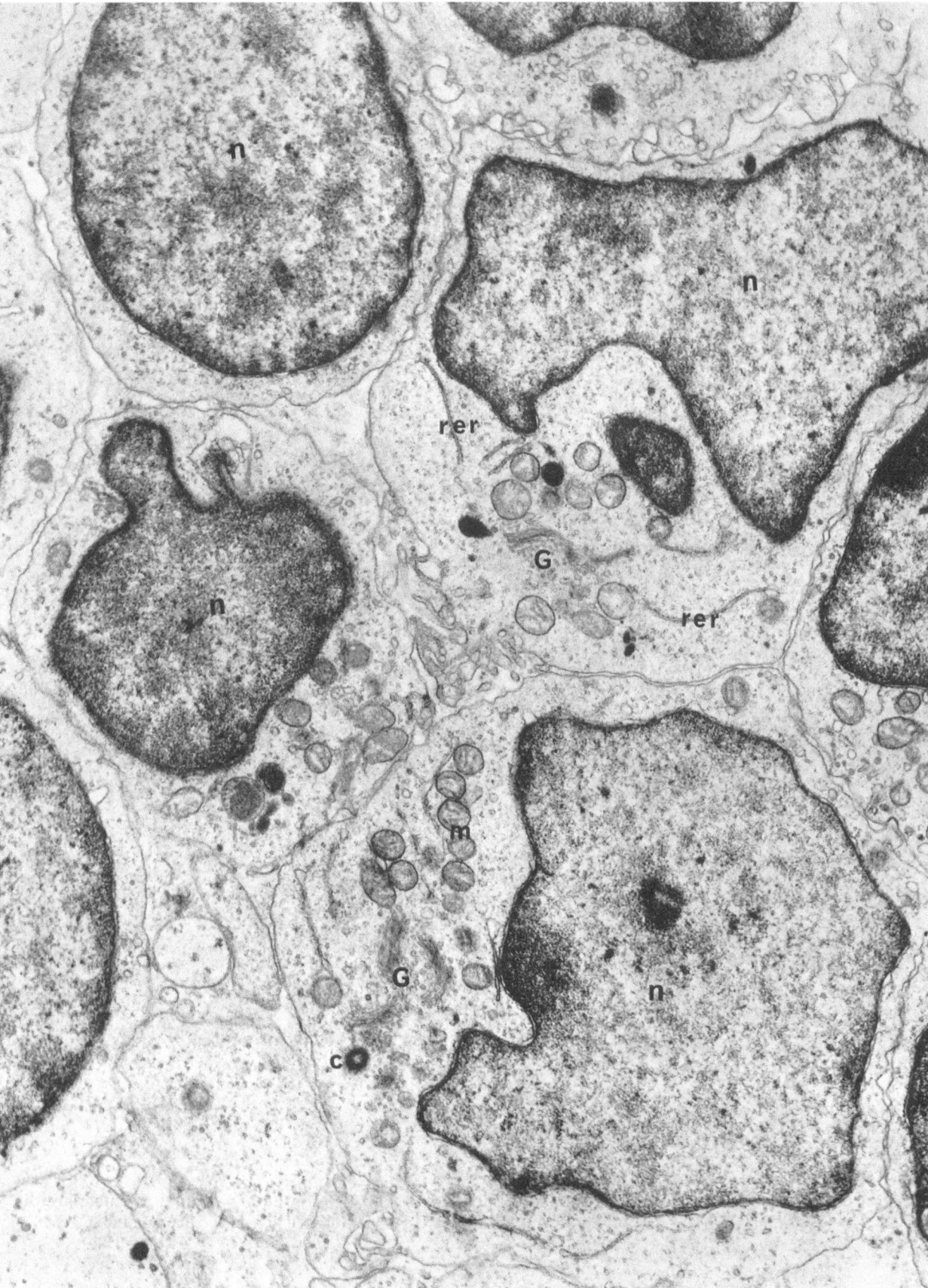

Fig. 140   Acute erythremia
The cell shows abundance of polysomes, a few dense granules without a visible limiting membrane and one lipid droplet. The vesicular Golgi region is inconspicuous.

Case 59. ×20,000

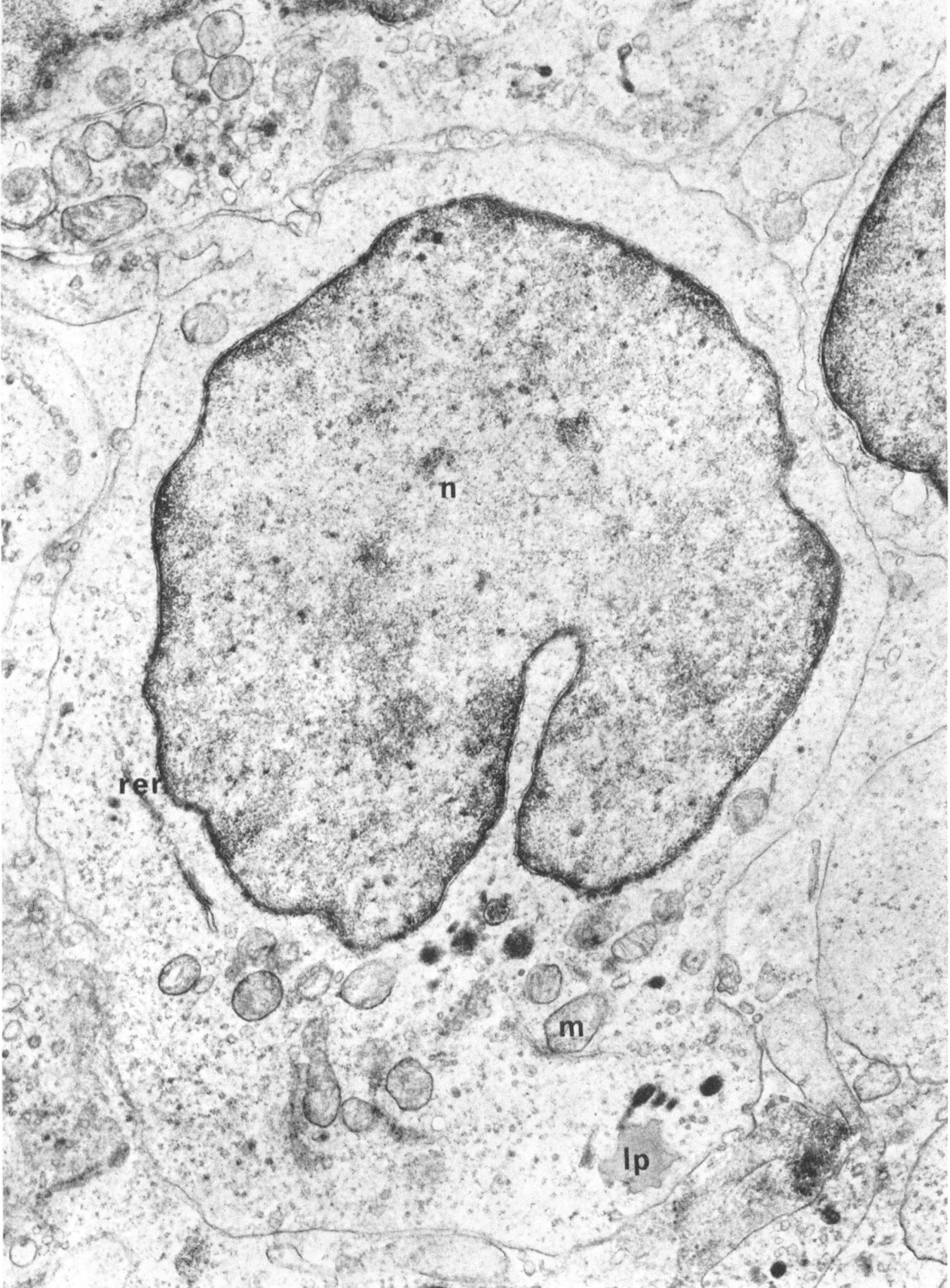

Letterer-Siwe's Reticulosis of Infants

Eosinophilic Granulomatosis of the Adult
(Lipoid Granulomatosis of Hand-Schüller-
Christian without Lipoid Storage)

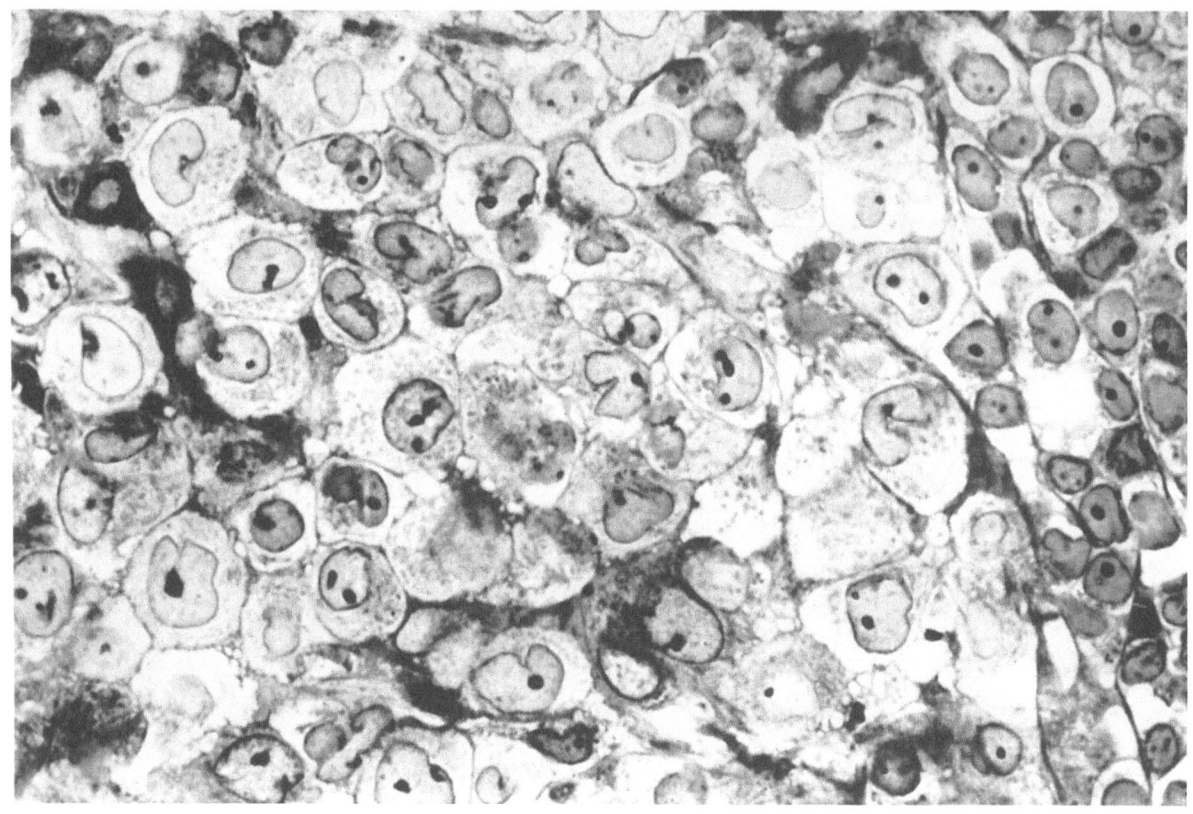

Fig. 141   Histiocytosis X of child (Letterer-Siwe's disease)

Note indented, frequently kidney-shaped nuclei with very sharp nuclear outlines. The abundant cytoplasm contains a massive accumulation of lysosomes.

Case 17. Osmium tetroxide fixation. Semithin section. Methylene blue-Azure II. ×1,230

Fig. 142   Histiocytosis X of child (Letterer-Siwe's disease)

Nuclear indentations are a prominent feature. The large nucleolus appears reticulated. Irregular mitochondria and abundant tubulovesicular endoplasmic reticulum, predominantly of the rough variety, are apparent in the cytoplasm. There is paucity of small electron-dense bodies.

Case 17. ×17,500

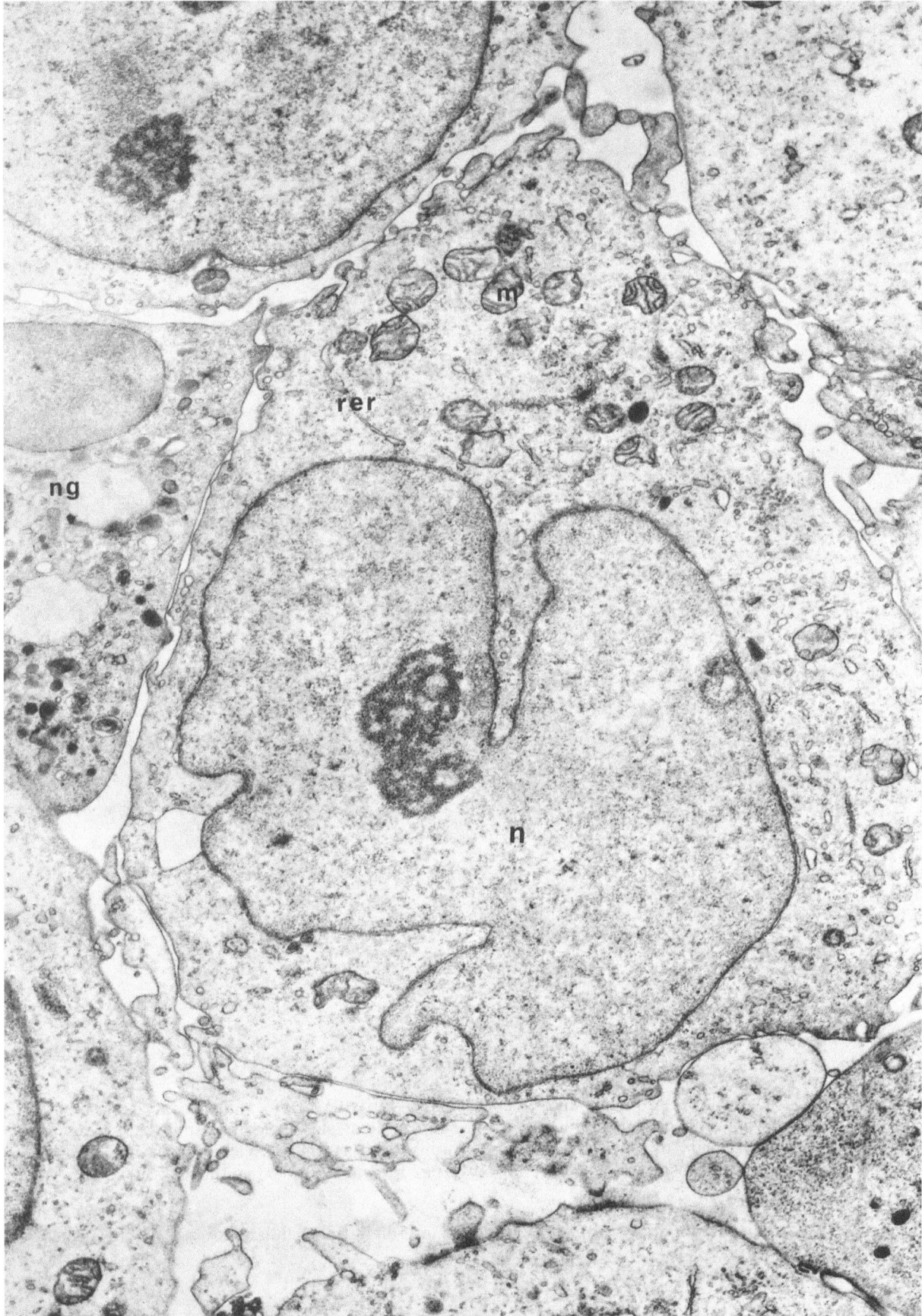

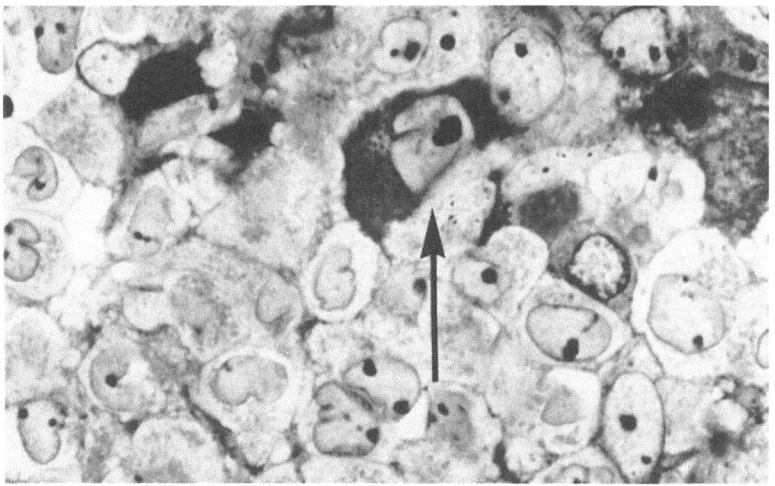

Fig. 143   Histiocytosis X of child (LETTERER-SIWE's disease)

This print illustrates another area of the section which is partially shown in Fig. 141. A giant cell with basophilic cytoplasm and a large nucleolus is apparent above the arrowhead.

Case 17. Osmium tetroxide fixation. Semithin section. Methylene blue-Azur II. ×1,230

Fig. 144   Histiocytosis X of child (LETTERER-SIWE's disease)

The cells bear a resemblance to epithelioid cells particularly in terms of the elaborate intertwining cellular membranes with prominent long slender cytoplasmic extensions. The multiplicity of Golgi regions is also reminiscent of epithelioid cells. Phagosome- and lysosome-like elements, however, are not usually found in typical epithelioid cells.

Case 17. ×17,500

Inset   This picture discloses the appearance of interdigitating cellular processes at higher magnification.

×30,000

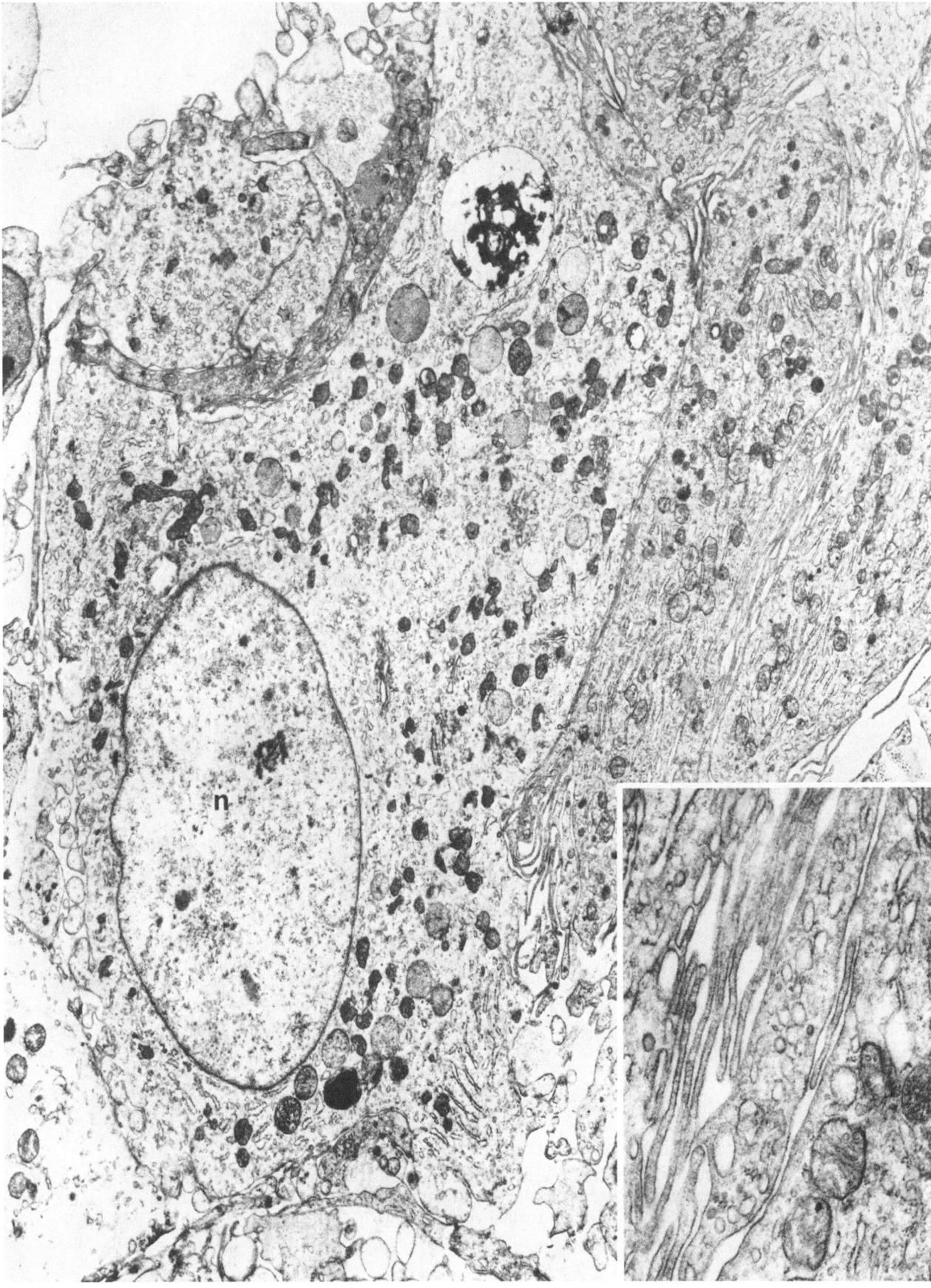

Fig. 145   Histiocytosis X of child (LETTERER-SIWE's disease)
Portion of giant cell with electron microscopic features reminiscent of LANGHANS' type giant
cell, particularly because of the prominent cellular interdigitations (bottom). The similarity
may be readily appreciated when compared with Figs. 51 and 52.

Case 17. ×20,000

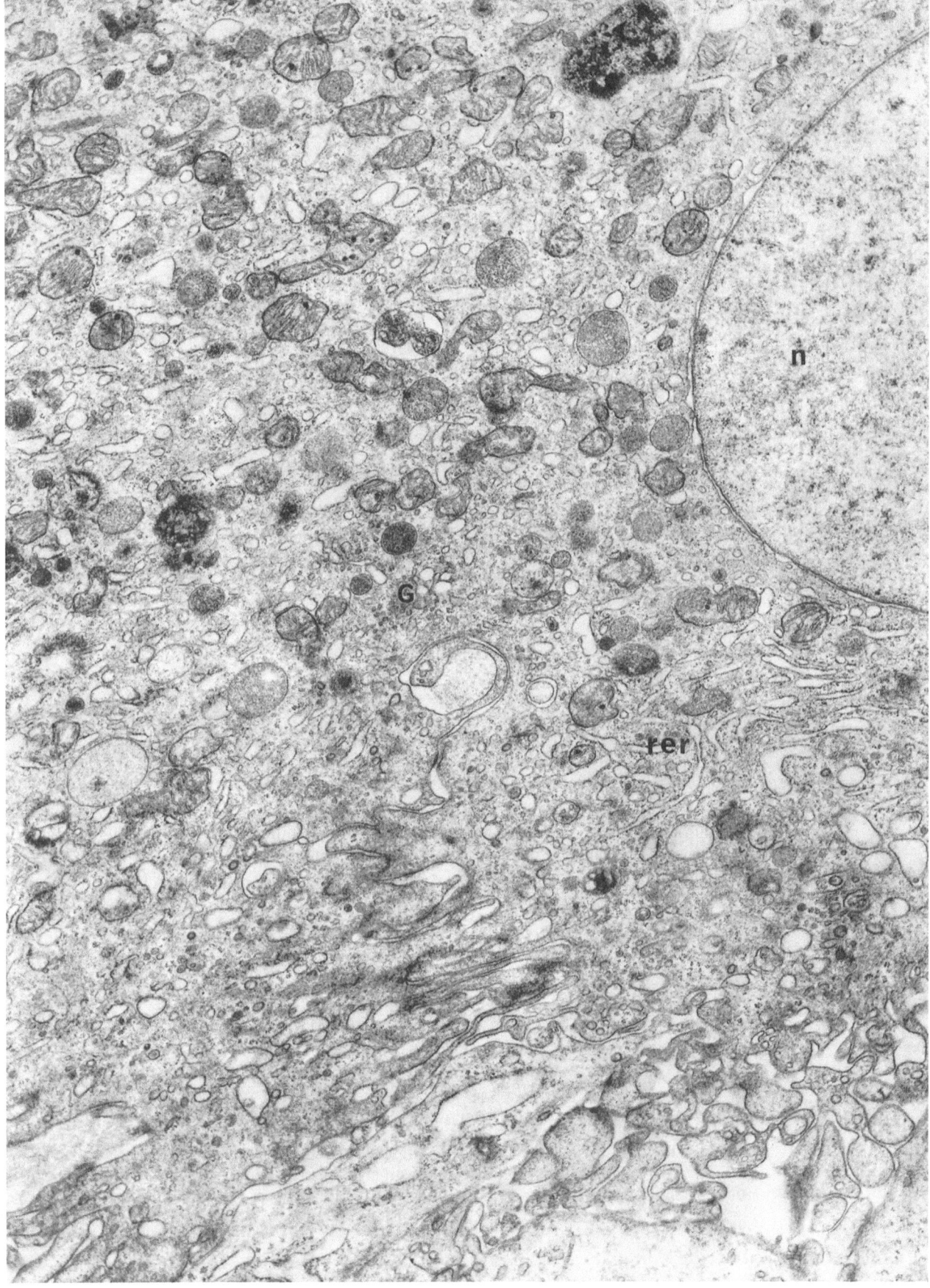

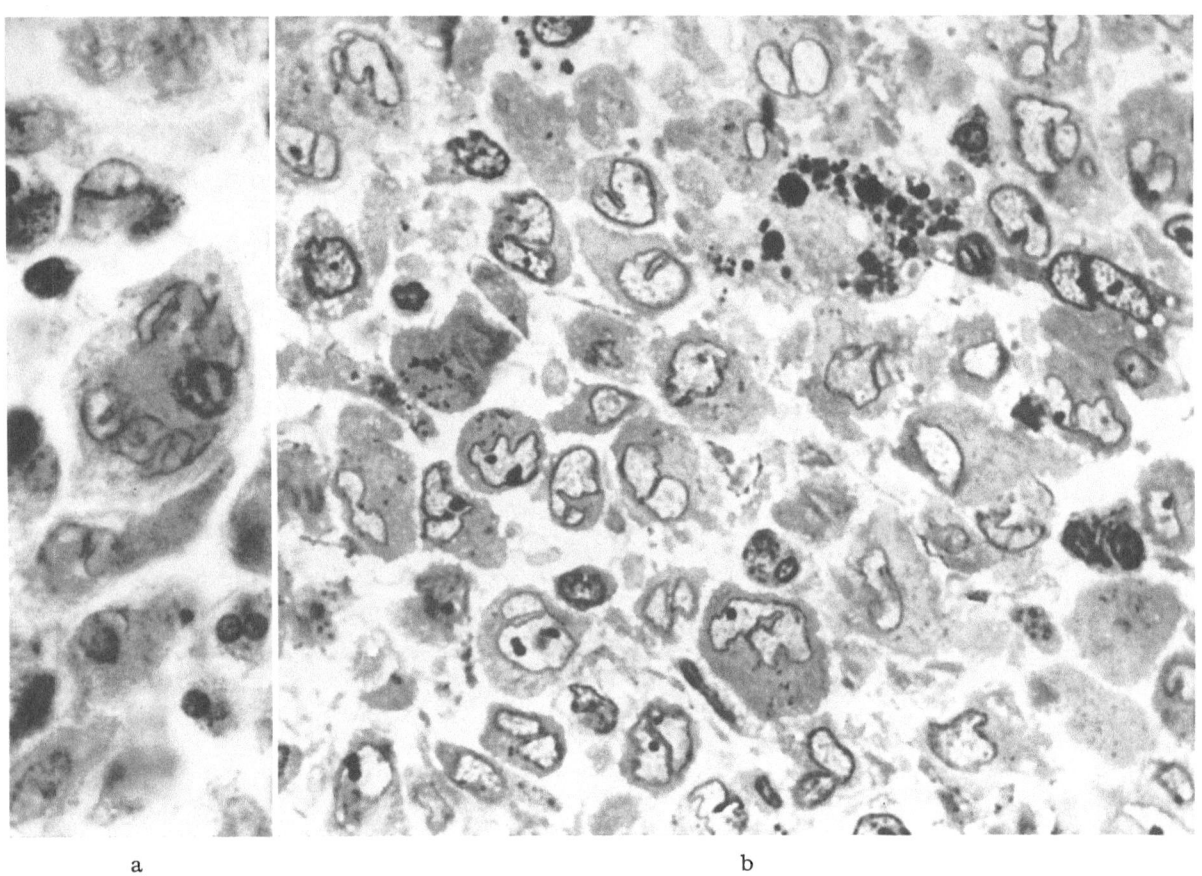

a
b

Fig. 146a and b   Histiocytosis X of adult

a) Multinucleated giant cell resembling osteoclast. Some "histiocytes" and numerous eosinophils are also seen.
Case 16. Formalin fixation. Giemsa-stained paraffin section. ×1,230

b) "Histiocytes" with abundant cytoplasm and pleomorphic nuclei. The nuclear membrane is frequently indented and folded. Occasional binucleated cells, numerous lysosomes and one cell rich in phagosomes are also seen.

Case 16. Osmium tetroxide fixation. Semithin section. Methylene blue-Azure II. ×1,230

Fig. 147   Histiocytosis X of adult

The cells shown in this print have morphologic features similar to epithelioid cells inasmuch as both are provided with villous cytoplasmic processes. Note irregular indentations of nucleus, long rod-shaped mitochondria and multiple small cavities of endoplasmic reticulum.

Case 16. ×17,500

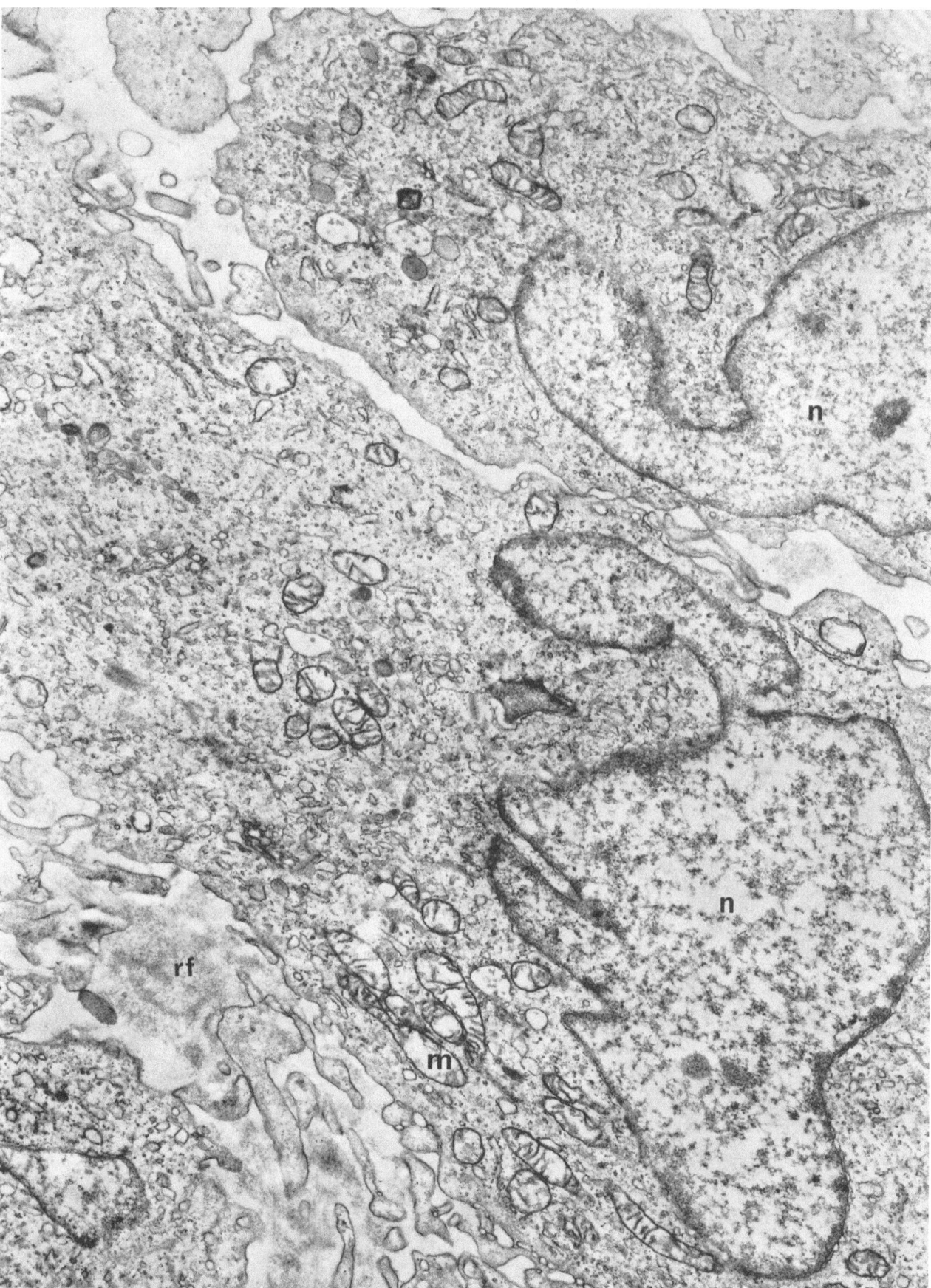

Lymphoepithelial Carcinoma
(Schmincke-Régaud)

Squamous Cell Carcinoma

Malignant Melanoma

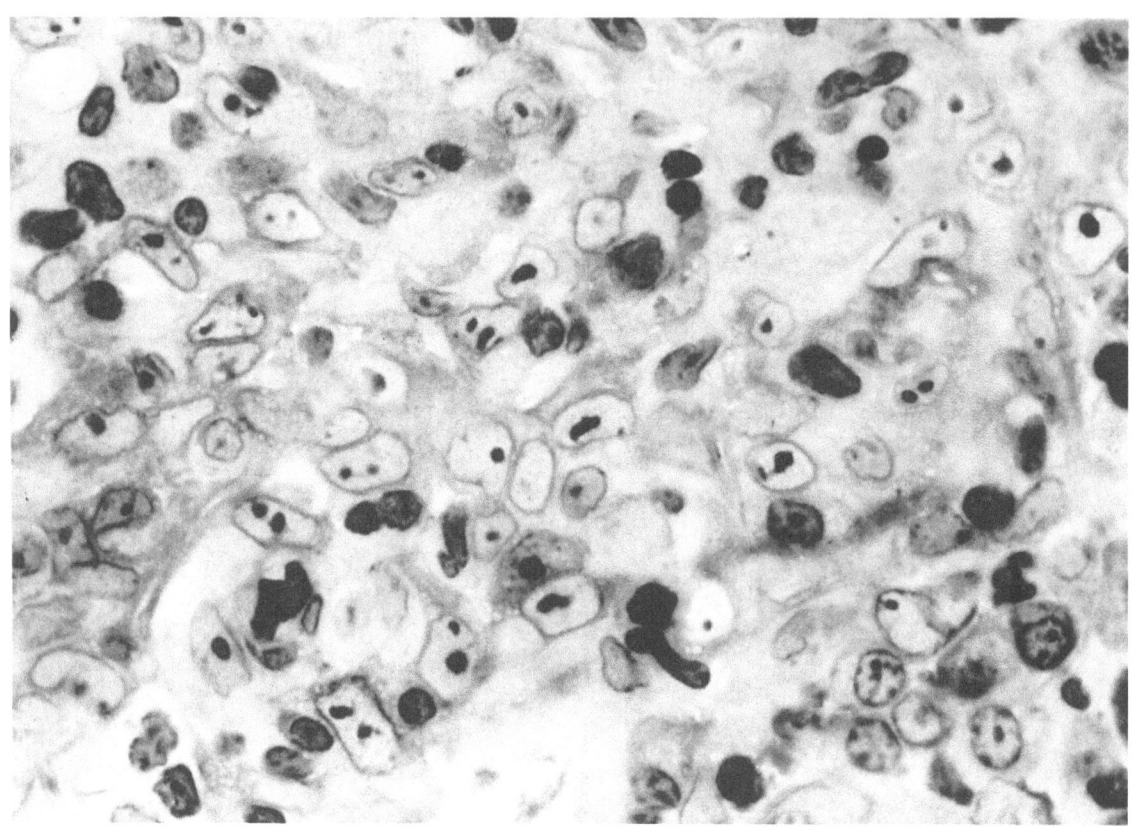

Fig. 148   Lymphoepithelial carcinoma (SCHMINCKE's tumor)
The few interspersed lymphocytes are of no diagnostic significance.

Case 51. Epipharynx. Osmium tetroxide fixation. Semithin section. Methylene blue-Azure II. ×1,230

Fig. 149   Lymphoepithelial carcinoma (SCHMINCKE's tumor)
Nuclei generally take a pale stain and contain large and slightly reticulated nucleoli. Some myelin figures and rough endoplasmic reticulum of varying prominence are encountered in the cytoplasm. Note the very characteristic desmosome (top) and tonofilaments in the cytoplasm.

Case 51. ×18,500

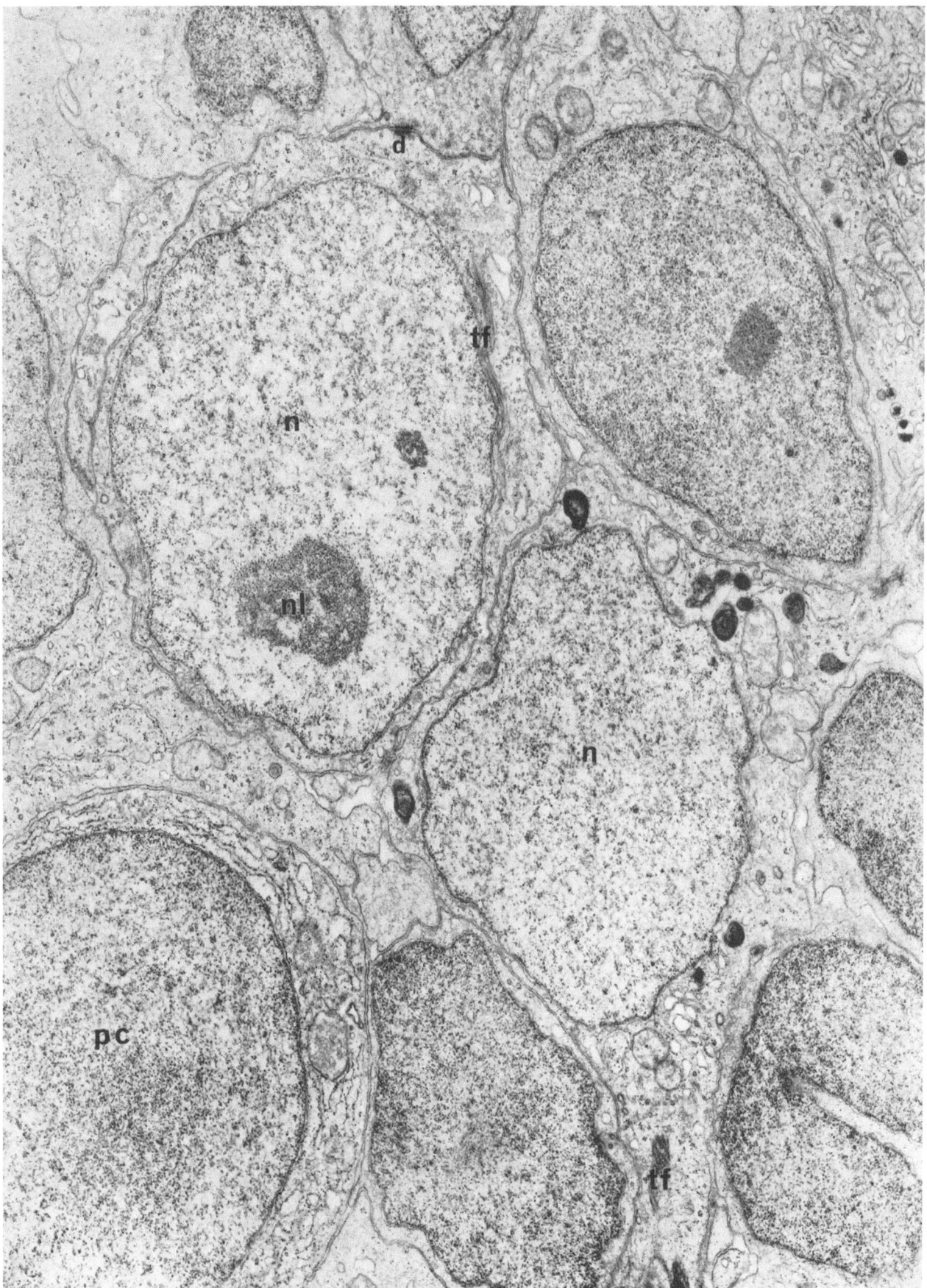

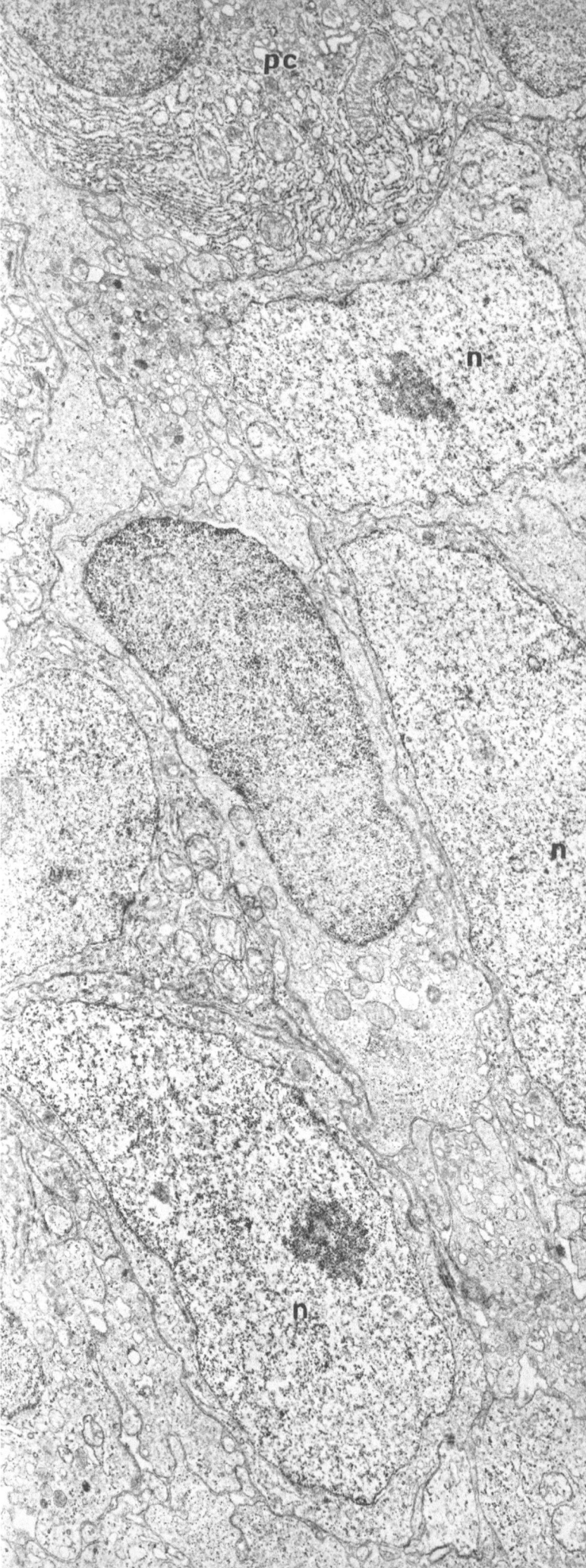

Fig. 150 Lymphoepithelial carcinoma
(SCHMINCKE'S tumor)
Mounted figure

The tumor cells are tightly packed
and connected by desmosomes (arrows). Tonofilaments are visible in
the cytoplasm. Normal lymphocytes
and plasma cells are interspersed
with tumor cells.

Case 51. ×5,500

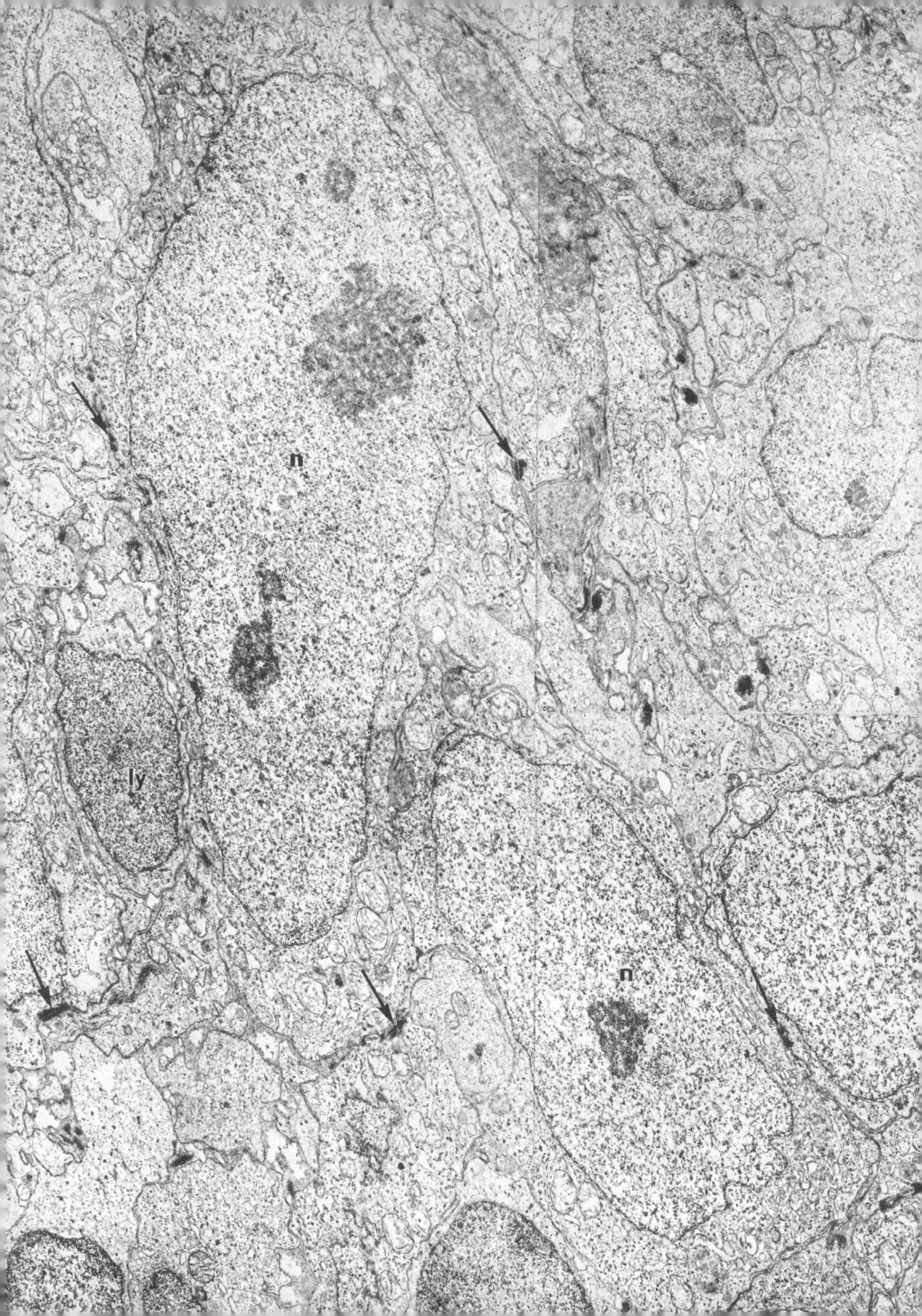

Fig. 151a and b   Lymphoepithelial carcinoma (SCHMINCKE's tumor)

Higher magnification highlights desmosomes and tonofilaments, but there are no inter-cellular bridges which are characteristic for squamous cell carcinoma.

Case 51. ×35,000

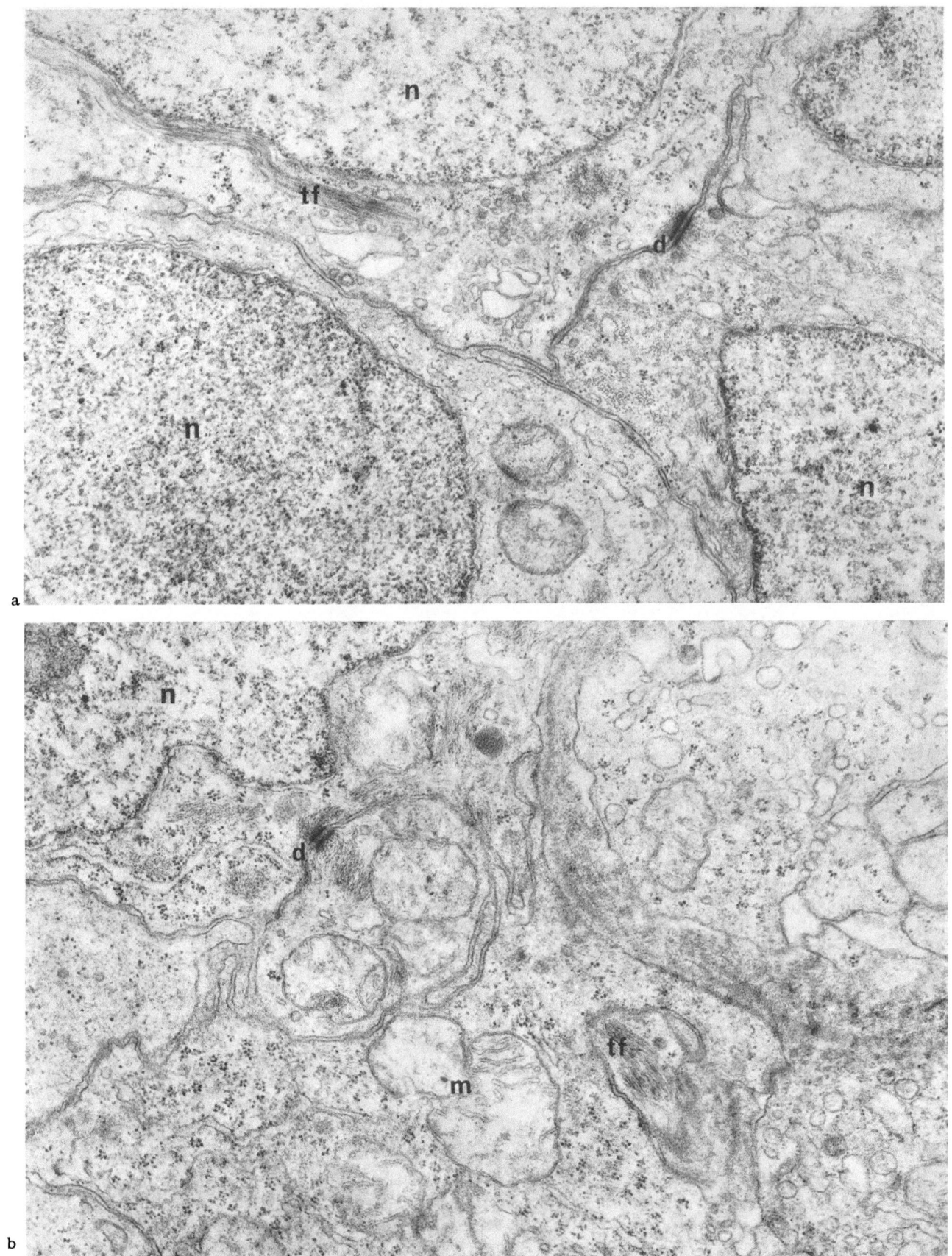

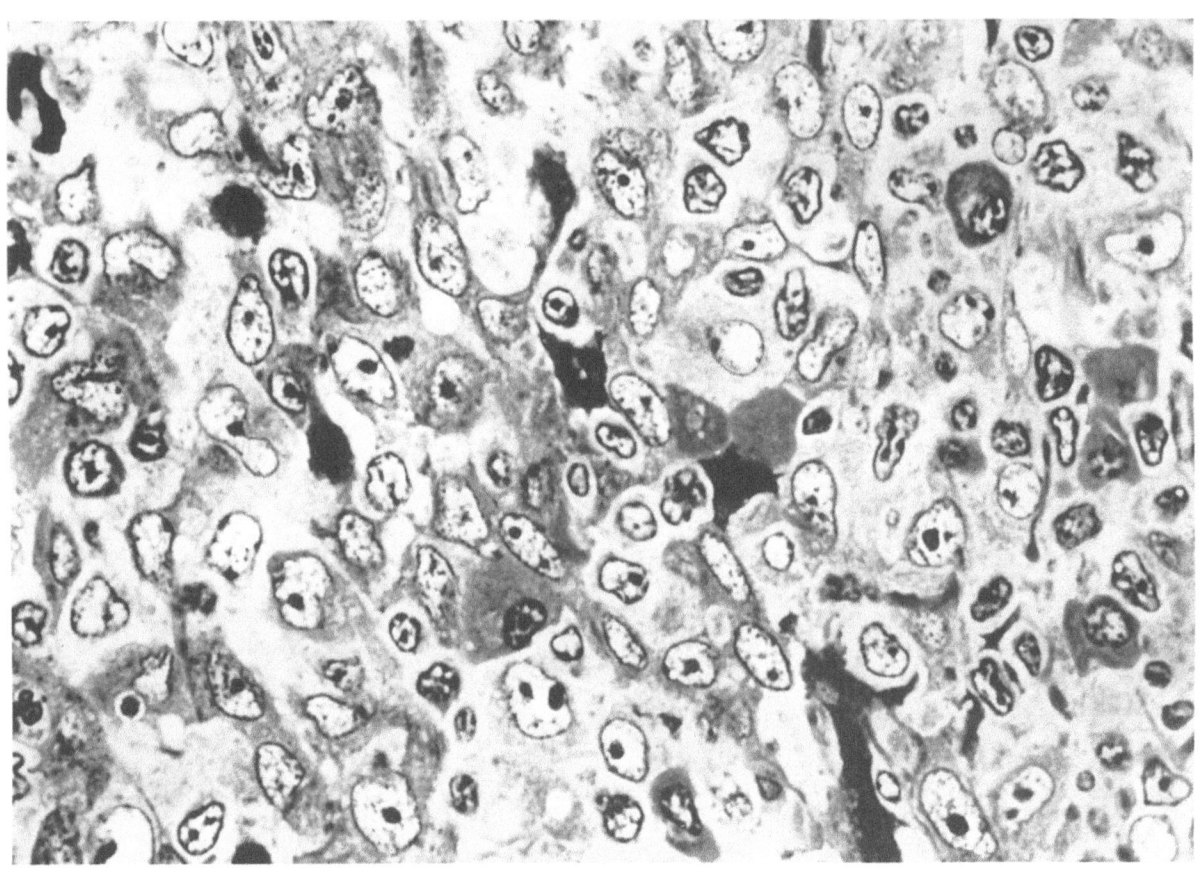

Fig. 152   Lymphoepithelial carcinoma (SCHMINCKE's tumor)
Lymph node metastasis of case depicted in Figs. 148—151.

Case 51. Initial fixation in unbuffered formalin, followed by osmium tetroxide fixation. Semithin section. Methylene blue-Azure II. ×1,230

Fig. 153   Lymphoepithelial carcinoma (SCHMINCKE's tumor)
Metastasis in cervical lymph node of case illustrated in Figs. 148—151. Desmosomes and tonofilaments are clearly visible, otherwise cellular details are somewhat obscured due to improper fixation in unbuffered formalin (routine surgical material).

Case 51. ×17,500

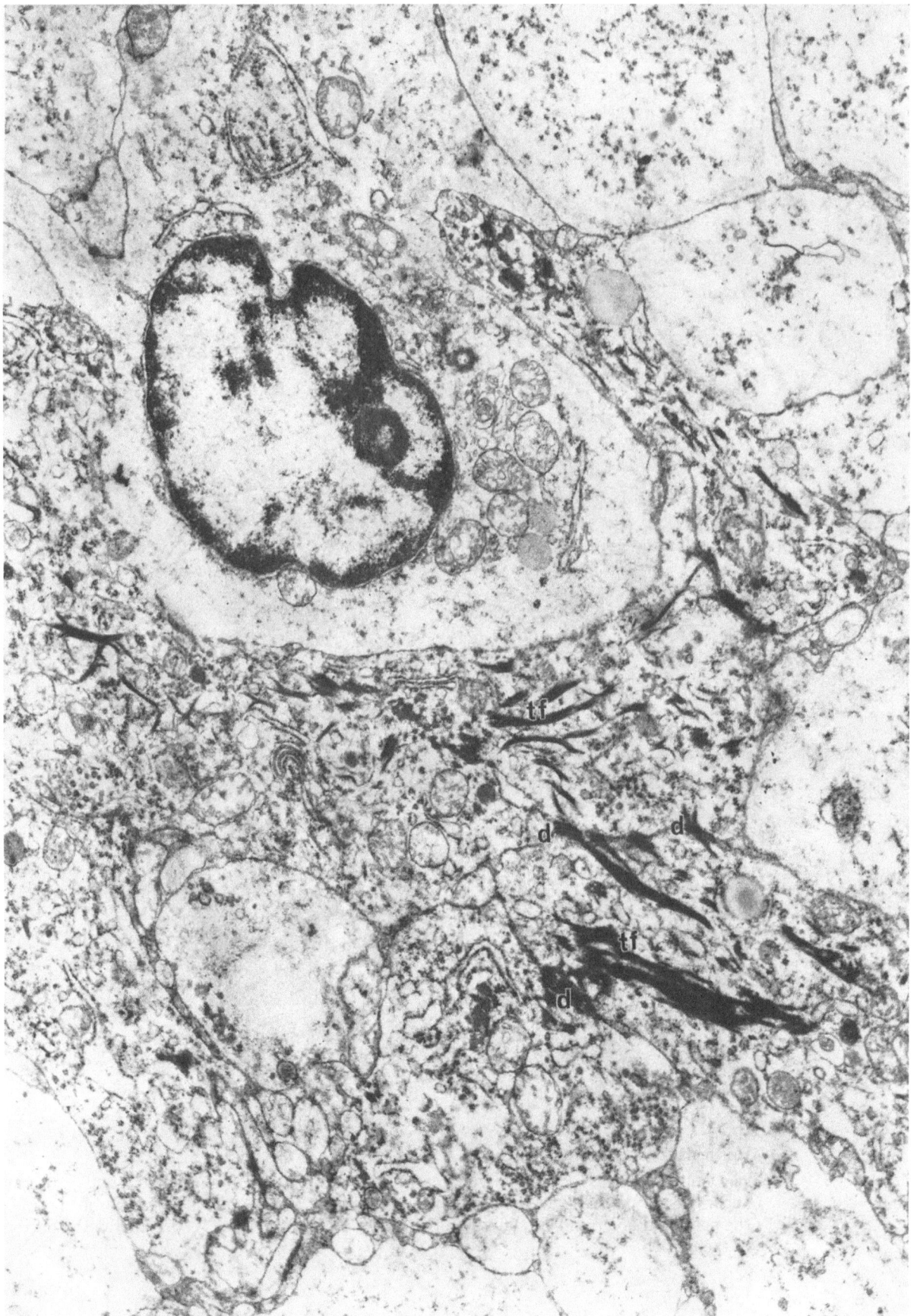

Fig. 154   Epithelial cells of normal tonsil

The epithelial cells occasionally extend deep into the lymphoreticular tissue, forming epithelial networks. They show distinct desmosomes and tonofibrils. The border between the epithelial layer and the lymphoreticular tissue is not distinct, especially at the bottom of the crypt.

×8,750

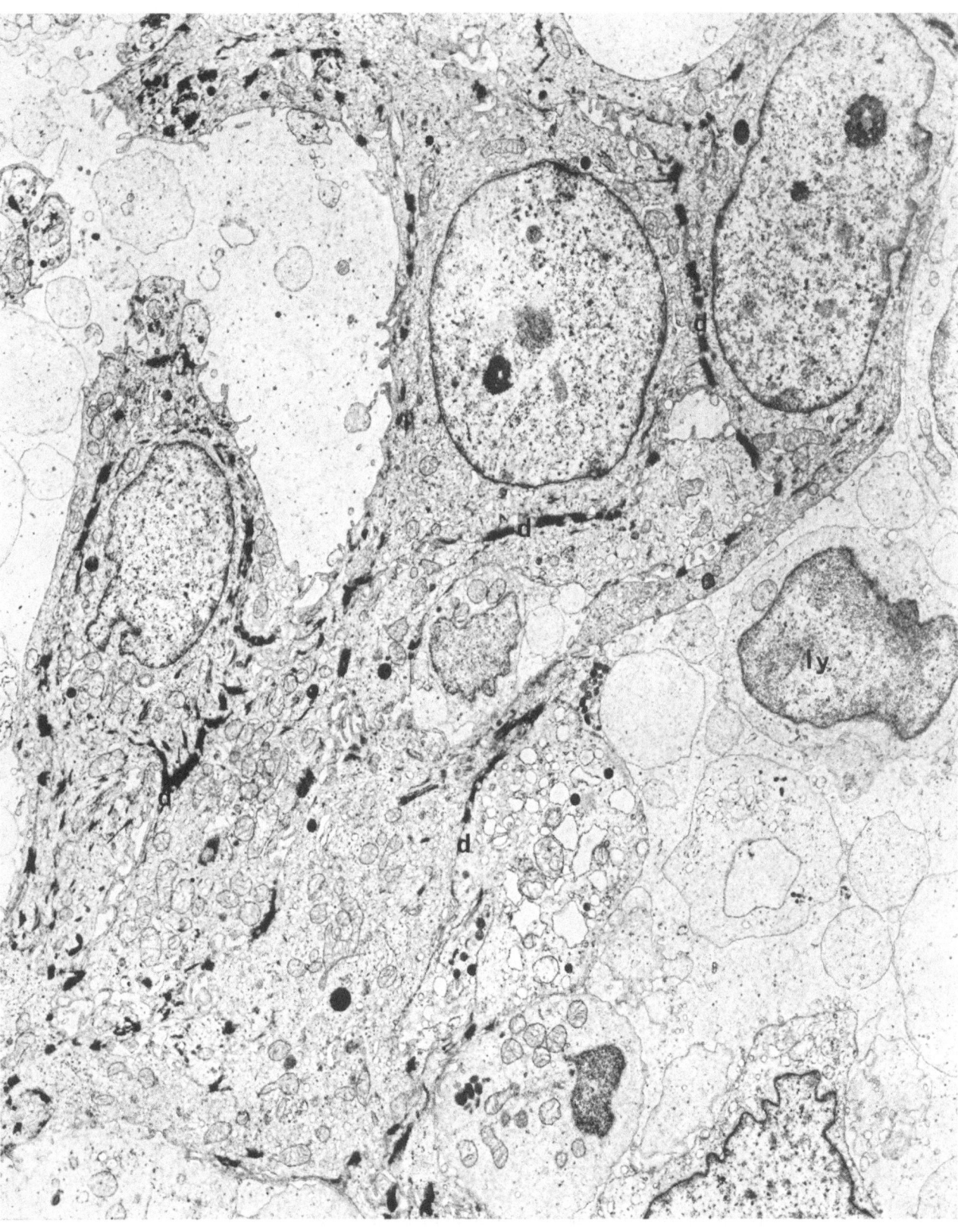

Fig. 155   Lymphoepithelial carcinoma of thymus
The tumor cells resemble those of SCHMINCKE's tumor. Tonofilaments and desmosomes are still recognizable (routine fixation in unbuffered formalin).

Case 6. ×8,750

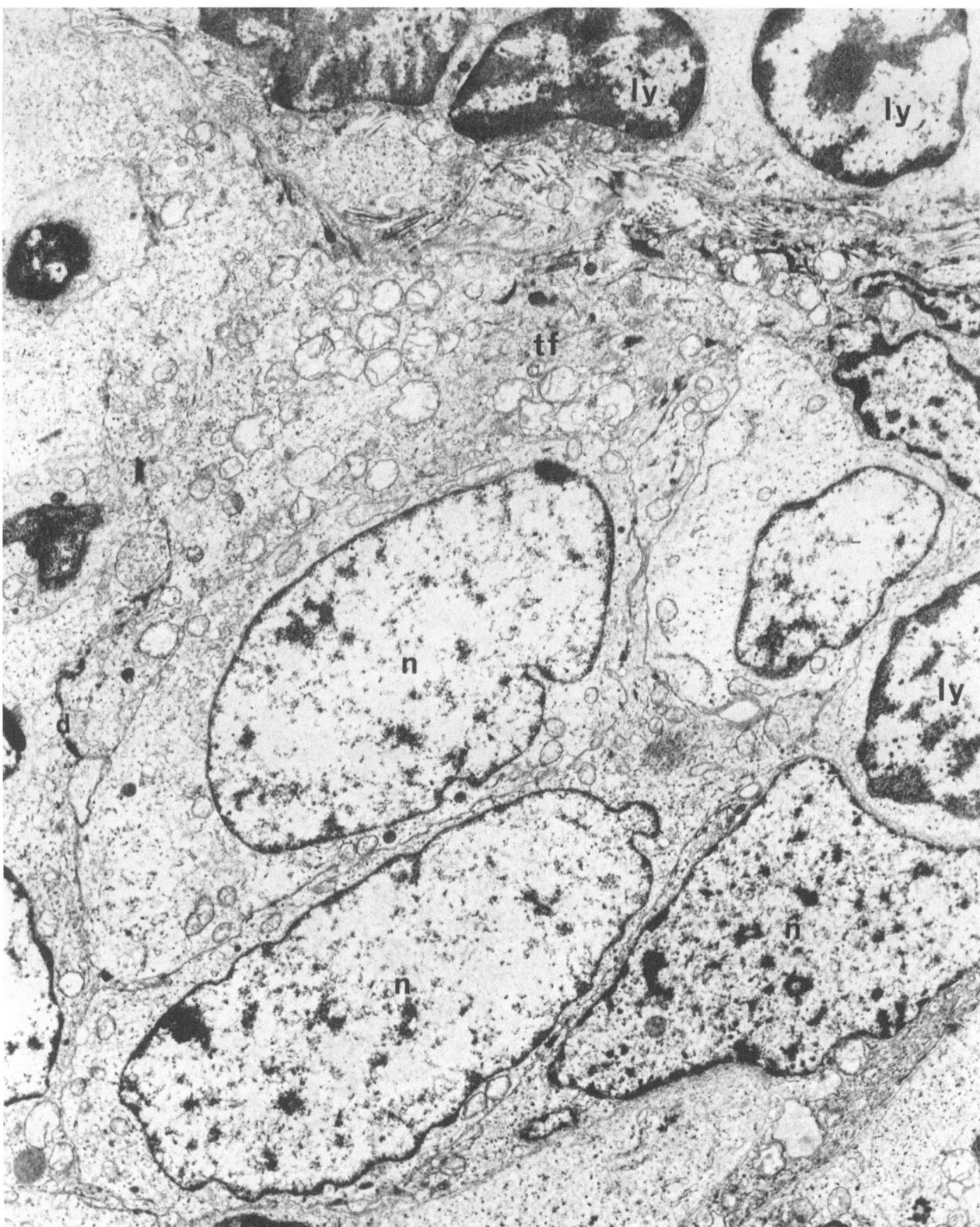

Fig. 156   Metastasis of squamous cell carcinoma (branchiogenic)

Tumor cells exhibiting intercellular bridges are united by desmosomes. In addition, irregular cytoplasmic processes project from the cellular surface. Tonofilaments are seen in the cytoplasm. Intercellular spaces are widened.

Case 7. ×17,500

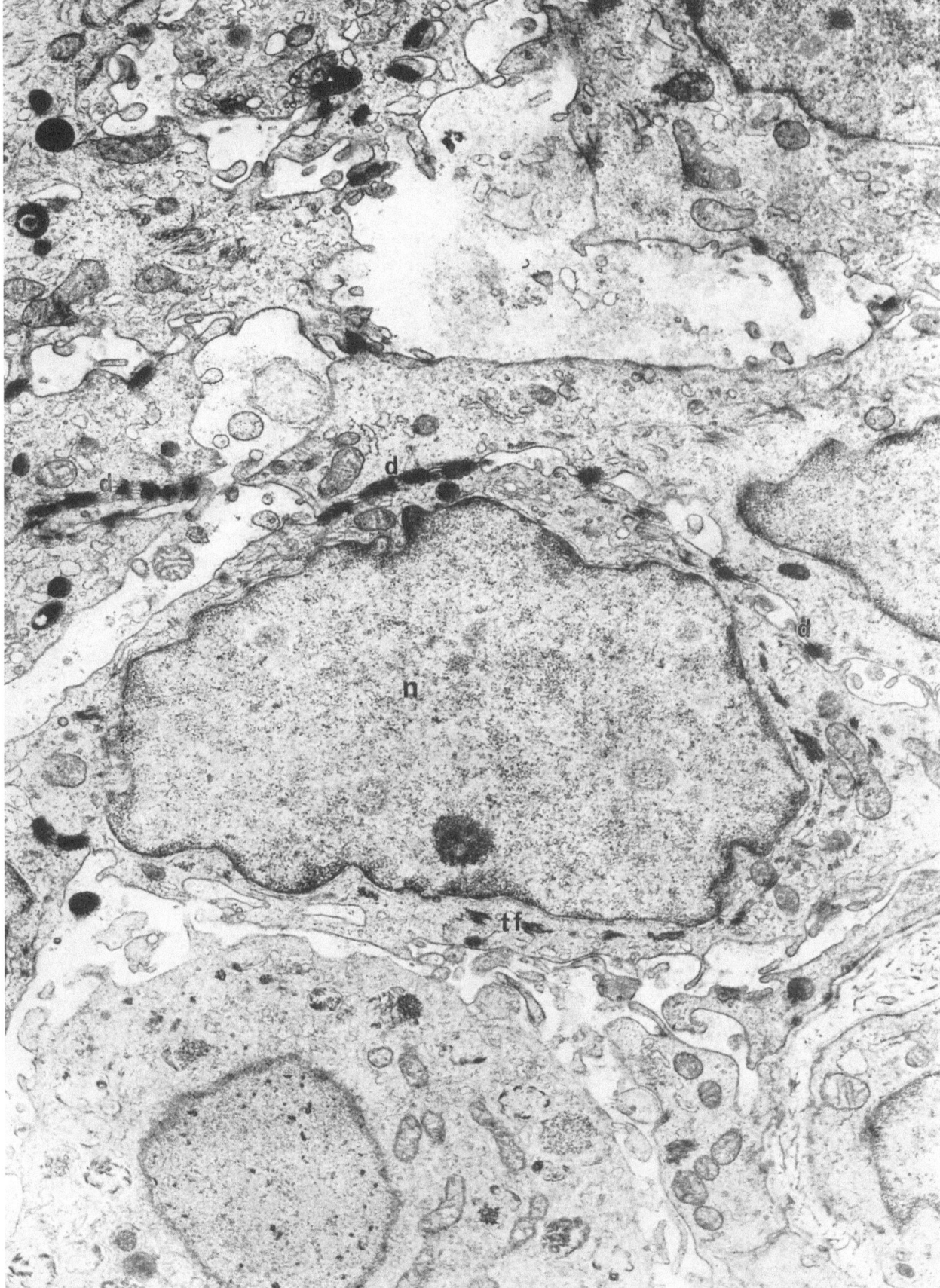

Fig. 157   Metastasis of squamous cell carcinoma (branchiogenic)

Well-differentiated carcinoma cell resembling normal prickle cells of squamous epithelium. Well-defined intercellular bridges with many desmosomal connections are seen. The cytoplasm contains tonofibrils.

Case 7. ×17,500

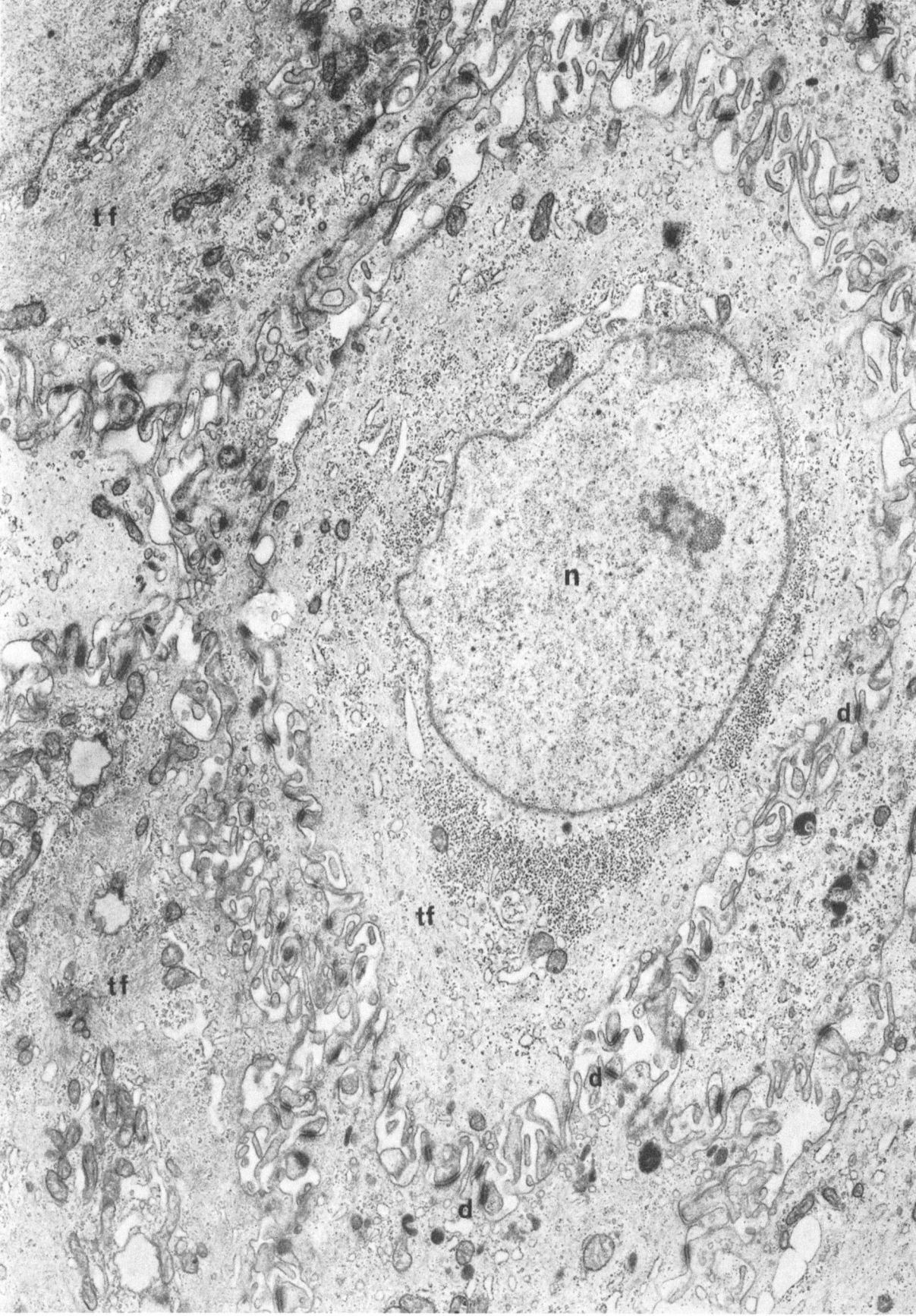

Fig. 158   Squamous cell carcinoma (branchiogenic)
Detailed view of desmosomal connections and slender cytoplasmic processes. Tonofilaments are also noted.
Case 7. ×50,000

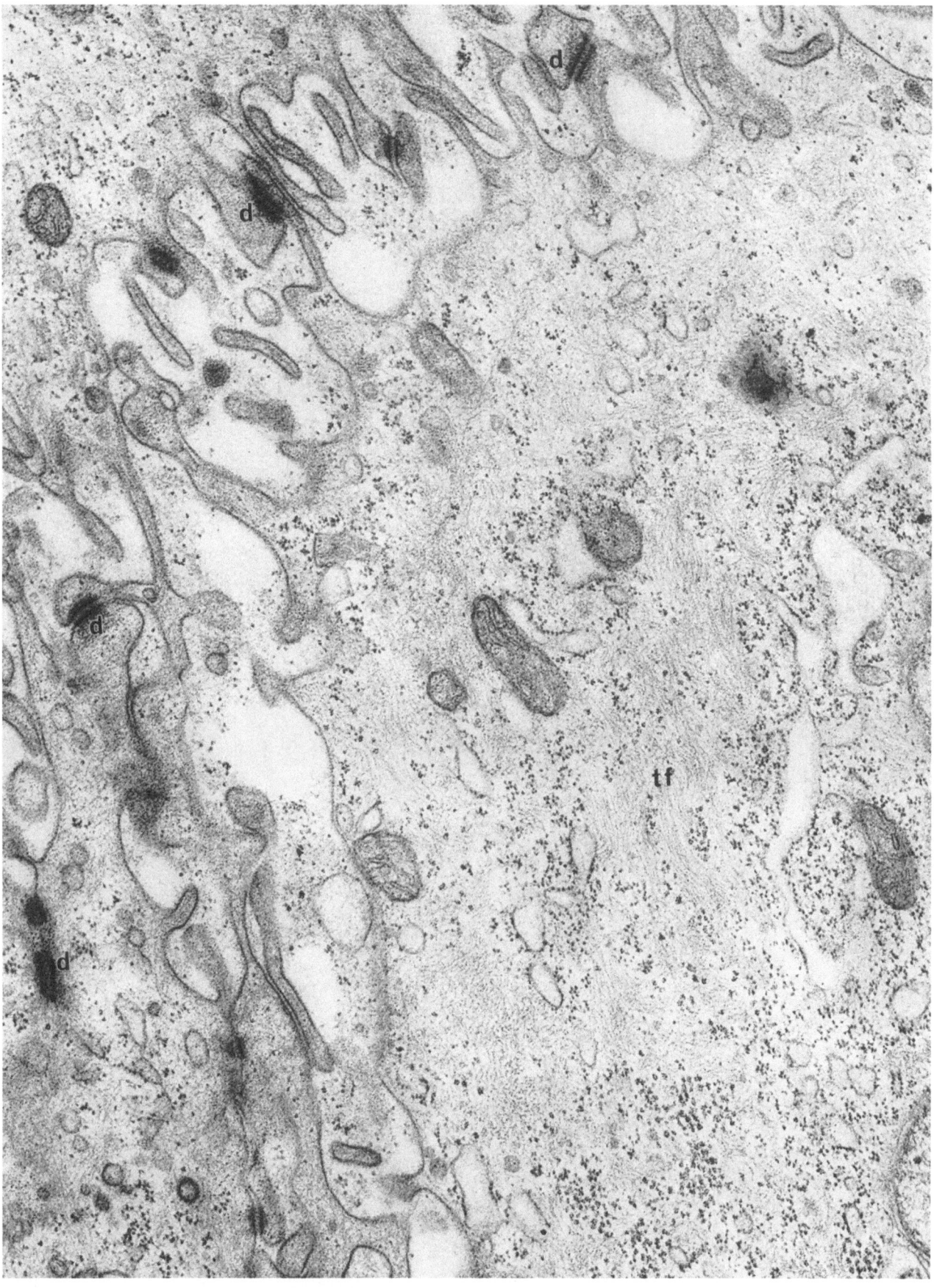

Fig. 159   Metastasis of malignant melanoma
             Fine electron dense granules (melanosomes) are prominently displayed.
             Case 42. ×8,750

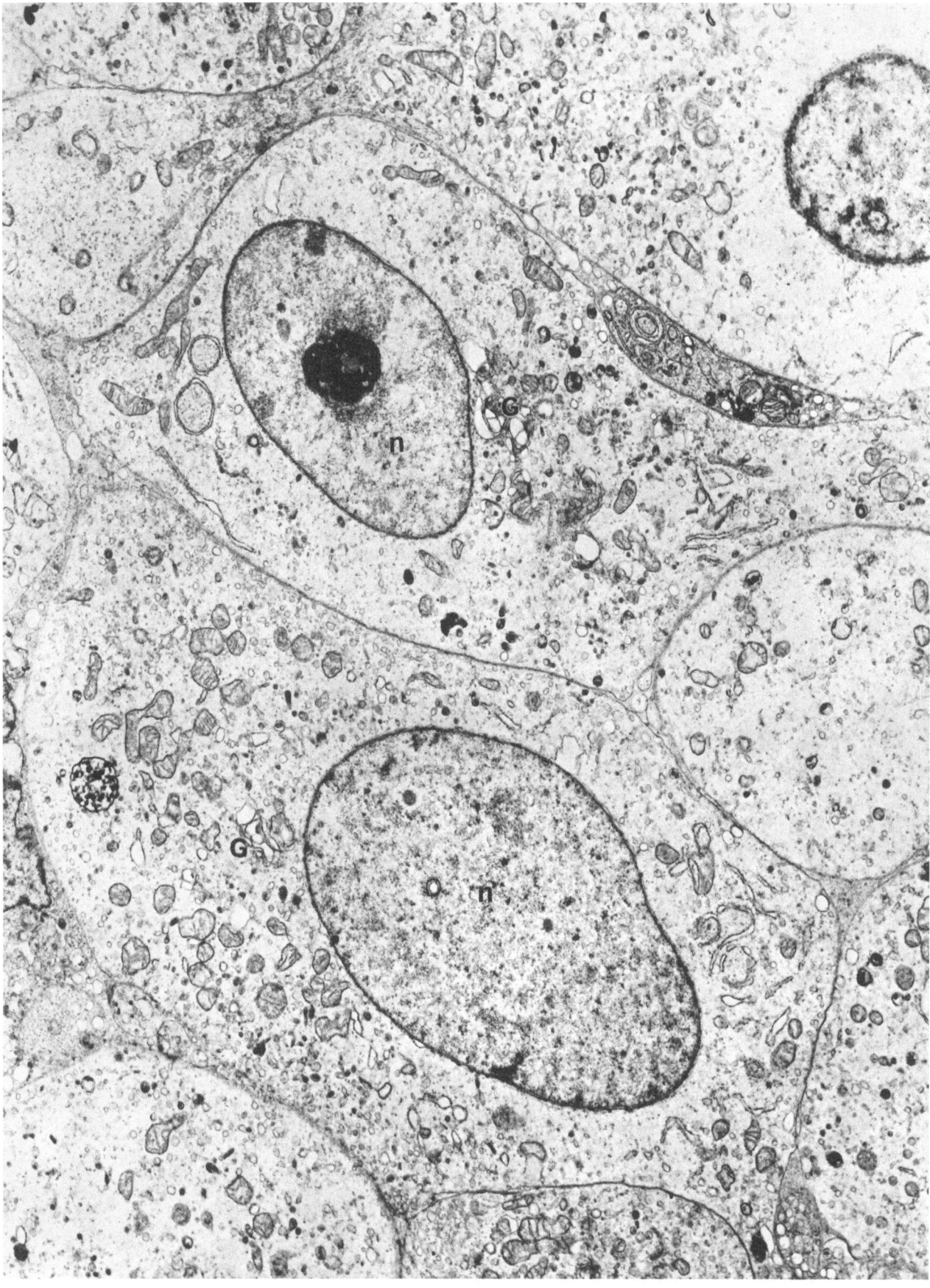

Fig. 160   Metastasis of malignant melanoma
Considerable numbers of melanosomes are encountered in the cytoplasm. Branching tubules and arches of endoplasmic reticulum can be discerned.

Case 42. ×17,500

Inset   Detailed view of melanosome. Note peculiar periodicity of structure.
×52,500

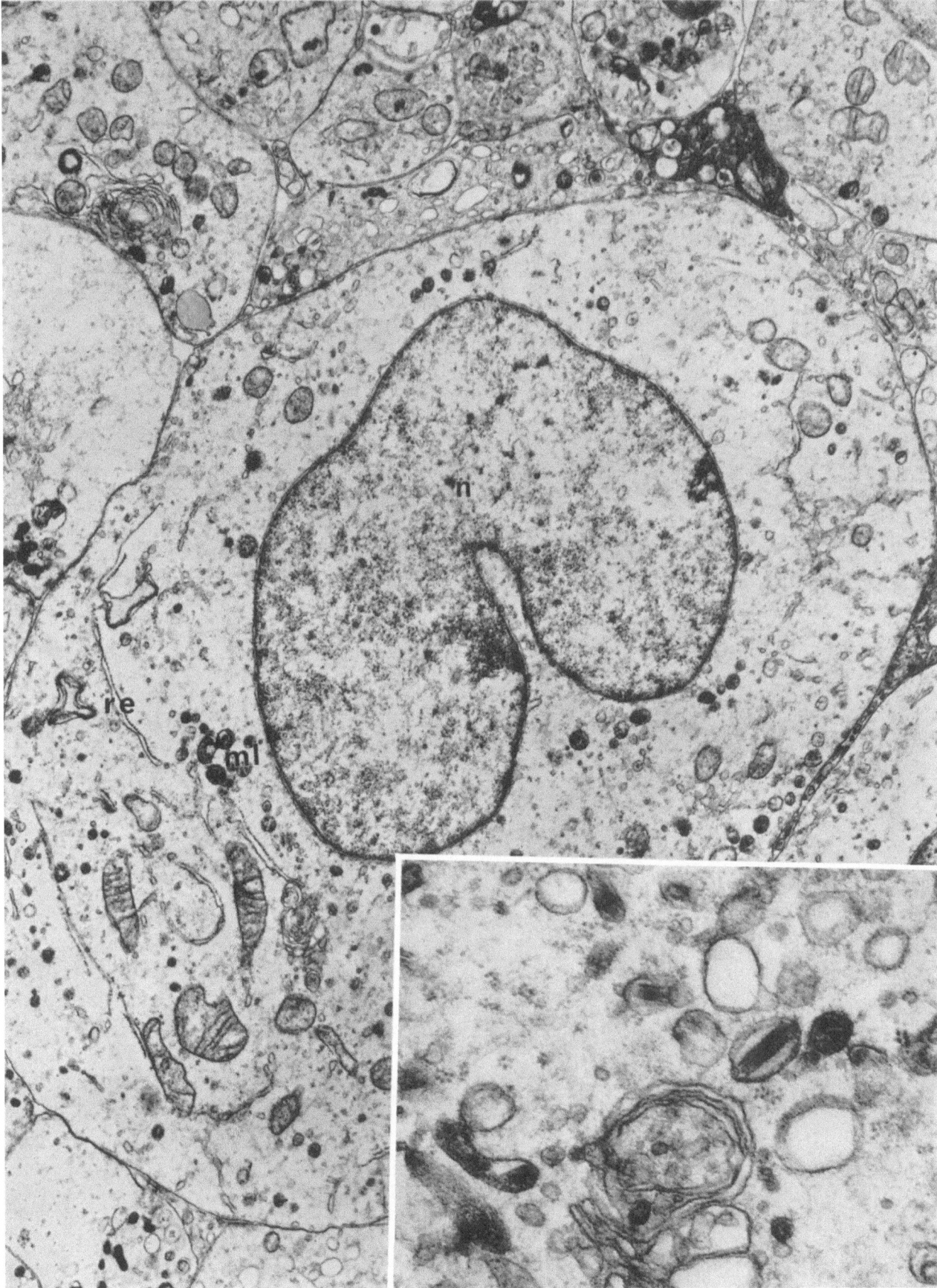

Index

Figure numbers are given in *italics*